neuroscience
for
neurologists

neuroscience for neurologists

edited by

patrick f. chinnery
The University of Newcastle upon Tyne, UK

ICP
Imperial College Press

Published by

Imperial College Press
57 Shelton Street
Covent Garden
London WC2H 9HE

Distributed by

World Scientific Publishing Co. Pte. Ltd.
5 Toh Tuck Link, Singapore 596224
USA office: 27 Warren Street, Suite 401-402, Hackensack, NJ 07601
UK office: 57 Shelton Street, Covent Garden, London WC2H 9HE

British Library Cataloguing-in-Publication Data
A catalogue record for this book is available from the British Library.

NEUROSCIENCE FOR NEUROLOGISTS

ISBN 1-86094-657-7

Printed in Singapore by Mainland Press

Contributors

S Bahn
Cambridge Centre for
Neuropsychiatric Research
Institute of Biotechnology
University of Cambridge
Cambridge, CB2 1QT UK

Oliver Bandmann
Academic Neurology Unit
Division of Genomic Medicine
University of Sheffield, UK

Academic Neurology Unit
E Floor Medical School
Beech Hill Road
Sheffield S10 2RX, UK
o.bandmann@sheffield.ac.uk

G O'Beirne
GE Healthcare, Whitchurch
Cardiff, CF14 7YT

Monika Bradl
Institute for Brain Research
Department of Neuroimmunology
Spitalgasse 4, A-1090 Vienna
Austria
monika.bradl@meduniwien.ac.at

Institute for Clinical
Neuroimmunology
Klinikum Grosshadern
D-81366 München, Germany

Patrick F Chinnery
Neurology, The Medical School
Newcastle upon Tyne, NE2 4HH
Tel: 0191 222 8334
Fax: 0191 222 8553
P.F.Chinnery@ncl.ac.uk

John Collinge
MRC Prion Unit and National
Prion Clinic
Institute of Neurology
and National Hospital for
Neurology and Neurosurgery
Queen Square
London WC1N 3BG

M Czosnyka
Academic Neurosurgical Unit
University of Cambridge Clinical
School, Cambridge, UK
Academic Neurosurgery, Box 167
Addenbrooke's Hospital
Cambridge CB2 2QQ, UK
MC141@medschl.cam.ac.uk

S M Hague
Academic Neurology Unit
Division of Genomic Medicine
University of Sheffield, UK
Academic Neurology Unit
E Floor Medical School
Beech Hill Road
Sheffield S10 2RX, UK

R Hohlfeld
Department of Neuroimmunology
Max-Planck-Institute of Neurobiology, Martinsried
Munich, Germany

H R Jäger
Lysholm Department of Neuroradiology, National Hospital of Neurology and Neurosurgery
Institute of Neurology
Queen Square, London WC1N3BG, UK

P Jezzard
Centre for Functional magnetic Resonance Imaging of the Brain (FMRIB)
Department of Clinical Neurology
University of Oxford
John Radcliffe Hospital
Oxford, UK

K Jurkat-Rott
Abteilung Angewandte Physiologie Universität Ulm
D-89081 Ulm, Germany

S Klaffke
Academic Neurology Unit
Division of Genomic Medicine
University of Sheffield, UK
Academic Neurology Unit,
E Floor Medical School
Beech Hill Road
Sheffield S10 2RX, UK

Peter G Kennedy
Division of Clinical Neurosciences
Institute of Neurological Sciences
Southern General Hospital
Department of Neurology
Glasgow G51 4TF
Tel: 0141 201 2474
Fax: 0141 201 2993
P.G.Kennedy@clinmed.gla.ac.uk

Frank Lehmann-Horn
Abteilung Angewandte Physiologie Universität Ulm
D-89081 Ulm, Germany
frank.lehmann-horn@medizin.uni-ulm.de

S C Lewis
Neurosciences Trials Unit
Division of Clinical Neurosciences
Western General Hospital
Crewe Road, Edinburgh EH4 2XU, UK
steff.lewis@ed.ac.uk

H S Markus
Centre for Clinical Neuroscience
St George's University of London
Cranmer Terrace
London SW17 0RE, UK
h.markus@sgul.ac.uk

P M Matthews
Centre for Functional Magnetic Resonance Imaging of the Brain
Department of Clinical Neurology
University of Oxford, Oxford, UK

Centre for Functional Magnetic Resonance Imaging of the Brain
John Radcliffe Hospital
Headington, Oxford OX3 9DU
UK
paul@fmrib.ox.ac.uk

C M Morris
Institute of Human Genetics
International Centre for Life
Central Parkway
Newcastle upon Tyne
NE1 4BZ, UK
c.m.morris@ncl.ac.uk

P R Orange
NextGen Sciences
Alconbury North Airfield
Huntigndon, Cambridgeshire
PE28 4DA

D P Pashby
Smiths Detection, Watford
Hertfordshire, WD23 2Bw

P Piccini
MRC Cyclotron Building
Imperial College School of Medicine
Hammersmith Hospital
Du Cane Road W12 0NN
London, UK
paola.piccini@csc.mrc.ac.uk

J D Pickard
Academic Neurosurgical Unit
University of Cambridge Clinical School, Cambridge, UK

Academic Neurosurgery, Box 167
Addenbrooke's Hospital
Cambridge CB2 2QQ, UK

J E Prime
Kudos Pharmaceuticals
Milton Road, Cambridge
CB4 0WG

S-M Pulst
Division of Neurology
Cedars-Sinai Medical Center
Departments of Medicine and
Neurobiology, David Geffen School of Medicine at UCLA
Los Angeles, CA, USA

8631 West Third Street, Suite 1145,
Los Angeles, CA 90048, USA
pulst@cshs.org

M M Ryan
Cambridge Centre for Neuropsychiatric Research
Institute of Biotechnology
University of Cambridge
Cambridge, CB2 1QT UK

Eric A Schon
Departments of Neurology and Genetics and Development
Columbia University College of Physicians and Surgeons, USA
eas3@columbia.edu

K Schmierer
Lysholm Department of Neuroradiology, National Hospital of Neurology and Neurosurgery
Institute of Neurology
Queen Square, London WC1N3BG, UK

Pamela J Shaw
Professor of Neurology, Academic Neurology Unit, University of Sheffield
Academic Neurology Unit, E Floor
School of Medicine and Biomedical Sciences, Beech Hill Road
Sheffield S10 2RX
Pamela.Shaw@sheffield.ac.uk

M Symms
Lysholm Department of Neuroradiology, National Hospital of Neurology and Neurosurgery
Institute of Neurology
Queen Square, London WC1N3BG, UK

Y F Tai
MRC Cyclotron Building
Imperial College School of Medicine
Hammersmith Hospital
Du Cane Road W12 0NN
London, UK

C P Warlow
Neurosciences Trials Unit
Division of Clinical Neurosciences
Western General Hospital
Crewe Road, Edinburgh EH4 2XU, UK

K E Wilson
Institute for Ageing and Health
Newcastle University
Newcastle General Hospital
Westgate Road
Newcastle upon Tyne
NE4 6BE

Alan F Wright
MRC Human Genetics Unit
Western General Hospital
Crewe Road, Edinburgh
EH4 2XU, UK
alan.wright@hgu.mrc.ac.uk

J G Whateley
GE Healthcare, Whitchurch
Cardiff, CF14 7YT

T A Yousry
Lysholm Department of
Neuroradiology, National
Hospital of Neurology
and Neurosurgery
Institute of Neurology
Queen Square, London WC1N3BG,
UK
t.yousry@ion.ucl.ac.uk

Preface

The last decade has seen major advances in our understanding of the basic scientific principles that underpin clinical neurology. Many of these advances have already had a major impact on routine clinical practice, and this is likely to continue in the future. Although this makes it an exciting time to practice neurology, it also presents new challenges. How can established general neurologists keep up to date with clinically relevant scientific advances, and how can the specialist remain competent outside his own field? What should trainee neurologists learn to prepare them for their future career?

Neuroscience for Neurologists is a single volume describing the scientific principles behind neurology. The chapters were initially commissioned as a series of review articles published in *The Journal of Neurology, Neurosurgery and Psychiatry*, and were written by experts in each field. When compiled into a single volume, the text takes the reader from the human genome, through gene expression, to molecular and cellular pathology, and subsequently to contemporary clinical investigations and clinical trials. The over riding aim is to provide succinct chapters that will be easily accessible by neurologists and trainees who have no expertise in the area.

Many of the chapters are co-written by a clinician and a basic scientist, and each has been subject to peer review — both by experts in the field and also by a practicing general neurologist. This has ensured that the chapters reflect contemporary scientific thinking which is presented in a way that is easy to digest. Diagrams and figures have been used to help explain difficult concepts, along with a glossary explaining the terminology, and web-links to further information for both physicians and patients. Finally, each we each chapter includes some speculation about the future, highlighting areas of potential growth and their relevance to routine clinical practice over the next decade.

I am very grateful to Professor Chris Kennard (Imperial College London) for his support putting this volume together, along with the staff at the *JNNP* office. Hopefully readers will find the book interesting, informative and easily accessible. The aim is to give confidence in areas that we do not consider our own, enhance and enrich our clinical practice, and thereby improve the care that we give to patients with neurological disease.

Patrick Chinnery
Newcastle upon Tyne, August 2006

Contents

1

Neurogenetics — Single Gene Disorders

*S-M Pulst**

The advent of molecular biology has changed the way in which neurological illnesses are classified, and the single genes causing a number of disorders have been identified. In addition, techniques such as linkage analysis and DNA sequencing have resulted in greater understanding of multigene diseases. This review covers some of the molecular tools and animal models used for genetic analysis and for DNA-based diagnosis, and a brief survey of information available on the Internet.

For centuries, neurology and the classification of neurological illnesses have relied mainly on the neurological phenotype. Subsequent classifications and subclassifications were aided by increased understanding provided by neuropathology and neurophysiology, and more recently by neuroimaging. Grouping disorders based on inheritance patterns did not necessarily result in an improved classification. Diseases such as the dominant ataxias or the dominant neuropathies share a similar inheritance

*Division of Neurology, Cedars-Sinai Medical Center, Departments of Medicine and Neurobiology, David Geffen School of Medicine at UCLA, Los Angeles, CA, USA. Dr S-M Pulst, 8631 West Third Street, Suite 1145, Los Angeles, CA 90048, USA; pulst@cshs.org

pattern, but phenotypes do not necessarily breed true within families, and overlap between families was significant. We now know that mutations in several different genes can cause these phenotypes. A truly novel understanding and grouping of disorders was only possible with the advent of linkage analysis and the eventual isolation of the causative disease genes.

The first neurological illness mapped to a specific chromosome was a disorder, now known as spinocerebellar ataxia type 1. Mapping was accomplished with polymorphic protein markers.[1] An inherited neuropathy, now known as CMT1B, was mapped to chromosome 1 by linkage to the Duffy blood group.[2] These chromosome assignments were contemporaneous with the suggestion that restriction fragment length polymorphisms (RFLPs) could be used to establish a genetic map of the human genome.[3] Another movement disorder, Huntington disease, was the first neurological illness to be genetically mapped using DNA based RFLP markers.[4]

In the following sections, we will highlight some of the theoretical approaches to neurogenetic diseases and introduce the reader to a small selection of the techniques used for genetic analysis and for DNA-based diagnosis. We will also provide a brief survey of information available on the worldwide web that can aid the clinician in phenotype-based

BOX 1 GLOSSARY OF GENETIC TERMS

- Allele: One of two (or more) forms of a gene; differing DNA sequence at a given locus.
- Exon: Coding part of a gene; those parts of a gene that are found in the mature messenger RNA (mRNA).
- Genotype: genetic constitution of an individual in contrast to the visible features (phenotype). Genotype at a given autosomal locus describes the two alleles (see also Fig. 1).
- Haplotype: string of alleles on a single chromosome.
- Homozygous: presence of identical alleles at a given locus.
- Heterozygous: Presence of two different alleles at a given locus.
- Intron: noncoding parts of a gene; introns are located between exons and are spliced out during formation of the mature mRNA.
- Polymorphism: DNA sequence variation between individuals in a given region of the genome.
- Splicing: process that removes introns (noncoding parts of a gene) from transcribed RNA; can also give rise to formation of different mRNA isoforms by selective splicing of specific exons.

diagnosis as well as the choice and interpretation of DNA-based test results. A glossary of the terms used can be seen in Box 1.

GENETIC LINKAGE ANALYSIS

Genetic linkage analysis has been one of the most important tools for the identification of disease genes. A recent brief review of these concepts for neurologists is provided elsewhere.[5] Genetic linkage refers to the observation that genes that are physically close on a chromosome are inherited together. With increasing physical distance between two genes, the probability of their separation during meiotic chiasma formation increases.

By comparing the inheritance pattern of the disease phenotype with that of DNA marker alleles, a chromosomal location for the causative gene can be assigned. A marker very close to the disease gene shows no recombination with the disease trait, but markers further distant or on other chromosomes show an inheritance pattern that is completely different. This is illustrated in Fig. 1 for three different markers in a pedigree with autosomal dominant inheritance.

The significance of observed linkage depends on the number of meioses in which the two loci remain linked. It is intuitively obvious that the observation of linkage in four meioses is less significant than the observation of linkage in 20 meioses. A measure for the likelihood of linkage is the logarithm of the odds (lod) score. The lod score Z is the logarithm of the odds that the loci are linked divided by the odds that the loci are unlinked.[6] The concept of genetic linkage can also be applied to polygenic diseases that are caused by changes in more than one gene and to complex diseases where several genetic and environmental factors interact.

LINKAGE DISEQUILIBRIUM AND ASSOCIATION

Linkage and association should not be confused with one another. Linkage relates to the physical location of genetic loci (or disease traits) and refers to their relationship. Association describes the concurrence of a specific allele with another trait, and thus refers to the relationship of alleles at a frequency greater than predicted by chance. Linkage disequilibrium refers to the occurrence of specific alleles at two loci with a frequency greater than that expected by chance.

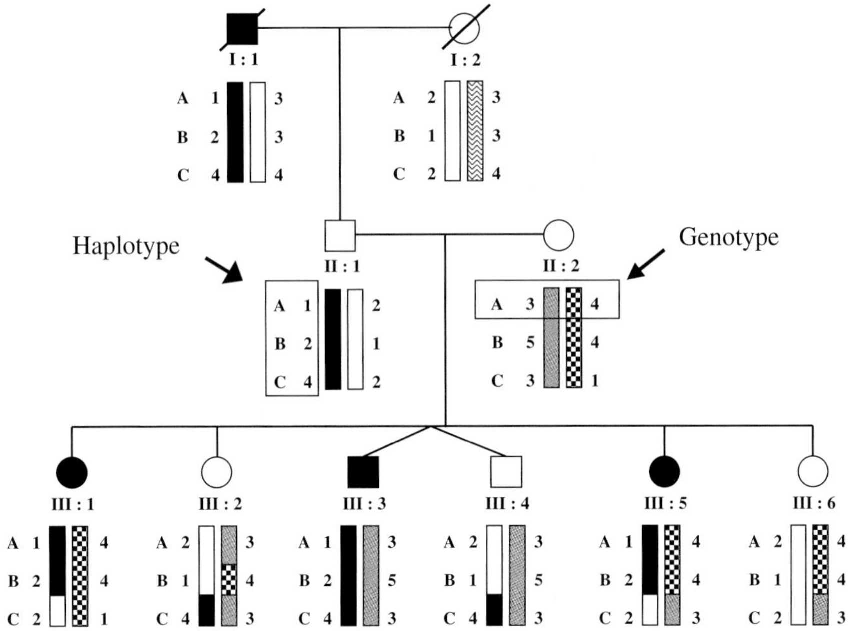

Fig. 1 Pedigree segregating an autosomal dominant trait. Standard pedigree symbols are used.[35] Squares, males; circles, females; solid symbols, affected individuals; a slash through a symbol indicates deceased status. Individuals III:3 and III:4 are fraternal twins; individual II:1 is non-penetrant. The segregation of alleles for three marker loci (A, B, and C) is shown. Vertical boxes provide a depiction of chromosome segments allowing easy visualisation of recombination events. A double recombination is shown for the maternally-inherited chromosome in individual III:2. The haplotype for markers A, B, C is boxed in individual II:1 (a1-b2-c4); the genotype for marker A is boxed in individual II:2 (a3/4).

If the alleles at locus A are a1 and a2 with frequencies of 70% and 30%, and alleles at the locus B are b1 and b2 with frequencies of 60% and 40%, the expected frequencies of haplotypes would be a1b1 = 0.42, a1b2 = 0.28, a2b1 = 0.18, and a2b2 = 0.12. Even if the two loci were closely linked, unrestricted recombination would result in allelic combinations that are close to the frequencies given above. When a particular combination occurs at a higher frequency, for example, a2b2 at a frequency of 25%, this is called linkage disequilibrium, and is a powerful tool for genetic mapping. When a disease mutation arises on a founder chromosome and not much time has elapsed since the mutational event, the disease mutation will be in linkage disequilibrium with alleles from loci close to the gene. At the population level, linkage disequilibrium can be used to investigate processes such as mutation, recombination, admixture, and selection.

In order to study association, allele frequencies in unrelated cases have to be determined and compared with the allele frequencies found in controls. For association studies, it is imperative to repeat the analyses in different patient populations to minimise effects attributable to population stratification, particularly when individuals with the disease belong to a genetically distinct subset of the population. The use of parents as controls can circumvent this problem (transmission disequilibrium test). The transmission of specific parental alleles to affected offspring is scored and statistically analysed.[7]

Association studies may point to genetic factors involved in the pathogenesis or susceptibility of a disease. Well known examples of genetic association studies are the involvement of certain apolipoprotein E (ApoE) alleles in Alzheimer's disease and of tau alleles in PSP.[8,9] PSP Association studies are very powerful for the detection of alleles that have a relatively small effect, but also require that the tested polymorphism be relatively close to the genetic variant that is responsible for the phenotypic variation. This often means that the polymorphism itself has a causative effect or is at least located within the gene of interest.

MUTATIONAL SPECTRUM

Mutations (heritable changes) and polymorphisms in DNA are the basis of genetic variation. Types of mutations range from changes of a single base pair change in a gene, to deletions or duplications of exons or entire genes. Some mutations involve large alterations of parts of a chromosome or involve an entire chromosome such as a trisomy for chromosome 21. Chromosomal abnormalities cause a significant proportion of genetic diseases. They are the leading cause of pregnancy loss and mental retardation. Structural chromosomal abnormalities usually affect several genes and cause phenotypes involving malformations and dysfunction of several organ systems. With the exception of chromosomal translocations, structural chromosomal abnormalities would be a rare cause of a phenotype dominated by a dysfunction in a single neurological system without dysmorphological features or involvement of other organ systems.

Mutations that affect single genes are most commonly located in the coding region of genes, at intron-exon boundaries, or in regulatory sequences. These changes can be due to changes of single base pairs, or insertions/deletions of one or multiple base pairs. Single base pair substitutions, also called point mutations, result when a single base pair is replaced by another. Many single base pair mutations, even when located in the coding region of a gene, may not change the amino acid sequence owing to the redundancy of the genetic code. These are often located in the third position of a codon, and are called silent substitutions. Although a base pair substitution may not change the amino acid sequence, it may nevertheless cause disease by introducing cryptic splice signals.

Most amino acid changes are not pathogenic. In particular, when the amino acid change is conservative (substitution with a similar amino acid), the resulting protein may represent a normal variant and have normal function. At times, it may be difficult to distinguish a normal variant from a sequence change causing disease, especially when the function of a protein is poorly understood.

Deletions or insertions of one or several base pairs of DNA represent another mutation type. If the change involves three base pairs or a multiple thereof, amino acids are added or deleted from the protein, but the reading frame and the remainder of the protein remain intact. Deletion of a single GAG codon in torsin is sufficient to cause dystonia type I (DYT1).[10] Deletions/insertions that are not a multiple of three will alter the reading frame and the resulting amino acid sequence downstream of the deletion/insertion. This change in the reading frame (frameshift mutations) usually results in a shortened polypeptide, because the frameshift will result in the recognition of a premature stop codon.

MOLECULAR GENETIC TOOLS

The array of tools and techniques used in molecular genetics could fill many book volumes.[11,12] For the purpose of this review, we will focus on two basic techniques that are regularly used for DNA-based diagnosis, the polymerase chain reaction (PCR) and DNA sequencing. Genetic tests are certainly not limited to the analysis of DNA; metabolic disorders have long been diagnosed by determination of enzyme activity.

Polymerase Chain Reaction

Molecular biology was revolutionised by the introduction of PCR.[13,14] In an ingeniously simple procedure, PCR combines the sequence specificity of restriction enzymes with amplification of DNA fragments, previously only possible by cloning of restriction fragments. Sequence specificity is provided by the annealing of oligonucleotide probes complementary to the DNA sequence of interest (also called a DNA oligonucleotide primer). Amplification is achieved by repeated rounds of oligonucleotide primed DNA synthesis (Fig. 2). The specificity of PCR is primarily determined by the annealing temperature. When the annealing temperature is lowered, the primers will anneal to DNA sequences that are not perfectly matched, and other fragments may be amplified.

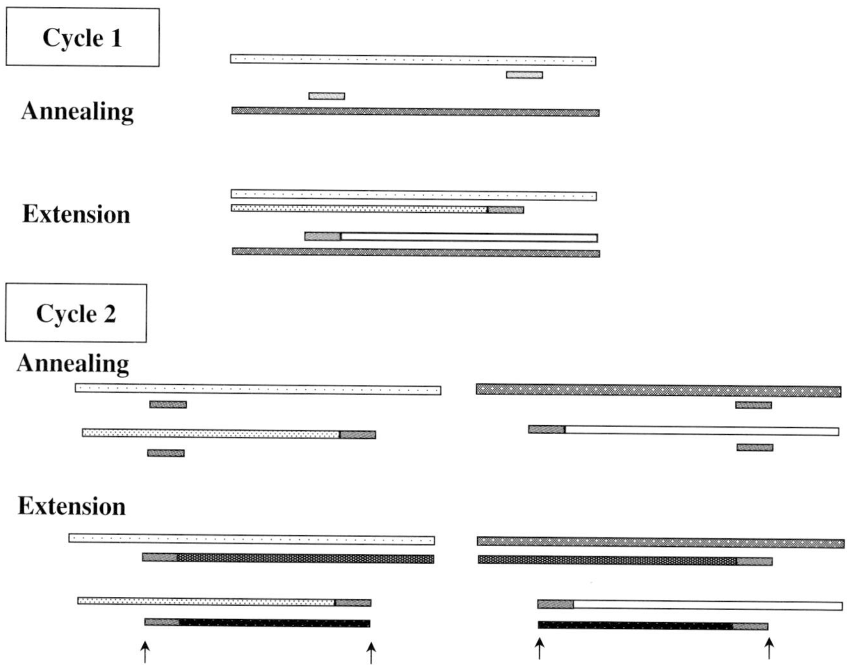

Fig. 2 Schematic of the polymerase chain reaction (PCR). Annealing of primers and extension of DNA fragments are shown for two cycles. Note that the DNA fragments shown between the arrows have a length defined by the spacing of the oligonucleotide primers. (Reproduced with permission.[5])

Almost all DNA-based testing involves PCR. Deletion or insertion mutations can be detected as a variation in amplicon length. This approach is typically used for disorders involving DNA repeats such as the polyglutamine (polyQ) disorders; for Friedreich ataxia; or for myotonic dystrophy. PCR is also used for mutation analysis that involves sequencing of the gene of interest.

Although PCR is a highly specific and sensitive method, it has technical limitations that need to be recognised when interpreting DNA test results. Any setting that results in non-amplification of the mutant allele may result in a false negative DNA test, because amplification from the wild type allele will proceed normally. Thus, non-amplification of the mutant allele may appear as homozygosity for the wild type allele. For example, elongation of the PCR product may be impeded by a secondary structure of the intervening DNA sequence or by large insertions. This is particularly relevant for mutations caused by large DNA repeats. If a mutation disrupts annealing of the oligonucleotide primer, an amplicon from the mutant allele is not generated, and thus will escape detection. Similarly, if a genomic deletion involves the entire length of DNA that is to be amplified (such as an entire exon) it will not be detected, because the homologous DNA on the other chromosome will generate an amplicon.

DNA Sequencing

Several sequencing methods exist. Only the chain termination method will be further described, because it is the method currently being used for automated sequencing. In this method, a DNA fragment is provided as a template for the synthesis of new DNA strands using a DNA polymerase.[15] The reaction is primed by a sequencing primer usually 17–22-bp in length that specifically binds to the region of interest. In addition to regular deoxynucleotide triphosphates (dNTPs), smaller amounts of dideoxynucleotide triphosphates (ddNTPs) are added to the reaction mix. Although ddNTPs are incorporated into the newly synthesised DNA chain, they cause abrupt termination of the chain due to lack of the 3′-hydroxyl group, preventing formation of the phosphodiester bond. Chain termination will occur randomly whenever a ddNTP is incorporated into the growing chain instead of dNTP. The length of the fragments can be determined using polyacrylamide gel or by capillary electrophoresis and the fragments visualised by labelling with radioisotopes or fluorophores.

Although sequence-based mutation detection may appear straightforward, technical difficulties can occur. DNA structure may interfere with the abundance of a particular fragment (seen as a weak band in a radioactive gel or a decreased peak in automated sequencing). As most sequencing for mutations occurs on genomic DNA (with two alleles present except for X chromosomal sequences in males), detection of sequence differences occurs in the heterozygous state. A weak peak marking a single base mutation may be masked by the presence of a normal sized peak from the normal allele. This can be circumvented by adjustments in sequence analysis software, sequencing in forward and reverse directions, and using different sequencing primers.

MOLECULAR GENETIC TESTING

DNA-based diagnosis falls into two broad categories: direct mutation detection and indirect detection of a mutated gene using the analysis of polymorphic genetic markers that flank the gene of interest or lie within it. When the disease gene has been identified and the types of mutations are limited, direct mutational analysis is possible. For some diseases, however, a large number of different mutations exist without mutational "hot spots," so that indirect testing is technically more feasible than direct mutational analysis. A similar situation arises when the chromosomal location of a disease gene is known, but the gene itself has not been identified. In the above two scenarios, DNA diagnosis can be performed using polymorphic DNA markers, which are used to track the disease chromosome in a given family. With ever improving sequencing technology, even genes with multiple nonrecurring mutations distributed over many exons, such as the *CACNA1A* gene, are now amenable to direct mutation analysis. Genetic diagnostic tests involving the sequence analysis of an entire gene (connexin 32 for CMTX, *MECP2* for Rett syndrome) are now commercially available.

Direct Mutation Detection

For diseases caused by a specific mutation, direct detection of the mutation is simple and inexpensive. The direct genetic test more closely resembles a conventional laboratory test. A blood sample is taken from a patient and the test is used to determine whether the individual carries a specific mutation in a given gene, with a yes/no answer. Barring technical problems or a sample switch resulting in false positive results, a positive test indicates that a symptomatic individual has the disease. A positive result is independent of the accuracy of the clinical assessment.

Examples of direct mutation testing include diseases caused by DNA repeats. The repeat is amplified by PCR and the allele with the expansion can be detected as a larger than normal fragment by gel electrophoresis. The same is true for other mutations that result in deletion or additions of DNA codons. Despite the fact that all mutation detection is in the end sequence based, various technologies and strategies are employed depending on the nature and frequency of mutations and the size of the gene to be analysed.

As with all laboratory tests, clinical judgement is the final arbiter. It is possible that a patient has symptomatic multiple sclerosis and an inherited ataxia as indicated by a DNA test for which she is pre-symptomatic. Two recent examples may illustrate this point. A 5-year-old boy presented with dysarthria and a very severe hypertrophic cardiomyopathy. On molecular analysis, he was found to have a double mutation in two distinct genes, the frataxin gene and the cardiac troponin T gene.[16] In a family segregating an autosomal dominant trait for tremor, one child developed a dystonic tremor. This proband was found to have Wilson disease, which obviously had significant implications for therapy.[17]

A negative test in a symptomatic individual needs to be interpreted, realising its limitations. Phenotype and gene test need to be correctly matched. If a patient with a dominant spinocerebellar ataxia (SCA) has a negative test for *SCA1* or *SCA3*, this does not mean that this individual does not have a dominant SCA, because the mutation could be in one of the other SCA loci. Alternatively, the mutation could be in the respective gene, but it was not assayed by the gene test that was performed. For example, a progressive ataxia can be caused by expansion of a polyglutamine repeat in the α_{1A} voltage dependent calcium channel subunit encoded by the *CACNL1A* gene on chromosome 19p. Specific missense mutations in this gene, which would not be detected by a PCR-based test for CAG repeat length, can also be associated with a progressive ataxia.[18,19]

Direct sequence analysis is increasingly used to analyse genes for point mutations. It is particularly useful when several different mutations are clustered in one exon and can be analysed by sequencing of a single PCR product. The availability of newer sequencing technologies will undoubtedly broaden the applicability of sequence analysis to larger genes with multiple different mutations.[20] Sequence analysis of entire genes, initially only available in research laboratories, is now also performed commercially.

As it is unknown where the mutation might occur, the quality of the sequence has to be high throughout the entire region that is being sequenced. Less than 100% sensitivity of mutation detection by sequencing has its causes in these technical limitations, but it can also be due to the presence of mutations outside the sequenced regions (such as intronic or promoter mutations or large scale deletions). Another cause for a false negative test may be the result of non-allelic heterogeneity (identical phenotypes caused by mutations in different genes).

The differentiation of non-disease causing DNA polymorphisms from pathogenic mutations may not be straightforward. This is illustrated by the fact that changes in the DRD2 receptor were found in patients with myoclonus dystonia, but it is not yet resolved whether these DNA changes are indeed disease-causing.[21] Similarly, some alleles may represent mutations associated with reduced penetrance. Examples include HD and *SCA2* alleles with intermediate expansion, or Friedreich alleles associated with very late onset. Widespread application of sequence analysis to disease genes will undoubtedly result in the detection of sequence variants that at the time of detection may be of unknown biological significance.

ANIMAL MODELS

Genetic technologies have provided the tools to create animal models resembling human diseases, even if the respective diseases do not naturally occur in these animals. In principle, these technologies can be applied to any animal, but the mouse (*Mus musculus*), zebra fish (*Danio rerio*), fruit fly (*Drosophila melanogaster*), and nematode OK; more precise than 'worm'? (*Cenorhabditis elegans*) are most commonly used. For phenotypes involving cognitive dysfunction, we will increasingly see the use of the rat.[22] Two main approaches are used to alter the genetic constitution of animals. The first involves insertion of a novel, often mutated gene into the germline, the second an alteration of endogenous genes by gene targeting or random mutagenesis.

Expression of Foreign Transgenes

Transgenes can be injected into fertilised oocytes or introduced into embryonic stem cells. The expression of transgenes in mice does not necessarily mirror the expression of a mutated allele in human disease conditions. Firstly, the transgene is expressed in addition to the two endogenous mouse alleles, often from multiple copies of the transgene. Secondly, unless the endogenous promoter is used, expression may be higher than under physiological conditions, may occur in different sets of cell types, and may not have the correct temporal profile. Even if the physiological promoter is used, the transgene construct may lack all the appropriate regulatory elements. Finally, by chance the transgene can integrate into an endogenous mouse gene and disrupt proper functioning of this gene (so-called insertional mutagenesis).

Despite these shortcomings, expression of transgenes with mutated human cDNAs can replicate important morphological and functional aspects of human movement disorders as shown by animal models of polyglutamine diseases.[23,24] It is important to compare phenotypes obtained by expression of the wild type gene with those obtained by expression of a mutant gene to discern effects of the mutation *v* those of overexpression of a normal protein.

Gene Targeting (Knockout, Knockin)

Using homologous recombination, a cloned gene or gene segment can be exchanged for the endogenous gene.[25,26] The cloned gene can contain inactivating mutations (knockout) or missense mutations (knockin). In this fashion, heterozygous animals can be generated that when mated can generate homozygous offspring. Knockout animals are models for human recessive diseases. Knockin mice are potentially closer models of human disease than mice expressing a transgene, because the proper endogenous promoter is used, and gene dosage is not disturbed.[27,28]

A problem with homozygous knockouts for many genes is that embryonic lethality results, because the respective genes have a function not only in adult tissues but also during embryogenesis. This can be circumvented by generating chimeric animals that carry a mixture of wild type and homozygous deficient cells. For instance, by using the Cre-loxP recombination system, mice containing alleles with inserted loxP sites can be mated with mice expressing a Cre recombinase transgene under the control of a tissue specific promoter so that function of a gene is only abolished in a specific tissue.[29]

WEB-BASED INFORMATION FOR GENETIC DIAGNOSIS AND TESTING

A large amount of information is contained in publicly available databases. Box 2 lists examples of different kinds of databases that in turn provide links to other relevant sites. For the clinician, a classic resource has existed in the form of *Mendelian Inheritance in Man*.[30] It is now available as a web based version called *Online Mendelian Inheritance in Man* (OMIM). OMIM is easily searchable by disease name, symptom or sign, gene symbol, or name of a protein. Its advantages are that it is comprehensive and very well updated. Several editors regularly perform literature searches and add information to disease categories. This is also one of its disadvantages, in that information for some entries is lengthy and not revised in its entirety. Entries identified by searches of the PubMed database are often linked to corresponding entries in OMIM.

BOX 2 SITES ON THE INTERNET WITH MOLECULAR GENETIC DATABASES

- http://www.Geneclinics.org — Online textbook of selected diseases and information on commercial and research laboratories
- http://www3.ncbi.nlm.nih.gov/Omim/ — *Online Mendelian Inheritance in Man*
- http://www.gene.ucl.ac.uk/hugo — Human Genome Organization
- http://www4.ncbi.nlm.nih.gov/PubMed/medline.html — Online search of the medical literature
- Http://www.neuro.wustl.edu/neuromuscular — Clinical and research data on neuromuscular syndromes including ataxias

(*Continued*)

(*Continued*)

- http://www.uwcm.ac.uk/uwcm/mg/hgmd0.html — Human Gene Mutation Database
- http://hgbase.interactiva.de — Human gene polymorphisms
- http://www.botany.uwc.ac.za/mirrors/MIT-bio/bio/7001main — MIT biology hypertext book
- http://www.biology.arizona.edu — The Biology Project (site for basics in biology and genetics)
- http://www.orpha.net — European site in several languages for rare diseases and orphan drugs
- http://avery.rutgers.edu/index.php?topic=welcome — A well illustrated genetics tutorial

A site primarily dedicated to neuromuscular diseases and inherited ataxias is maintained by the neurology department at Washington University. This site has useful tabular information for classifications and diagnosis based on typical phenotypic features. The site is not primarily focused on genetic causes of neuromuscular disorders and ataxia syndromes, and this feature makes the site useful for the evaluation of patients without a clear family history.

The GeneTests web site (formerly called GeneClinics) is a site maintained by the University of Washington through funding from the National Institutes of Health (USA). It provides information on genetic diseases and associated tests including academic and commercial laboratories. It has four main sections: GeneReviews, a laboratory directory, a clinic directory, and a section with educational materials that includes a Microsoft PowerPoint presentation. GeneReviews is an on-line review of currently 195 genetic syndromes and diseases, many of which are neurological. Disease reviews in the site are well-organised, and updated on a frequent basis. It now provides several new features that facilitate the choice of a particular genetic test for the diagnosis of disease and links to genetic testing laboratories and specialty clinics for genetic diseases. Using the site search engine, a particular disease entry can be identified with links to "testing," "research," and "reviews." The diagnosis section has a clinical and a testing subsection. If applicable, the testing section provides a list of different molecular tests that can be used for the diagnosis of a particular disorder.

In the author's opinion, GeneTests is rapidly becoming the most relevant online resource for clinicians to choose a genetic test and to aid in the interpretation.

OUTLOOK

Neuroscience has undergone an inexorable transition toward a molecular discipline.[31] The availability of the human genome sequence will further accelerate this process. The conservation of many genes involved in specific cellular pathways throughout evolution will unite neuroscientists working in different model systems.[32] Similarly, many human disease genes are conserved in lower species.[33] This provides the opportunity to model the normal or mutant functions of these genes, making use of the unique advantages of each species. Genome comparisons will help to identify those DNA sequences that represent gene control regions as well as DNA domains important in alternative splicing, and will lead to novel ways to analyse development and differentiation of the brain. Comparisons of human and chimpanzee genomes, which differ by only 1%, will provide novel directions for cognitive neurosciences.

How will this impact be felt in the practice of neurology? Neuroscience and neurology have been transformed by the ability to chromosomally map and subsequently isolate disease genes. Both of these steps will be drastically accelerated. Once a gene has been mapped to a chromosomal region, the previously time-consuming tasks of identifying and isolating genes in the candidate region are now greatly shortened. Novel mapping techniques and large scale sequencing can now be applied to the study of complex diseases.

The collection of DNAs from participants in clinical trials will become commonplace with the goal of genetic stratification, identification of risk alleles, and pharmacogenomic analysis. Once drugs tested in trials are approved for clinical use, the genomic data obtained during the clinical trial will probably follow the drug into neurological practice and will aid in patient selection and drug choice. Preventive medication choices will be tailored to a disease risk profile specific for the individual patient. It is likely that the decision matrix will be computer-aided, but if counselling for Mendelian disorders is any guide, the physician–patient interaction will remain central to this decision-making process in the end.

Advances on the technological front will have to be paralleled by major educational efforts for physicians and patients with regard to the genome project and genetic testing. These tasks have traditionally been assumed by medical geneticists and genetic counsellors. Unless major changes in the number of these professionals are made in the immediate future, it is probable that neurologists will have to deliver the majority of neurogenetic care. Curricula for medical students and neurology residents will need to reflect these changes.

In addition to dealing with single causative alleles that greatly increase susceptibility to a disorder, the neurologist will need to interpret the effect of alleles that each may have only a relatively small effect on a phenotype and often act in concert with environmental stressors. Both patients and physicians will have to adjust to dealing with this changed scenario. To keep the physician–patient interaction free and unencumbered, privacy of genetic data has to be guaranteed at the local and national levels. The use of information obtained through DNA testing by other family members, employers, insurance companies, and governmental agencies requires societal discussion. The realisation that we all carry deleterious alleles should foster compassion, and knowledge of these alleles could ideally lead to enlightened choices for lifestyles and medical interventions.

ACKNOWLEDGEMENTS

This review is based in part on previously published material.[5,34] The author thanks Ms C Moran for assistance with references. Work for this review was supported by the Rose Moss Foundation, the Carmen and Louis Warschaw Endowment for Neurology, and F.R.I.E.N.D.s of Neurology.

REFERENCES

1. Jackson JF, Currier RD, Teresaki PI, *et al*. Spinocerebellar ataxia and HLA linkage: Risk prediction by HLA typing. *N Engl J Med* 1997;**296**:1138–1141.
2. Bird TD, Ott J, Giblett ER. Evidence for linkage of Charcot-Marie-Tooth neuropathy to Duffy locus on chromosome 1. *Am J Hum Genet* 1982;**34**:388–394.
3. Botstein D, White RL, Skolnick M, *et al*. Construction of a genetic linkage map in man using restriction fragment length polymorphisms. *Am J Hum Genet* 1980;**32**:314–331.

4. Gusella JF, Wexler NS, Conneally PM, *et al.* A polymorphic DNA marker genetically linked to Huntington's disease. *Nature* 1983;**306**:234–238.
5. Pulst SM. Introduction to medical genetics and methods of dna testing. In: Pulst SM, ed. *Genetics of movement disorders*. San Diego: Academic Press, 2003:1–18.
6. Ott J. *Analysis of human genetic linkage*. Baltimore: Johns Hopkins University Press, 3rd edition, 1999.
7. George V, Laud PW. A Bayesian approach to the transmission/disequilibrium test for binary traits. *Genet Epidemiol* 2002;**22**:41–51.
8. Bachman DL, Green RC, Benke KS, *et al.* Comparison of Alzheimer's disease risk factors in white and African American families. *Neurology* 2003;**60**: 1372–1374.
9. Pastor P, Ezquerra M, Tolosa E, *et al.* Further extension of the H1 haplotype associated with progressive supranuclear palsy. *Mov Disord* 2002;**17**:550–556.
10. Ozelius LJ, Hewett JW, Page CE, *et al.* The early-onset torsion dystonia gene (DYT1) encodes an ATP-binding protein. *Nat Genet* 1997;**17**:40–48.
11. Ausubel FM, Brent R, Kingston RE, *et al.* Short Protocols in Molecular Biology, John Wiley and Sons, 4th Edition,1999.
12. Strachan T, Read AP. Human molecular genetics Bios Scientific Publishers Ltd, Oxford, 1999.
13. Saiki RK, Scharf SJ, Faloona F, *et al.* Enzymatic amplification of beta-globin sequences and restriction site analysis for diagnosis of sickle cell anemia. *Science* 1985;**230**:1350–1354.
14. Mullis KB. The unusual origin of the polymerase chain reaction. *Sci Am* 1990;**262**:56–65.
15. Sanger F, Nicklen S, Coulson AR. DNA sequencing with chain-terminating inhibitors. *Proc Natl Acad Sci USA* 1977;**74**:5463–5467.
16. Cuda G, Mussari A, Concolino D, *et al.* Co-existence of frataxin and cardiac troponin T gene mutations a child with Friedreich Ataxia and familial hypertrophic cardiomyopathy. *Hum Mutat* 2002;**19**:309–310.
17. Nicholl DJ, Ferenci P, Polli C, *et al.* Wilson's disease presenting in a family with an apparent dominant history of tremor. *J Neurol Neurosurg Psychiatry* 2001;**70**:514–516.
18. Alonso I, Barros J, Tuna A, *et al.* Phenotypes of spinocerebellar ataxia type 6 and familial hemiplegic migraine caused by a unique CACNA1A missense mutation in patients from a large family. *Arch Neurol* 2003;**60**:610–614.
19. Jen J, Yue Q, Nelson SF, *et al.* A novel nonsense mutation in CACNA1A causes episodic ataxia and hemiplegia. *Neurology* 1999;**53**:34–37.
20. Hacia JG, Brody LD, Chee MS, *et al.* Detection of heterozygous mutations in BRCA1 using high density oligonucleotide arrays and two-color fluorescence analysis. *Nat Genet* 1996;**14**:441–447.

21. Klein C, Liu L, Doheny D, *et al*. Epsilon-sarcoglycan mutations found in combination with other dystonia gene mutations. *Ann Neurol* 2002;**52**:675–679.
22. Von Horsten S, Schmitt I, Nguyen HP, *et al*. Transgenic rat model of Huntington's disease. *Hum Mol Genet* 2003;**12**:617–624.
23. Burright EN, Clark HB, Servadio A, *et al*. SCA1 transgenic mice; a model for neurodegeneration caused by an expanded CAG trinucleotide repeat. *Cell* 1995;**82**:937–948.
24. Huynh DP, Figueroa K, Hoang N, *et al*. Nuclear localization or inclusion body formation of ataxin-2 are not necessary for SCA2 pathogenesis in mouse or human. *Nat Genet* 2000;**26**:44–50.
25. Thomas KR, Capecchi MR. Introduction of homologous DNA sequences into mammalian cells induces mutation in the cognate gene. *Nature* 1986;**324**:34–38.
26. Doetschman T, Gregg RG, Maeda N, *et al*. Targeted correction of a mutant HPRT gene in mouse embryonic stem cells. *Nature* 1987;**330**:576–578.
27. Miranda CJ, Santos MM, Ohshima K, *et al*. Frataxin knockin mouse. *FEBS Lett* 2002;**512**:291–297.
28. Wheeler VC, White JK, Gutekunst CA, *et al*. Long glutamine tracts cause nuclear localization of a novel form of huntingtin in medium spiny striatal neurons in HdhQ92 and HdhQ111 knock-in mice. *Hum Mol Genet* 2000;**9**:503–513.
29. Puccio H, Simon D, Crossee M, *et al*. Mouse models for Friedreich ataxia exhibit cardiomyopathy, sensory nerve defect and Fe-S enzyme deficiency followed by intramitochondrial iron deposits. *Nat Genet* 2001;**27**:181–186.
30. McKusick VA. *Mendelian inheritance in a man; catalogs of autosomal dominant, autosomal recessive, and x-linked phenotypes*. 1 ed. Baltimore: Johns Hopkins University Press, 1966.
31. Cowan WM, Kandel ER. Prospects for neurology and psychiatry. *JAMA* 2001;**285**:594–600.
32. Aravind L, Dixit VM, Koonin EV. Apoptotic molecular machinery: vastly increased complexity in vertebrates revealed by genome comparisons. *Science* 2001;**291**:1279–1284.
33. Satterfield TF, Jackson SM, Pallanck LJ. A drosophila homolog of the polyglutamine disease gene SCA: Is a dosage-sensitive regulator of actin filament formation. *Genetics* 2002;**162**:1687–1702.
34. Pulst SM. Genomes, neuroscience, and neurology. *Arch Neurol* 2001;**58**: 1755–1757.
35. Bennett RL, Steinhaus KA, Uhrich SB, *et al*. Recommendations for standardized human pedigree nomenclature. Pedigree standardization task force of the national society of genetic counselors. *Am J Hum Genet* 1995;**56**:745–752.

2

Neurogenetics II: Complex Disorders

*A F Wright**

The genetic analysis of common neurological disorders will be a difficult and protracted endeavour. Genetics is only one of many disciplines that will be required, but it has already thrown considerable light on the aetiology of several major neurological disorders through the analysis of rare inherited subgroups. Identification of individual susceptibility genes with variants of smaller effect will be a hard and difficult road, but there is no sharp demarcation between large and small genetic effects, so that many new and important insights will emerge using existing and new technologies. The availability of improved neuroimaging, better animal models of disease, and new genetic tools, such as high-throughput gene chips, expression microarrays, and proteomics, are extending the range of traditional genetic mapping tools. Finally, an understanding of the genetic and epigenetic mechanisms that restrain the differentiation and integration of human neural stem cells into mature neuronal networks could have an enormous impact on clinical practice. These approaches will be illustrated in the

*MRC Human Genetics Unit, Western General Hospital, Crewe Road, Edinburgh EH4 2XU, UK; alan.wright@hgu.mrc.ac.uk

context of Alzheimer disease, Parkinson disease, and synucleinopathies, tauopathies, amyotrophic lateral sclerosis, and stroke.

The sequencing of the human genome signalled a major shift in the Human Genome Project from gene discovery in monogenic disorders towards the "post genome challenge" of gene characterisation and the genetic analysis of complex disorders. This change was largely driven by the increasing facility of gene identification, which led to the identification of >1200 predominantly Mendelian disease genes. An important conceptual development was the common disease/common variant (CD/CV) hypothesis as a model for complex disorders[1,2] (see Appendix A, Human genetic variation). This model proposed that the genetic basis of common, genetically complex disorders is principally due to genetic variants that are common in the population. In contrast, the common disease/rare variant (CD/RV) model argued that in complex disorders there is a significant contribution from rare variants, which include most of those with the most significant individual effects.[3,4] Although the debate continues, the heritability of a complex trait almost certainly results from both common and rare variants. One estimate, based on more than two decades of research on such traits in the experimentally tractable organism *Drosophila melanogaster*, suggested that between one third and two thirds of the typical variation in a complex trait, with at least some effect on reproductive fitness, results from rare variants with adverse effects.[4] The remainder is due to common variants, many of them with opposite effects on different traits (some beneficial, others detrimental) allowing them to be maintained in the population. The motivation for finding common variants is currently greater than for finding rare variants, for three main reasons. First, they provide potential mechanistic insights; second, they are easier to identify than rare variants; third, and most important, they may be of public health importance and allow identification of subpopulations at increased risk of disease.[5]

The success in finding both common and rare genetic variants influencing susceptibility to Alzheimer disease (see below) shows that both CD/CV and CD/RV models are "correct", but it is a matter of debate as to which will provide the most useful insights. This is perhaps the biggest issue at stake, since the majority of complex traits are polygenic — resulting from the combined action of many different genes, in combination with often proportionately greater environmental effects. In addition, recent evidence suggests that interaction effects — gene-gene and gene-environment — are common, even in experimental organisms where genotype and environment are well controlled.[6]

The methods currently being used to unravel the genetics of common neurological disorders, such as Alzheimer disease and stroke, are essentially the same as those used in the early phase of the Human Genome Project, namely low resolution genetic mapping by linkage analysis in families with multiple affected individuals, followed by high resolution mapping using case-control association studies. However, increasing emphasis is being placed on the latter, fuelled by technological advances using single nucleotide polymorphism (SNP) chips (see below). However, the large scale use of candidate gene association studies has led to a serious problem, with many unreplicated and, in many cases, spurious associations being published. As an example, out of 127 candidate gene associations with Alzheimer disease reported in a single year, only three were found to have been replicated in three or more independent studies.[8]

A number of principles which have emerged to guide researchers through the maze of complex genetic disorders are discussed below.

GUIDING PRINCIPLES

Large Sample Sizes

Most individual genetic effects on complex traits or diseases are small, emphasising the need for large sample sizes to reliably detect them.[9] Very few genes are capable of exerting large effects, but many genes can exert small marginal effects. A widely accepted model for the distribution of effect sizes of genetic variants influencing complex traits is an L shaped distribution — many genes with variants showing small and peripheral effects on disease (both rare and common) and a smaller number with variants showing moderate to large effects (which tend to be rare).[4] The effect of individual variants will therefore often be obscured by those of other genes and by large environmental and interaction effects.

Quantitatively Varying Intermediate Disease Endpoints

Quantitative traits (QTs) which influence disease risk are used whenever possible to increase study power. In a recent review, it was commented that "studies using a single clinical endpoint are akin to a shot at the moon", and compare unfavourably with studies focusing on genetically and physiologically simpler intermediate traits.[5] All individuals with QT information are informative in genetic mapping studies, in contrast to studies focusing on disease, where most of the power comes from the comparatively few affected individuals. It has been difficult to find useful QTs in neurological disorders, compared with cardiovascular or metabolic diseases. The use of disease age of onset or severity, plasma amyloid β_{42} in Alzheimer disease, well validated questionnaires, and structural brain imaging may facilitate this process.

Ascertainment Strategies

It is relatively easy to study "typical" patients with disease, but other ascertainment schemes are more powerful. Families of individuals with complex disorders do not generally have multiple affected members, since the incidence in relatives declines exponentially with decreasing relationship to the proband, as expected under a polygenic model. The identification of individuals at the extremes of the QT distribution is helpful in contributing to study power. Extreme individuals may show large genetic effects, without necessarily developing overt disease (for example, because they lack other risk factors). Screening of large samples may therefore be required to detect such extreme individuals. For example, a study of personality traits targeted 88 000 individuals to fill in a postal questionnaire, which identified over 34 000 sib pairs including many with extreme or contrasting trait values. A genetic linkage analysis of extreme or discordant sib pairs led to the identification of several significant linkage peaks.[10] Similarly, the ascertainment of rare individuals with early onset Parkinson disease was necessary for the identification of a major gene (*DJ-1*) causing this disorder.[11]

Genetic Linkage and Case-Control Association Designs

These two methods form the core of the genetic mapping effort. Linkage analysis is carried out using extended or small nuclear families (for example, affected sib pairs) (Fig. 1). The term "genetic linkage" refers to the finding of an association between disease and genetic marker within a series of families, each containing affected individuals, after carrying out a

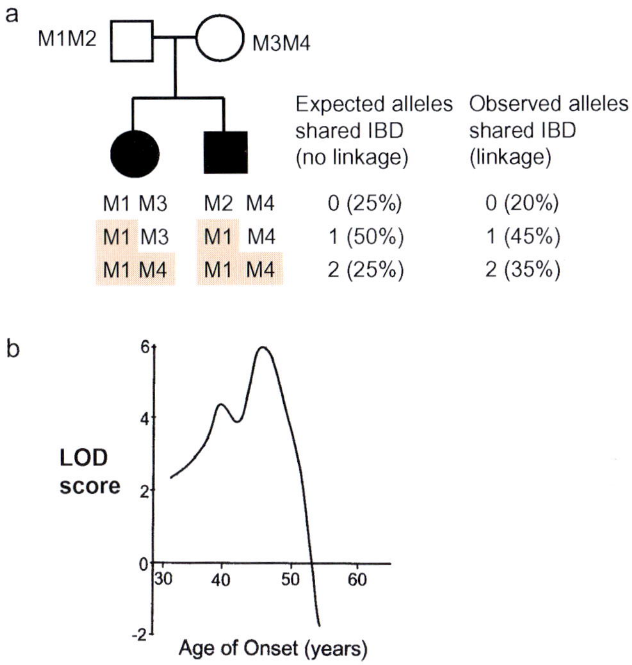

Fig. 1 The principles underlying genetic linkage analysis of a common neurological disorder. (a) Affected sib pairs (ASPs) are first genotyped for several hundred genetic markers. For late onset disorders the parents are usually unavailable, but allele sharing between ASPs can still be inferred. In the absence of linkage, the extent of genetic marker allele (M1–4) sharing between ASPs is zero, one, or two alleles shared (identical by descent or IBD) as a result of common ancestry. This is expected to occur 25% (no sharing), 50% (one allele shared), or 25% (both alleles shared) of the time, under the null hypothesis of no linkage (shared alleles are shaded in grey). In the presence of linkage, there is an increase in allele sharing over the null hypothesis, as shown. This method is robust when the precise mode of inheritance is unknown. Large numbers of ASPs (for example, 500–1000) are often required to accumulate significant evidence of linkage to a common disorder. Affected individuals are shown as filled boxes (males) or circles (females) and unaffected individuals are unfilled. (b) Linkage to a late onset disorder may vary according to age of onset. These LOD score data are from Hall *et al.*[62] who reported stronger evidence for linkage to the *BRCA1* gene on chromosome 17 in early compared with late onset familial breast cancer families. An LOD score just over 3 is significant for a complex disorder. The cumulative LOD scores are shown for all families in which the mean age of onset is less than or equal to the age shown on the x axis.

whole genome scan. The latter involves genotyping many "genetic markers" — variant sites unrelated to gene function which show common variation in the general population — situated at regular intervals throughout the genome (see Appendix B, Genetic linkage and association analyses). If successful, genetic linkage can identify a large genomic region, often containing several hundred genes, in which the disease gene is sought. Linkage disequilibrium (LD) implies a failure of markers that are close to one another to recombine freely, even over long periods of time, and hence these markers are associated with each other in the population. The presence of LD between SNP markers makes it possible to infer the location of a disease gene that is in LD with a genotyped SNP. Fine mapping is carried out using the more familiar case-control association study design (Fig. 2) in which excess marker sharing is sought within cases compared to controls, following a more dense marker genotyping effort within the identified region. In fine mapping a broad region of genetic linkage, often containing about 100 genes, is narrowed by carrying out dense SNP marker genotyping across the region in cases and controls. This identifies small shared ancestral regions that are associated either with cases or controls. Since the common ancestor is remote, genomic regions that are shared IBD (shown in black in Fig. 2) become progressively smaller over successive generations as a result of recombination. The number of genes in the identified region of association now contains a finite number of candidate genes which can be analysed for sequence variation.

Choice of Study Population

Modern urban populations are often extremely diverse and are far from ideal for gene mapping studies because of genetic heterogeneity.[12] However, there is a trade off between obtaining large well characterised study cohorts, which are generally available in urban contexts, and smaller but more homogeneous cohorts from less diverse population groups. The Icelandic population was chosen to study complex diseases to minimise both genetic and environmental heterogeneity, which led to the discovery of several susceptibility genes, including the *PDE4D* gene in stroke (see below).[13]

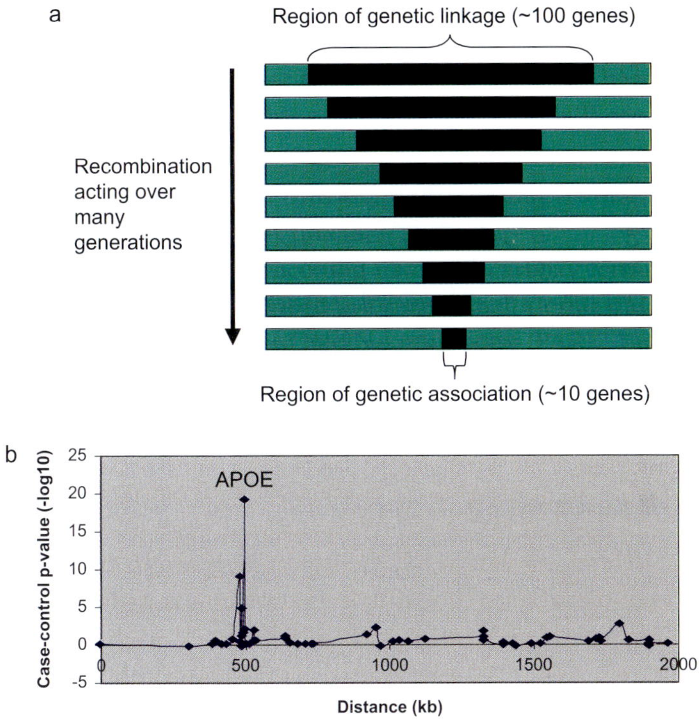

Fig. 2 (a) Genetic fine mapping using association studies. Genomic regions shared IBD are shown in black. (b) Allelic association between Alzheimer disease and SNPs across the region flanking the *APOE* gene on chromosome 19.[63] Distance in kilobases (kb) is from *APOE*. As cases are only very distantly related, the region of shared genome is greatest within ~40 kb of *APOE* and falls off steeply beyond this, impling a very dense genome scan would have been required to identify *APOE*. The association was originally identified by genetic linkage to a broad region on chromosome 19q followed by candidate gene association studies. (c) Linkage disequilibrium (LD) map of the *DPP10* gene associated with asthma[64] showing regions that are associated or in LD with each other as a result of non-random association between pairs of markers. The chromosomal region runs from left to right on the x axis at the bottom of the figure. The strength of association to asthma (red) and the QT immunoglobulin E levels ($\log_e$ IgE) (yellow) is plotted as −log(P) against position. The markers showing strongest association correspond to the highest peaks. The graph is superimposed on the distribution of LD between markers (measured as D′), which are colour coded and plotted at the marker locations with red (high LD) and dark blue (low LD) at opposite ends of the scale. The four initial exons of the causal *DPP10* gene are shown as white bars.

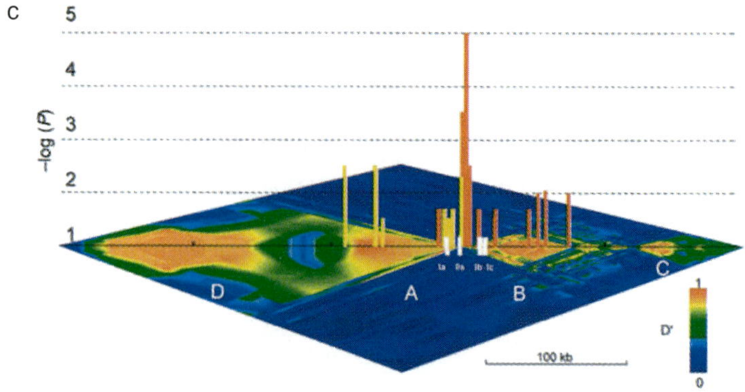

Fig. 2 (*Continued*)

Choice of Research Strategy

The research strategy should be specifically designed to answer the question posed. If the aim is to identify common variants with predominantly small genetic effects on a categorical endpoint, such as disease, a broadly based candidate gene screening approach may be appropriate, using common genetic variants (SNPs) (see Appendix A, Human genetic variation) and a hierarchical case-control strategy. For example, a moderate number of cases (for example, $n = 500$) and matched controls could be systematically screened for association between disease and candidate gene SNPs (at an appropriate density per gene). Candidate genes could be selected on the basis of (a) expression within a tissue of interest (for example, hippocampus or substantia nigra), (b) functional criteria, such as membership of a known disease pathway, or (c) localisation to an implicated chromosomal region, on the basis of previous genetic linkage studies. Positive associations could then be followed up using an independent and preferably larger cohort, to eliminate false positives. Alternatively, if the goal is to identify rarer genetic variants of intermediate effect, the strategy could be quite different. A genetic linkage analysis using a large set of families segregating for a QT or disease would be an appropriate initial strategy, as used to identify the chromosomal locations of the *APOE*, αT-catenin, and *GST01* genes (see below). Fine mapping could follow using a case-control association study, and a dense set of SNPs confined to the implicated region(s) (Fig. 2).

Nature of Disease Susceptibility Variants

Susceptibility genes in complex diseases are often expressed in a wide range of tissues and may contain only subtle variants or combinations of variants, some or all of which lie outside protein coding sites. This makes identification of susceptibility variants difficult. Overall, about 5% of the human genome has functional significance and so is potentially involved in disease.[14] About 1.5% of the genome contains the protein or RNA coding regions of the 20 000–30 000 human genes, in which lie an estimated 20 000 coding or cSNPs.[5] These represent an important initial target for whole genome association studies. Firstly, they are more likely to influence disease than non-coding SNPs and, secondly, a genome scan could be carried out using substantially fewer markers than the estimated 600 000–1 000 000 non-coding SNPs required to provide coverage of the entire genome.[5] A further 1% of the human genome lies within genes and is transcribed but is not translated into protein. Finally, an additional 2.5% of the genome lies outside of the genes altogether but is conserved across species, suggesting that these regions also have functional importance. Proving that such subtle non-coding variants influence a complex disease is difficult. In monogenic disorders, the situation is quite different, with 99% of mutations occurring in protein coding or splice sites, and only 1% within non-coding regulatory regions.[15] The best evidence that a gene influences disease susceptibility comes from the identification of several different genetic variants within its coding or splice sites in different affected (or extreme QT value) individuals, coupled with the demonstration that variants affect gene function and show relevant tissue expression.

APPLICATIONS TO CLINICAL NEUROLOGY

Alzheimer Disease

Alzheimer disease provides an excellent paradigm for the genetic basis of a complex disorder, with contributions from both common modifier genes and rare variants of large effect.[16] Heritability estimates in Alzheimer disease are in the region of 60%,[17] suggesting that genetic variation plays a significant role in the disease process. However, the major insights into disease mechanisms to date have come from mutations in genes that are so rare that they make essentially no contribution to the heritability of the disease as a whole.

One of the best paradigms for the CD/CV hypothesis was the discovery of common variants in the *APOE* gene which influence susceptibility to Alzheimer disease. There are three common *APOE* alleles (E2, E3, E4) in human populations, resulting from differences at two amino acid residues (residues 112, 158).[18] Associations between the E4 allele, which is present in about one third of Caucasians, and Alzheimer disease have been widely confirmed, but associations have also been found with several other disorders — the Lewy body variant of Alzheimer disease, Parkinson disease, susceptibility to herpes simplex virus infection, poor recovery from head injury, intracerebral haemorrhage, and elective cardiac bypass surgery.[19] A protective effect of the E2 allele in Alzheimer disease has also been reported. APOE is the primary cholesterol transporter in the brain and is a component of both amyloid (senile) plaques and neurofibrillary tangles. The mechanism for the effects of APOE isoforms on brain damage and dementia is unclear, although transgenic ApoE deficient mice ($Apoe^{-/-}$) engineered to express a human *APOE* E4 allele showed age related spatial learning and memory defects, in contrast to $Apoe^{-/-}$ controls or mice carrying the E3 allele.[20] Lipid carrying apoE3 binds amyloid β ($A\beta$) peptide, the major constituent of amyloid plaques, with 20-fold higher affinity than lipidated apoE4, which may enhance the clearance of $A\beta$.[21] The close relationship between APOE and Alzheimer disease risk is highlighted by the finding that transgenic mice overexpressing familial Alzheimer disease mutations on an $Apoe^{-/-}$ null background show very little $A\beta$ amyloid deposition, compared with those on a normal (wildtype) $Apoe^{+/+}$ background.[22] This suggests that *APOE* is essential for $A\beta$ deposition in transgenic models of familial Alzheimer disease (FAD). It remains unclear whether this effect is mediated by increased formation or decreased clearance of $A\beta$ amyloid.

The effect of the *APOE* E4 allele is dosage dependent, so that carriers of a single E4 copy have a twofold increased risk of Alzheimer disease compared with a fivefold risk for homozygotes with two copies. The E4 allele appears to be a disease modifier, exerting its effect on disease risk by influencing age of onset in both Alzheimer disease and Parkinson disease, rather than disease risk per se. Despite the relatively large effects of these variants, the use of *APOE* genotype information in disease prediction remains limited, since its diagnostic sensitivity is only 0.65 and specificity

0.68, compared with clinical diagnosis, which has a reported sensitivity of 0.93 and specificity of 0.55.[23]

A number of Alzheimer disease modifier loci have recently been proposed, none of which have yet been consistently replicated, but they illustrate some of the approaches taken and difficulties encountered. The glutathione-S-transferase (*GST01*) gene was proposed to be a determinant of age of onset, here used as a QT, in both Alzheimer disease and Parkinson disease.[24] GST01 is widely expressed and is thought to be concerned with the biotransformation of compounds such as free radicals and interleukin-1β. The gene was identified by narrowing the number of genes in the large region of chromosome 10 implicated by linkage analysis from several hundred genes to only four, on the basis that only these genes showed altered expression in the hippocampus of Alzheimer disease compared with control subjects. This is an interesting but potentially misleading assumption. Using a case-control strategy, and large sample sizes, the authors found a significant association with one of the three genes, *GST01*.[24] One of the common variants analysed, SNP7, was associated with the substitution of aspartic acid for alanine at residue 140 (Ala140Asp) in the *GST01* product. However, since about 90% of the population carry one or two copies of this early onset "risk" allele (Ala140), it remains unclear how much of the original linkage signal is explained by this (and the associated SNP9) variant, or how useful the resultant mechanistic insights will be.

The identification of another proposed genetic modifier in Alzheimer disease followed the discovery of an association between the insulin degrading enzyme (*IDE*) gene and Alzheimer disease itself,[25,26] age at onset in both Alzheimer and Parkinson disease,[27] and plasma amyloid $A\beta_{42}$, sometimes used as a QT risk factor for Alzheimer disease.[28–30] The $A\beta_{42}$ peptide is a secreted cleavage product of the amyloid β protein precursor (APP), which is strongly expressed in brain and cerebral spinal fluid (CSF). $A\beta_{42}$ is present in CSF at 50 times its concentration in plasma, but, in a longitudinal study, individuals who developed Alzheimer disease showed higher levels of plasma $A\beta_{42}$, suggesting its use as a surrogate for brain $A\beta_{42}$ production. Plasma $A\beta_{42}$ is elevated in individuals with familial late onset Alzheimer disease, in early onset FAD, and in Down syndrome (since the *APP* gene is carried on chromosome 21). It remains unclear which variants in or close to the *IDE* gene are directly concerned with Alzheimer disease risk, age of onset, and plasma $A\beta_{42}$ levels. *IDE* is an interesting candidate gene since it has been shown to regulate $A\beta_{42}$ levels in brain neurons and microglial cells.[29,30] Increased degradation of $A\beta_{42}$ by transgenic mice overexpressing IDE or another $A\beta$-degrading protease, neprilysin, slows $A\beta_{42}$ deposition and reduces Alzheimer-like pathology in mouse models of FAD.[31]

The most significant advances in the genetics of Alzheimer disease and Parkinson disease to date have come not from the identification of the common variants discussed above, but from the study of genes which have virtually no role in common forms of these disorders. Mutations in three genes account for about half of all cases of FAD,[32] which is an extremely rare disease, with fewer than 200 confirmed FAD families worldwide, compared with an estimated 4–5 million Alzheimer disease individuals in the USA alone.[33] FAD is clinically and pathologically indistinguishable from Alzheimer disease except for age of onset. The most common cause is a mutation in the presenilin-1 (*PS1*) gene, which is found in about half of all FAD families. Mutations in the related presenilin-2 (*PS2*) gene and in the *APP* gene account for <1% and <5% of FAD families, respectively.[32] Mutations in all three genes give rise to increased $A\beta_{42}$ formation since the presenilins form part of a protein complex concerned with the processing and release of the neurotoxic $A\beta_{42}$ peptide from APP.[16] Mutations in the *APP* and *PS1* genes give rise to a fully penetrant autosomal dominant disorder with onset in the age range 35–55 years, while *PS2* mutations are more variable, often showing later onset (age range 40–85 years) and occasional non-penetrance.

The importance of these rare mutations lies in the identification of a pathogenetic pathway, involving the endoproteolytic cleavage of the transmembrane APP protein by the enzymes BACE1 and the γ-secretase complex.[16] The common factor in Alzheimer disease arising from Down syndrome and mutations in the *APP*, *PS1*, and *PS2* genes is an excess production of the neurotoxic $A\beta_{42}$ peptide or an increased ratio of $A\beta_{42}$ to the less toxic $A\beta_{40}$ peptide. Paradoxically, the pathogenetic sequence in the transition from old age through mild cognitive impairment to Alzheimer disease emphases the role of neurofibrillary degeneration (NFD), associated with paired helical filament (PHF)-tau deposition, rather than amyloid plaque formation.[34,35] Amyloid deposits are deposited randomly throughout the entire cerebral cortex, and tend to appear subsequent to NFD and PHF-tau deposits in any one region. NFD progresses hierarchically along specific neuronal pathways (starting in the trans-entorhinal cortex and progressing to the temporal cortex), suggesting a specific vulnerability in these pathways. It has been suggested that this vulnerability may be enhanced in the presence of increased $A\beta_{42}$ formation, which can result from genetic mutations or environmental events such as head injury or stroke. There is an apparent progression in the extent of both NFD and amyloid deposits from normal ageing to Alzheimer disease. For example, in one study, 100% of individuals over age 75 showed NFD in the hippocampus, often in the absence of amyloid plaques or dementia, whereas those with Alzheimer disease (by definition) also have both significant neuronal loss and amyloid plaques.[34]

TAUOPATHIES

The discovery of mutations in the *Tau* gene in a subset of patients with fronto-temporal lobe dementia (FTD) linked to chromosome 17 (FTDP-17) throws further light on Alzheimer disease mechanisms.[36] FTD is an early onset (<65 years) disorder associated with prominent frontal lobe symptoms, such as behavioural disinhibition, with fronto-temporal atrophy due to neuronal loss, spongiform degeneration, and gliosis, sometimes extending to the substantia nigra (SN), amygdala, and spinal cord. Clinical presentation can be accordingly varied. There are no amyloid or Lewy bodies and a small proportion of patients have *Tau* gene mutations.[37] Tau is a phosphoprotein expressed in peripheral and central nervous systems, predominantly in neurons, where it is associated with axons and concerned with the microtubule binding and assembly that is necessary for axoplasmic transport.[37] Hyperphosphorylated Tau deposits are associated with PHF and the NFD found in Alzheimer disease. In FTDP-17, both loss of function mutations and mis-expression of the *Tau* gene, which is normally processed into different isoforms, are found. The precise disease sequence and mechanism remains unclear, but amyloid $A\beta_{42}$ overexpression appears to exacerbate Tau pathology. One possibility is that APP mis-processing in Alzheimer disease leads to post-translational modification of the Tau protein and subsequent neurodegeneration. The observation that amyloid deposition follows rather than precedes Tau mis-processing could however also be explained by the proposal that $A\beta_{42}$ neurotoxicity results from formation of the more toxic soluble protofibrils rather than the later appearing insoluble fibrillar aggregates.[38]

Parkinson Disease and Synucleinopathies

The presence of neuronal loss and insoluble aggregates of α-synuclein, called Lewy bodies, in the SN are the major pathological features of Parkinson disease.[39] Surprisingly, the prevalence of SN Lewy bodies in the general population is ten times greater than the prevalence of Parkinson disease, but there appears to be a threshold, so that those with SN neuronal loss exceeding about 60% show symptoms of Parkinson disease. This may be because in disorders of protein aggregation, the characteristic aggregates are actually protective but when present in large numbers are indicative of a more sinister underlying process or extent of disease. Post mortem studies show that SN cell loss in the normal population follows an exponential distribution, with 4.4% of cells lost per decade.[40] In contrast, cell loss in Parkinson disease appears to occur ten times faster, at a rate of 45% per decade, with onset about 4–5 years before symptomatic disease.[40] Lewy bodies are also a prominent feature in other neurological disorders — dementia with Lewy bodies, multiple system atrophy, Down syndrome, and neurodegeneration with brain iron accumulation I.[41] Ten genes have been mapped by genetic linkage to rare monogenic forms of familial Parkinson disease (FPD), four of which have been isolated: the α-synuclein (*SNCA*), ubiquitin C-terminal hydrolase like 1 (*UCH-L1*), parkin (*PRKN*), and *DJ-1* genes.[42] These have again provided mechanistic insights into common forms of Parkinson disease. Firstly, mutations in the α-synuclein gene result in early onset autosomal dominant FPD.[43] Autosomal dominant FPD families showing triplication or duplication of the *SNCA* gene present FPD symptoms in the fourth and fifth decades respectively, implying that overexpression even of normal α-synuclein is sufficient to cause disease. Genetic variability in the *SNCA* promoter region was associated with increased risk of sporadic Parkinson disease. This is consistent with the possibility that, like overexpression of $A\beta_{42}$ in Alzheimer disease and Down syndrome, increased formation of normal α-synuclein can be disease causing.

Mutation in the *PRKN* gene causes juvenile or early adult (<45 years) onset autosomal recessive PD.[44] Complete loss of parkin due to homozygous deletion of the *PRKN* gene is associated with severe loss of dopaminergic neurons in the SN and locus coeruleus but a notable absence of Lewy bodies. Some amino acid changing (missense) mutations in *PRKN* do show both Lewy bodies and abnormal tau deposits (NFD), suggesting a possible gain of function. One explanation is that since parkin is an E3 ubiquitin ligase, it is a component of the ubiquitin proteasome system, which may be required to produce Lewy bodies. The ubiquitin protease system is involved with the degradation of misfolded proteins, some of which — such as α-synuclein and perhaps some types of mutant parkin itself — can give rise to aggregation and neurodegeneration. The importance of this pathway is reinforced by the finding of mutations in the *UCHL1* gene, coding for ubiquitin carboxy-terminal hydrolase L1, one of the most abundant proteins in the brain, in rare autosomal dominant FPD families.[45] The UCHL1 enzyme is found in Lewy bodies and is also concerned with protein degradation. *UCHL1* mutations lead to accumulation of α-synuclein in cells and may influence susceptibility to Parkinson disease by altering the balance of ubiquitin hydrolase and ligase activities, both of which are present in UCHL1, impairing the degradation of α-synuclein.[46]

DJ-1 is another component of the ubiquitin/proteasome protein degradation pathway which is mutated in a rare autosomal recessive form of early onset Parkinson disease.[11] The gene was identified by genetic linkage analysis in a large inbred Dutch community in which the mutant gene appeared to be more common as a result of a founder effect and cultural isolation of this population. Since both Parkin and DJ-1 are components of the ubiquitin proteasome pathway, and are concerned with the degradation of fibrillogenic proteins within the SN, these rare genes have again identified an important pathogenetic pathway in all forms of Parkinson disease, despite making essentially no contribution to heritability in the common form.

Amyotrophic Lateral Sclerosis (ALS)

ALS is a progressive disease associated with degeneration of motor neurons in the brain stem and spinal cord. Surviving neurons contain inclusions of neurofilament components and ubiquitin. It is generally sporadic but rare familial forms of ALS occur in about 10% of patients, about 20% of which are associated with missense mutations in the cytoplasmic enzyme Cu/Zn superoxide dismutase 1 (SOD1), which is also present in the inclusions.[47,48]

It is unclear whether the disease results from a gain of function, such as protofibril toxicity, or loss of function and oxidative stress. SOD1 catalyses the dismutation of the superoxide radical to form hydrogen peroxide and oxygen. One possibility is that an oxidising environment (due to reduced SOD1 activity) causes protein instability, aggregation, and neurotoxicity, since mutant SOD1 aggregates have been seen under such conditions.

Cerebrovascular Disease and Stroke

Stroke is a heterogeneous group of ischaemic and, less commonly, haemorrhagic disorders, which are associated with atherosclerosis of large blood vessels or occlusion of small penetrating arteries in the brain. All forms of stroke share common risk factors, including hypertension, hyperlipidaemia, diabetes, and smoking. Family history is an independent risk factor, suggesting that genetic factors may contribute to susceptibility.[49] Genetic linkage analysis of Icelandic families segregating for stroke provided the initial evidence for a susceptibility gene on chromosome 5. Fine mapping was carried out in a case-control study of 864 affected individuals from the Icelandic population and 908 controls, using 98 markers spanning the implicated chromosomal region. A broad definition of stroke was employed, including both cardiogenic and carotid stroke, and common variants within the phosphodiesterase 4D (*PDE4D*) gene were found to be associated.[13] The highest risk haplotype (present in 9% of controls) conferred a twofold relative risk. A protective haplotype (present in 21% of controls) was also identified, with a relative risk of 0.7. However, none of the associated variants were present in protein coding or gene splicing regions, suggesting that the identified and/or associated variants affect gene regulation (such as expression level) rather than having a direct functional effect on the protein. Some protein isoforms associated with the risk haplotype may be expressed at a lower level in patients than in controls. The *PDE4D* risk haplotype has an effect that is largely independent of known risk factors. The *PDE4D* gene encodes a cyclic nucleotide phosphodiesterase which degrades cyclic AMP and regulates signal transduction in a wide variety of cells. One possibility is that *PDE4D* variants cause low cyclic AMP levels, increasing the tendency for proliferation and migration of vascular smooth muscle cells, although similar effects in the immune system are also possible. These findings and their pathogenic significance remain to be confirmed and elucidated.

A similar approach led to the identification of another gene, *ALOX5AP*, coding for 5-lipoxygenase activating protein, in which certain common haplotypes double the risk of both stroke and myocardial infarction.[50] The initial finding was a suggestive linkage to a region of chromosome 13 in a series of 296 Icelandic families with multiple affected members. A case-control association study was carried out using a high density of markers across the implicated region (containing 40 known genes) which led to the identification of the *ALOX5AP* susceptibility gene. This was confirmed in a UK population, although the associated haplotype was different. The individual or combination of variants associated with disease risk remain to be identified. ALOX5AP and 5-lipoxygenase together convert unesterified arachidonic acid to the leukotriene LTA4, which is further converted to LTB4 or LTC4.[50] These are important proinflammatory mediators which are active in macrophages and leukocytes invading atherosclerotic lesions.

NEW TECHNOLOGIES

Increasing access to powerful new technologies will facilitate the discovery of genetic influences in neurological disorders. Perhaps the most important ones are those concerned with refining the clinical phenotype, such as brain imaging techniques, and developing quantitative intermediate disease endpoints. The goal of reliably defining simpler phenotypes than disease itself, such as carotid intima media thickness, instead of more complex and categorical traits such as stroke, is particularly important. Other enabling technologies are allowing high throughput analysis of genes and their products in health and disease, which is beginning to influence neurological research. The new technologies are discussed below.

Microarrays

High density arrays of DNA sequences, such as SNP alleles or expressed gene sequences (cDNA), can be immobilised on miniaturised grids (chips), in order to perform large scale screening experiments.[51] For example, the messenger RNA (mRNA) from both normal and diseased neurological tissues can be extracted, converted to DNA (cDNA) and labelled prior to hybridisation to the chip, in order to identify genes that are differentially expressed in disease. Alternatively, genomic DNA from an individual could be labelled and hybridised to an SNP chip containing tens or hundreds of thousands of SNP variants, to search for a disease association. Finally, if a candidate gene for a disease has been mapped to a specific genomic region containing a few hundred genes, it may be useful to know which genes from that region are expressed in the diseased region using microarrays.

This technology has been used to investigate neurological disorders.[52] In one study, cDNA microarrays containing 18 000 genes were hybridised to cDNA from hippocampal CA1 neurons with or without neurofibrillary tangles in Alzheimer and control brains.[53] Similarly, prefrontal cortex from schizophrenic versus control brains was screened using arrays containing 7000 genes to detect differences in gene expression, which showed decreased expression of genes regulating presynaptic function.[54] It is important to confirm changes in gene expression shown by microarray using other methods, such as immunohistochemistry, *in situ* hybridisation, or reverse transcription polymerase chain reaction. A final example is the use of microarrays in the transcriptional analysis of brain plaques from multiple sclerosis (MS) samples compared with control brain samples.[55] This type of study identified osteopontin (*OPN*) gene expression exclusively in MS plaques, which led to the proposal that this proinflammatory molecule is expressed by infiltrating T lymphocytes, microglia, and macrophages, and promotes damage to the myelin sheath as a result of an autoimmune process. Polymorphisms in OPN also appear to influence the disease course.[56,57]

Proteomics

Gene expression profiles provide little information on genetic variation and may give misleading information on the function or expression of their protein products. The proteome, which is the sum of all expressed proteins in a tissue or cell, is regulated at different levels, including synthesis, degradation, and a wide variety of post-translational modifications, such as phosphorylation. The abundance of the mRNA coding for a specific protein may be poorly correlated with protein abundance. However, the variety and different physico-chemical properties of proteins complicates the "protein chip" approach, although the entire yeast proteome has now been arrayed on a chip. Instead, the techniques of two-dimensional gel electrophoresis, in-gel digestion, and peptide identification by microsequencing or mass spectrometry, are together enabling the high throughput analysis and identification of unknown proteins dissected from healthy or diseased tissues. Two-dimensional gel electrophoresis allows the separation of several hundred proteins by molecular size and net charge while techniques such as MALDI-TOF or tandem (q-TOF) mass spectrometry facilitate their identification.[58] For example, over 300 proteins were identified from subcellular fractions of human frontal cortex using such an approach.[58] Current limitations include the difficulty of analysing hydrophobic proteins, such as membrane receptors, and the identification of post-translational modifications in a high throughput manner. These techniques however have the potential for refining the analysis of cells and tissues in neurological disorders. Firstly, they can provide critical information on the structure and function of specific proteins, such as disease related post-translational modifications. Secondly, they can provide an overview of the collective changes occurring within a brain region which can help to subdivide and refine molecular subtypes of disease.

Neural Stem Cells

There is considerable interest in the possibility of inducing resident human neural stem cells, that are known to be present in the subependymal zone and hippocampus, to differentiate into and replace neurons damaged by ischaemia, trauma, or neurodegeneration.[59–61] This property is retained in the brains of some simpler non-mammalian vertebrates but appears to have been progressively lost with the evolution of increasing brain complexity from amphibians through to rodents and primates. The precise number of human neural stem cells is unknown but $<1\%$ of human subependymal cells display the Ki-67 marker that is associated with a capacity for cell division.[60] In human bone marrow, only about 1 in 10^6 cells show the properties of haematopoietic stem cells. Human neural stem cells display glial astrocyte but not neuronal markers, although they are able to generate both neuronal and glial cells in culture. It therefore appears that there is an inherent resistance of such cells to undergo neurogenesis *in vivo*, perhaps because of the need to retain the complex neuronal networks built up by experience and learning. The goal of replacing cells from the temporal or parietal association cortex which are lost in Alzheimer disease therefore currently seems remote. The more limited goal of understanding the restraints on neural differentiation that limit the neurogenic potential of subependymal neural stem cells *in vivo* compared with *in vitro* may well be achievable. This knowledge could ultimately lead to replacement of specific motor or sensory neurons serving less advanced brain functions.

CONCLUSIONS

The neurogenetic analysis of complex disorders will be a difficult and protracted endeavour. Genetics is only one of many disciplines that will be required to elucidate disorders like epilepsy and dementia. However, it is a very powerful tool for dissecting such complex phenotypes. Historically, the power of the genetic approach has come from the analysis of relatively simple and rare Mendelian disorders which resemble complex traits or diseases and elucidate key disease mechanisms and pathways. This is well illustrated by the analysis of genes responsible for early onset forms of Alzheimer disease and Parkinson disease. The identification of individual susceptibility genes with variants of smaller effect will be a hard and difficult road, but there is no sharp demarcation between large and small genetic effects so that many new and important insights will inevitably emerge using existing technologies. The increased availability of animal models of inherited neurological diseases, and of high throughput gene based technologies, such as microarrays and proteomic analyses, extend the range of traditional genetic tools, such as gene mapping. Finally, an understanding of the genetic and epigenetic mechanisms that restrain the differentiation and integration of human neural stem cells into mature neuronal networks could have an enormous impact on clinical practice.

APPENDIX A

HUMAN GENETIC VARIATION

Humans are on average 99.9% identical, with one variant base every 1300 base pairs.[7] Most of the genetic differences between any two individuals consist of SNPs, which are single base changes present in at least 2% of the population (allele frequency >0.01). There are probably over 10 million SNPs and an almost unlimited number of rare variants in the human population.[7] Most common variants are extremely ancient, pre-dating the divergence of human racial groups >100 000 years ago. They survive in the human genome because the majority are "neutral" in their effects on reproductive fitness. They therefore confer no reproductive advantage or disadvantage. Some common variants have arisen or become common within more recent times (for example, <10 000 years) as a result of selection for some favourable characteristic. In contrast, genetic variants with intermediate or large effects on disease are predicted to be at low population frequency, since they tend to have adverse effects both on disease related traits and on reproductive fitness (which are usually correlated).[4] Collectively, however, there are many more rare variants than common ones in the human population and these are the ones with large functional effects that contribute most to human Mendelian diseases. It remains to be seen to what extent these rather than common variants provide most insights into common disorders.

APPENDIX B

GENETIC LINKAGE AND ASSOCIATION ANALYSES

A genetic linkage analysis (Fig. 1a) aims to identify a gene of moderate effect by scanning the genome with several hundred evenly spaced genetic markers to find one or more that segregates with the trait or disease. An association is first sought between each marker and the trait or disease within each family. The probability of the observed data, assuming either linkage or the null hypothesis of no linkage, are summarised in a LOD score table or graph (Fig. 1b). In some late onset disorders, the LOD score declines with age of onset, indicating that other factors, such as polygenic or environmental influences, obscure the effect of single genes (Fig. 1b). Significant evidence of linkage can occur either by chance or because genetic marker and susceptibility gene are adjacent to one another on the same chromosome (true genetic linkage). Different families may have different mutations, but in linkage analysis it is assumed that these occur predominantly within a single gene, and account for much of the variation in disease susceptibility.

A case-control association study compares the frequency of a single SNP marker or more usually a combination of SNPs on a single chromosome (SNP haplotype) in cases and controls (Fig. 2). An excess of marker alleles or haplotypes in cases compared with controls may occur by chance or as a result of genetic association. A true association occurs when apparently unrelated individuals share a region of the genome as a result of distant common ancestry (Fig. 2a). In order to identify such regions, a high density of genetic markers is required, which is often restricted to the vicinity of a linkage peak (Fig. 2b). Association can occur between a disease or QT and genetic marker even if the genetic variant(s) conferring disease susceptibility is not tested directly, provided it is associated with adjacent (tested) markers, due to common ancestry (linkage disequilibrium) (Fig. 2c). Regions of association between SNP markers are being defined in the HapMap project, which aims to determine the most efficient combinations and density of marker SNPs for disease gene mapping. The aim is to use sufficient well chosen SNPs so that any untested but disease associated SNP will still be detectable in an association study, as a result of its association with adjacent (tested) SNPs (Fig. 2c).[5]

Genetic association methods work well for fine mapping within a (linkage) defined region, but their use in screening the entire genome for disease susceptibility genes requires very high marker densities — in the region of hundreds of thousands of SNPs, since only a small segment of genome is shared between distantly related individuals (Fig. 2a). This generates many false positive associations. A second problem is the underlying assumption in association studies that a significant fraction of the variation in disease susceptibility results from not only a single gene, but a single variant within a single gene, making it more restrictive than the linkage approach. It is however a powerful approach for identifying common, small effect variants in large population samples, for example using candidate genes.

REFERENCES

1. Lander ES, Schork NJ. Genetic dissection of complex traits. *Science* 1994; **265**:2037–2048.
2. Reich DE, Lander ES. On the allelic spectrum of human disease. *Trends Genet* 2001; **17**:502–510.
3. Pritchard JK. Are rare variants responsible for susceptibility to complex diseases? *Am J Hum Genet* 2001; **69**:124–137.

4. Wright A, Charlesworth B, Rudan I, *et al.* A polygenic basis for late-onset disease. *Trends Genet* 2003; **19**:97–106.
5. Carlson CS, Eberle MA, Kruglyak L, *et al.* Mapping complex disease loci in whole-genome association studies. *Nature* 2004; **429**:446–452.
6. Dilda CL, Mackay TF. The genetic architecture of Drosophila sensory bristle number. *Genetics* 2002; **162**:1655–1674.
7. Kruglyak L, Nickerson DA. Variation is the spice of life. *Nat Genet* 2001; **27**: 234–236.
8. Bertram L, Tanzi RE. Alzheimer's disease: one disorder, too many genes? *Hum Mol Genet* 2004; **13**(spec no 1):R135–R141.
9. Dahlman I, Eaves IA, Kosoy R, *et al.* Parameters for reliable results in genetic association studies in common disease. *Nat Genet* 2002; **30**:149–150.
10. Fullerton J, Cubin M, Tiwari H, *et al.* Linkage analysis of extremely discordant and concordant sibling pairs identifies quantitative-trait loci that influence variation in the human personality trait neuroticism. *Am J Hum Genet* 2003; **72**:879–890.
11. Bonifati V, Rizzu P, van Baren MJ, *et al.* Mutations in the DJ-1 gene associated with autosomal recessive early-onset parkinsonism. *Science* 2003; **299**:256–259.
12. Wright AF, Carothers AD, Pirastu M. Population choice in mapping genes for complex diseases. *Nat Genet* 1999; **23**:397–404.
13. Gretarsdottir S, Thorleifsson G, Reynisdottir ST, *et al.* The gene encoding phosphodiesterase 4D confers risk of ischemic stroke. *Nat Genet* 2003; **35**:131–138.
14. Bentley DR. Genomes for medicine. *Nature* 2004; **429**:440–445.
15. Botstein D, Risch N. Discovering genotypes underlying human phenotypes: past successes for mendelian disease, future approaches for complex disease. *Nat Genet* 2003; **33**(suppl):228–237.
16. Hardy J, Selkoe DJ. The amyloid hypothesis of Alzheimer's disease: progress and problems on the road to therapeutics. *Science* 2002; **297**:353–356.
17. Bergem AL, Engedal K, Kringlen E. The role of heredity in late-onset Alzheimer disease and vascular dementia. A twin study. *Arch Gen Psychiatry* 1997; **54**: 264–270.
18. Mahley RW, Rall SC Jr. Apolipoprotein E: far more than a lipid transport protein. *Annu Rev Genomics Hum Genet* 2000; **1**:507–537.
19. Mayeux R. Epidemiology of neurodegeneration. *Annu Rev Neurosci* 2003; **26**: 81–104.
20. Raber J, Wong D, Yu GQ, *et al.* Apolipoprotein E and cognitive performance. *Nature* 2000; **404**:352–354.
21. LaDu MJ, Pederson TM, Frail DE, *et al.* Purification of apolipoprotein E attenuates isoform-specific binding to beta-amyloid. *J Biol Chem* 1995; **270**: 9039–9042.

22. Bales KR, Verina T, Dodel RC, *et al.* Lack of apolipoprotein E dramatically reduces amyloid beta-peptide deposition. *Nat Genet* 1997; **17**:263–264.
23. Mayeux R, Saunders AM, Shea S, *et al.* Utility of the apolipoprotein E genotype in the diagnosis of Alzheimer's disease. Alzheimer's Disease Centers Consortium on Apolipoprotein E and Alzheimer's Disease. *N Engl J Med* 1998; **338**: 506–511.
24. Li YJ, Oliveira SA, Xu P, *et al.* Glutathione S-transferase omega-1 modifies age-at-onset of Alzheimer disease and Parkinson disease. *Hum Mol Genet* 2003; **12**:3259–3267.
25. Prince JA, Feuk L, Gu HF, *et al.* Genetic variation in a haplotype block spanning IDE influences Alzheimer disease. *Hum Mutat* 2003; **22**:363–371.
26. Ertekin-Taner N, Allen M, Fadale D, *et al.* Genetic variants in a haplotype block spanning IDE are significantly associated with plasma Abeta42 levels and risk for Alzheimer disease. *Hum Mutat* 2004; **23**:334–342.
27. Blomqvist ME, Silburn PA, Buchanan DD, *et al.* Sequence variation in the proximity of IDE may impact age at onset of both Parkinson disease and Alzheimer disease. *Neurogenetics* 2004; **5**:115–119.
28. Ertekin-Taner N, Ronald J, Asahara H, *et al.* Fine mapping of the alpha-T catenin gene to a quantitative trait locus on chromosome 10 in late-onset Alzheimer's disease pedigrees. *Hum Mol Genet* 2003; **12**:3133–3143.
29. Farris W, Mansourian S, Chang Y, *et al.* Insulin-degrading enzyme regulates the levels of insulin, amyloid beta-protein, and the beta-amyloid precursor protein intracellular domain *in vivo*. *Proc Natl Acad Sci USA* 2003; **100**:4162–4167.
30. Farris W, Mansourian S, Leissring MA, *et al.* Partial loss-of-function mutations in insulin-degrading enzyme that induce diabetes also impair degradation of amyloid beta-protein. *Am J Pathol* 2004; **164**:1425–1434.
31. Leissring MA, Farris W, Chang AY, *et al.* Enhanced proteolysis of beta-amyloid in APP transgenic mice prevents plaque formation, secondary pathology, and premature death. *Neuron* 2003; **40**:1087–1093.
32. St George-Hyslop PH. Molecular genetics of Alzheimer's disease. *Biol Psychiatry* 2000; **47**:183–199.
33. Hebert LE, Scherr PA, Bienias JL, *et al.* Alzheimer disease in the US population: prevalence estimates using the 2000 census. *Arch Neurol* 2003; **60**:1119–1122.
34. Delacourte A, David JP, Sergeant N, *et al.* The biochemical pathway of neurofibrillary degeneration in aging and Alzheimer's disease. *Neurology* 1999; **52**:1158–1165.
35. Nagy Z, Esiri MM, Jobst KA, *et al.* Relative roles of plaques and tangles in the dementia of Alzheimer's disease: correlations using three sets of neuropathological criteria. *Dementia* 1995; **6**:21–31.

36. Hutton M, Lendon CL, Rizzu P, *et al.* Association of missense and 5′-splice-site mutations in tau with the inherited dementia FTDP-17. *Nature* 1998; **393**: 702–705.

37. Rosso SM, van Swieten JC. New developments in frontotemporal dementia and parkinsonism linked to chromosome 17. *Curr Opin Neurol* 2002; **15**:423–428.

38. Caughey B, Lansbury PT. Protofibrils, pores, fibrils, and neurodegeneration: separating the responsible protein aggregates from the innocent bystanders. *Annu Rev Neurosci* 2003; **26**:267–298.

39. Goldberg MS, Lansbury PT Jr. Is there a cause-and-effect relationship between alpha-synuclein fibrillization and Parkinson's disease? *Nat Cell Biol* 2000; **2**:E115–E119.

40. Fearnley JM, Lees AJ. Ageing and Parkinson's disease: substantia nigra regional selectivity. *Brain* 1991; **114**:2283–2301.

41. Galvin JE, Lee VM, Trojanowski JQ. Synucleinopathies: clinical and pathological implications. *Arch Neurol* 2001; **58**:186–190.

42. Maries E, Dass B, Collier TJ, *et al.* The role of alpha-synuclein in Parkinson's disease: insights from animal models. *Nat Rev Neurosci* 2003; **4**:727–738.

43. Polymeropoulos MH, Lavedan C, Leroy E, *et al.* Mutation in the alpha-synuclein gene identified in families with Parkinson's disease. *Science* 1997; **276**:2045–2047.

44. Kitada T, Asakawa S, Hattori N, *et al.* Mutations in the parkin gene cause autosomal recessive juvenile parkinsonism. *Nature* 1998; **392**:605–608.

45. Leroy E, Boyer R, Auburger G, *et al.* The ubiquitin pathway in Parkinson's disease. *Nature* 1998; **395**:451–452.

46. Liu Y, Fallon L, Lashuel HA, *et al.* The UCH-L1 gene encodes two opposing enzymatic activities that affect alpha-synuclein degradation and Parkinson's disease susceptibility. *Cell* 2002; **111**:209–218.

47. Rosen DR, Siddique T, Patterson D, *et al.* Mutations in Cu/Zn superoxide dismutase gene are associated with familial amyotrophic lateral sclerosis. *Nature* 1993; **362**:59–62.

48. Bruijn LI, Becher MW, Lee MK, *et al.* ALS-linked SOD1 mutant G85R mediates damage to astrocytes and promotes rapidly progressive disease with SOD1-containing inclusions. *Neuron* 1997; **18**:327–338.

49. Humphries SE, Morgan L. Genetic risk factors for stroke and carotid atherosclerosis: insights into pathophysiology from candidate gene approaches. *Lancet Neurol* 2004; **3**:227–235.

50. Helgadottir A, Manolescu A, Thorleifsson G, *et al.* The gene encoding 5-lipoxygenase activating protein confers risk of myocardial infarction and stroke. *Nat Genet* 2004; **36**:233–239.

51. Butte A. The use and analysis of microarray data. *Nat Rev Drug Discov* 2002; **1**:951–960.
52. Lockhart DJ, Barlow C. Expressing what's on your mind: DNA arrays and the brain. *Nat Rev Neurosci* 2001; **2**:63–68.
53. Ginsberg SD, Hemby SE, Lee VM, *et al.* Expression profile of transcripts in Alzheimer's disease tangle-bearing CA1 neurons. *Ann Neurol* 2000; **48**:77–87.
54. Mirnics K, Middleton FA, Marquez A, *et al.* Molecular characterization of schizophrenia viewed by microarray analysis of gene expression in prefrontal cortex. *Neuron* 2000; **28**:53–67.
55. Steinman L, Zamvil S. Transcriptional analysis of targets in multiple sclerosis. *Nat Rev Immunol* 2003; **3**:483–492.
56. Caillier S, Barcellos LF, Baranzini SE, *et al.* Multiple Sclerosis Genetics Group. Osteopontin polymorphisms and disease course in multiple sclerosis.*Genes Immun* 2003; **4**:312–315.
57. Niino M, Kikuchi S, Fukazawa T, *et al.* Genetic polymorphisms of osteopontin in association with multiple sclerosis in Japanese patients. *J Neuroimmunol* 2003; **136**:125–129.
58. Lubec G, Krapfenbauer K, Fountoulakis M. Proteomics in brain research: potentials and limitations. *Prog Neurobiol* 2003; **69**:193–211.
59. Eriksson PS, Perfilieva E, Bjork-Eriksson T, *et al.* Neurogenesis in the adult human hippocampus. *Nat Med* 1998; **4**:1313–1317.
60. Sanai N, Tramontin AD, Quinones-Hinojosa A, *et al.* Unique astrocyte ribbon in adult human brain contains neural stem cells but lacks chain migration. *Nature* 2004; **427**:740–744.
61. Rakic P. Neuroscience: immigration denied. *Nature* 2004; **427**:685–686.
62. Hall JM, Lee MK, Newman B, *et al.* Linkage of early-onset familial breast cancer to chromosome 17q21. *Science* 1990; **250**:1684–1689.
63. Martin ER, Gilbert JR, Lai EH, *et al.* Analysis of association at single nucleotide polymorphisms in the APOE region. *Genomics* 2000; **63**:7–12.
64. Allen M, Heinzmann A, Noguchi E, *et al.* Positional cloning of a novel gene influencing asthma from chromosome 2q14. *Nat Genet* 2003; **35**:258–263.

3

Functional Genomics and Proteomics: Application in Neurosciences

*K E Wilson, M M Ryan, J E Prime, D P Pashby, P R Orange, G O'Beirne, J G Whateley, S Bahn, C M Morris**

The sequencing of the complete genome for many organisms, including man, has opened the door to the systematic understanding of how complex structures such as the brain integrate and function, not only in health but also in disease. This blueprint, however, means that the piecemeal analysis regimes of the past are being rapidly superseded by new methods that analyse not just tens of genes or proteins at any one time, but thousands, if not the entire repertoire of a cell population or tissue under investigation. Using the most appropriate method of analysis to maximise the available data therefore becomes vital if a complete picture is to be obtained of how a system or individual cell is affected by a treatment or disease. This review examines what methods are currently available for the large scale analysis of gene and protein expression, and what are their limitations.

*Institute of Human Genetics, International Centre for Life, Central Parkway, Newcastle upon Tyne NE1 4BZ, UK; c.m.morris@ncl.ac.uk

"The Blind Man and the Elephant"
And so these men of Indostan
Disputed loud and long,
Each in his own opinion ~ Exceeding stiff and strong,
Though each was partly in the right ~ And all were in the wrong!
John Godfrey Saxe

Many of the currently used methods for the analysis of function in a cell or tissue rely on measuring one single analyte at a time, with the analysis being based on a prior hypothesis of mechanisms of action or biology. By measuring only one analyte, only a limited picture is gained of the complexities of an often very dynamic situation. Rather like the tale of the six blind men and the elephant, by observing only one facet at a time, a distorted view of the situation can be achieved, and the real answer to the problem remains hidden. With the advent of whole genome maps for several key organisms including man, and the use of methods for profiling all the changes that occur in cells or tissues, it is now possible to stand back and take an unbiased view of biological processes. This is particularly the case for neuropsychiatric disease, where the complexities of the brain, its multitude of connections, and the dynamic interplay between neurones and glia make an unbiased view of disease processes particularly important.[1] Two emerging fields are now coming into general use where the global approach to gaining insight into brain function can be applied. Functional genomics is the generic term used to describe methods that analyse the various genes expressed by a cell or tissue, while proteomics aims to define the protein complement. These technologies now allow an almost complete examination of gene and protein expression using single techniques. Our aim in this article is to provide an overview of the specialised methodologies and to point out the potential pitfalls and limitations of these techniques, with particular reference to investigations on human postmortem brain samples.

FUNCTIONAL GENOMICS

Cellular function is mediated through gene expression involving the production of messenger RNA. Several methods can be employed to profile gene expression in neurological and psychiatric disorders, including differential display and microarrays, ideally coupled with real time quantitative polymerase chain reaction (Q-PCR) cross validation. These technologies can be used independently or in parallel, where they investigate mRNA transcripts quantitatively by amplification of RNA from disease and control samples, and then detection of specific complementary DNA (cDNA) or antisense RNA (aRNA) species. It is hoped that pathologically relevant pathways and disease mechanisms involved in neurological and psychiatric disorders can be revealed with these techniques.

RNA PREPARATION

Essential to any gene expression studies is RNA of the highest quality.[2] While this is easily possible for RNA isolated from cells in tissue culture, or from animal tissues, it could be a problem for human studies where — certainly for brain derived RNA — there will be potential delays in obtaining the tissue sample, and for postmortem brain RNA this can be lengthy and also be affected by the agonal state.[3] It is, however, clear that high quality RNA can be obtained from postmortem brain and, moreover, useful results can be obtained in gene expression studies.[1]

Standard methods of RNA extraction from tissues or cells are based on rapid extraction into guanidine-phenol containing solutions providing total RNA.[4] Further purification into mRNA is not usually necessary for gene expression studies, as mRNA is easily worked on by preparing cDNA in a standard reverse transcription reaction. Column purification can also assist in removing potential contaminants such as extraction buffers, DNA and so on, or in concentrating small RNA samples.[5,6] The standard assessment of RNA quality is the use of the A260/A280 absorbance ratio, which be greater than 1.8; additionally, RNA should be visually examined by agarose gel electrophoresis or ideally in microfluidic systems,[6] which employ much smaller samples and provide a better visual representation of RNA quality. On visual inspection good quality RNA normally shows twice the amount of 28S compared with 18S ribosomal RNA bands.

Production of cDNA is a prerequisite for any functional genomics study. High temperature reverse transcriptase (RT) can improve the yields of cDNA produced, by melting secondary RNA structure, and for low levels of RNA, cDNA yields can be improved by use of nucleotide binding proteins.[7] The final assessment of RNA quality, however, can be more readily obtained by polymerase chain reaction (PCR) based assessment. Amplification of various housekeeping transcripts such as β actin, GAPDH, and so on can determine transcript length and amount. Using standard PCR, the first of these measures will give an indication of the amount of degradation of the transcripts in the mRNA pool, partly degraded RNA samples having shorter transcripts owing to the absence of the 5′ portions of the RNA, and subsequently reduced quantities of long transcripts are identified compared with optimally prepared RNA. Similar results are also obtained from using real time PCR with reduced transcript levels indicating either RNA degradation or the presence of RT or PCR inhibitors.

DIFFERENTIAL DISPLAY

Initial approaches to differential gene expression include subtractive cloning,[8] serial analysis of gene expression (SAGE),[9] and subtractive hybridisation techniques.[10] Differential display is a less laborious and more widely used approach,[11] and has yielded promising results, particularly with an adapted indexing based differential display PCR (DD-PCR) technique.[12] Standard differential display techniques involve conversion of RNA to cDNA, which is then split into a series of reactions involving a specific reverse primer and a random forward primer. Amplification of only a limited subset of the RNA present produces a specific RNA ladder for a given primer combination and tissue. By comparing the RNA ladder between test and control, it is possible to identify transcripts that are differentially expressed. For indexing based DD-PCR, cDNAs from disease and control tissue are digested with class II restriction enzymes (for example, *Bbv*I), followed by amplification of the internal fragments by adaptor primer PCR and visualisation by non-denaturing polyacrylamide gel electrophoresis, thus producing a representation of the RNA transcripts present in the samples. The fragments are cloned and identified by comparison with database entries. The main advantage of the DD-PCR technique is its sensitivity in detecting low expression fragments and the identification of unknown differentially expressed genes, particularly intragenic fragments. Its main disadvantage is a low throughput of samples. This technique and its modified forms have been successfully applied to investigating neurological and psychiatric disorders[13–15] — for example, in Alzheimer's disease, downregulation of genes involved in synaptic formation and organisation,[13] tau phosphorylation,[13] or protein targeting.[15]

SAGE

Serial analysis of gene expression (SAGE) uses the principle that a 10–14 base pair (bp) stretch of RNA can identify a transcript, and these short sequences can be ligated before PCR. SAGE combines cDNA library production with high throughput sequencing,[9] where double stranded cDNA is prepared using poly-dT priming and digested with a restriction enzyme which cuts at relatively high frequency (for example, NlaIII). Ligation of an adapter to the cDNA is then done, the adapter having a type IIS restriction enzyme site (such as BsmFI) which allows the cDNA to be cut again, but at a site up to 20 bp away from the recognition sequence. Following digestion, the short fragments are then blunt ended and ligated, and the subsequent products amplified by adaptor specific PCR. These PCR products can then be cleaved with the original restriction enzyme and ligated to give concatamers of these short fragments which are then cloned. These clones are composed of the various mRNA species initially present, and the frequency of each individual sequence will be proportional to the frequency of the mRNA in the starting material.[9] Sequencing and analysis of these clones provides a measure of gene expression in the tissue, and comparison of results between test and control allows comparison of gene expression. Several major projects have been based on the use of SAGE, most notably involving analysis of cancer gene expression (see cgap.nci.nih.gov/SAGE/). Drawbacks of SAGE are, however, the large scale sequencing that is required to build up sufficient expression information, and the bioinformatics support required to interpret the sequencing results, making it only suitable for larger laboratories. The use of poly-dT to prime synthesis may also reduce representation of certain mRNA species.

MICROARRAYS

It is probably fair to say that differential display and other methods have been superseded by microarrays as the technique of choice to investigate differential gene expression in large sample sets. Depending on the array used, microarrays can now give an almost global picture of gene expression status in one experiment. The key value of current microarray technology lies in its use as a high throughput, initial screen to identify potential disease related genes that can then be cross validated using more accurate semiquantitative methods, such as real time PCR.[16]

Microarrays encompass various different technologies that have developed over the past few years. Two principal methods exist: cDNA arrays and oligonucleotide arrays. cDNA arrays were first developed where the genes of interest as cDNA clones or PCR products are printed onto membranes (macroarrays) or microscope slides (microarrays) using a robotic arraying device. To compare the expression of the gene in the different samples, mRNA isolated from each of these samples is labelled with radioactive isotopes (^{33}P) or with different fluorescent dyes such as Cy3 (green) and Cy5 (red), by reverse transcription. Radioactively labelled samples are scanned using a phosphoimager, and results are compared after standard normalisation procedures — for example, housekeeping controls. For fluorescent labelling, the two pools of labelled cDNA probes (test and control labelled with a different CyDye) are mixed and hybridised to a microarray. After hybridisation, measurements are made with a high resolution laser scanner that illuminates each DNA spot (at two wavelengths) and measures the fluorescent intensity of each dye separately. A ratio measurement of the absolute and relative abundance of each specific gene in both samples is obtained. Numerous studies of neuropsychiatric diseases such as Alzheimer's disease, Parkinson's disease, and schizophrenia have produced promising results using these techniques.[17–21]

Oligonucleotide probe arrays are designed by printing the gene of interest directly onto a glass slide as a single stranded polynucleotide, or by direct synthesis of oligonucleotides on the substrate. These arrays are almost exclusively probed using fluorescently labelled cDNA or aRNA, and have become a popular choice as they have superior specificity over cDNA arrays.[22] Additionally, the cost of commercially available as well as custom designed oligonucleotide arrays has decreased recently. Several oligonucleotide arrays representing part or most of the genes in the human genome have been designed and used to investigate gene expression in neurological and psychiatric disorders. Companies such as Affymetrix, MWG Biotech and others have developed high throughput screening procedures, which include optimised protocols that minimise the variation generated within experiments. The current Affymetrix array for instance represents approximately 39 000 transcripts including 33 000 fully annotated genes (Fig. 1) on two GeneChips (HG-U133A and B chips). Each gene is represented by

Fig. 1 Typical image derived from an Affymetrix microarray probed with CyeDye labelled antisense mRNA. Each point on the array represents one unique oligonucleotide that has been directly synthesised on the microarray surface, with each array capable of representing over 30 000 transcripts. Image courtesy of Dr Heiko Peters, Institute of Human Genetics, Newcastle.

a Probe set which consists of 11 probe pairs typically designed with a bias towards the 3′ end of each gene. These probe pairs are distributed across the array and consist of a perfect match and a mismatch oligonucleotide (each averaging 25 base pairs in length). The mismatch oligonucleotide contains a single base pair mismatch in the centre of the probe and is used to quantify and subtract non-specific hybridisation and background signals. Each chip also includes *E coli* internal spike control probes (bioB, bioC, bioD, cre) to determine hybridisation efficiency. To investigate gene expression, total RNA is used to prepare complementary aRNA from cDNA, which is then fragmented and hybridised to the array. Following washing, the array is then scanned and data are analysed by specific software. The data can be further mined using a number of software tools, such as GeneSpring or Bioconductor.

A major consideration for microarray based gene expression profiling is data analysis. Given the ability to analyse thousands of genes at once, the potential to identify false positives is large. In order to overcome this, several approaches have been applied to maximise the likely hood of identifying true changes without losing significant effects owing to overstringent statistical analysis. In general, once the array background and background hybridisation to non-specific genes (often plant genes for mammalian arrays) has been subtracted from data, data are normalised to the mean expression of all genes on an array, or are normalised to housekeeping genes, for example GAPDH. As a general rule, genes showing a fold difference of ± 1.5 in mean expression levels between the test and control, and where $p = 0.005$, are considered to be differentially expressed, and to be further analysed by a more stringent method such as Q-PCR. All software packages for array analysis provide these analysis routines, with each package being tailored, often to individual arrays.

Various microarray experiments and other methodologies using RNA from postmortem human brain tissue have been published, including studies exploring expression profiles of complex neuropsychiatric disorders such as schizophrenia, bipolar affective disorder, and Alzheimer's disease.[19,20,23–25] For example, in order to understand the some of the early biochemical changes in Parkinson's disease, microarray based gene expression profiling has been applied to animal models of the disease, rather than studying end stage disease material where there may be changes in gene expression because of common pathological responses or neurone loss.[26] Here, using a membrane array containing over 1000 expressed genes, the investigators identified changes in striatal gene expression associated with transcription factors (for example, downregulated Nur-77) indicating a coordinated change in gene expression, changes in cell–cell communication (such as upregulated synapsin 1A), and altered kinase dependent cell signalling (downregulation of c-src).[26]

There are several potential limitations to array based methods of expression profiling. A limitation of array based methods in general is that low abundance transcripts are often not detected, simply because of limiting amounts of target RNA. However, linear amplification methods may allow improved sensitivity in detecting low abundance transcripts.[27] As with any determination of transcript levels there is the problem that RNA levels do not always correlate with protein levels owing to the post-transcriptional regulation that affects many RNA species.[28] Macroarrays, as an array method, almost exclusively use radiolabelled cDNA for interrogation, and only one membrane can be interrogated at a time, making experimental variation more likely. For macroarrays, another limitation is the number of probe sequences that are available on any one array which is usually no more than a few thousand; this means that for any RNA sample several macroarrays need to be hybridised, and on a sample with limited RNA this may prove difficult without RNA amplification. Macroarrays can, however, prove to be an advantage where a relatively focused approach to a problem is being investigated, for instance by using an array which only has genes involved in apoptosis printed onto it. Here a limited "hypothesis driven" array experiment can prove to be extremely powerful, providing a comprehensive overview of only the system under investigation (see, for example, Weinreb *et al.*[29]). Macroarrays are, however, highly suited to small laboratories as the equipment requirements are slight.

REAL TIME Q-PCR

High throughput gene profiling techniques can only be regarded, at best, as being semiquantitative as they only determine relative expression levels, and have only a limited dynamic range. Methods are therefore required that can provide validation of expression results derived from gene expression profiling, and which can provide semiquantitative data for levels of gene expression such as northern blotting or Q-PCR. Q-PCR is now considered the method of choice for validating gene expression.[30] Q-PCR uses fluorescence detection of the PCR product by combining a thermal cycler with a fluorescent spectrophotometer.[31] Two basic assay systems are available. As the PCR reaction proceeds and double stranded DNA is generated, this is detected by either binding a fluorescent dye to the double stranded DNA or release of a fluorescent reporter molecule. A relatively inexpensive and rapid method, it also provides a high degree of sensitivity, allowing the determination of low abundance genes that may not be detected by microarray.[32] With most Q-PCR methods, levels of specific transcripts are related to levels of specific housekeeping transcripts to provide a means of normalisation, thereby providing accurate relative quantitation (Fig. 2).

The most inexpensive Q-PCR method in relies on DNA intercalating dyes, perhaps the most popular of these being SYBR green I (SYBR green I

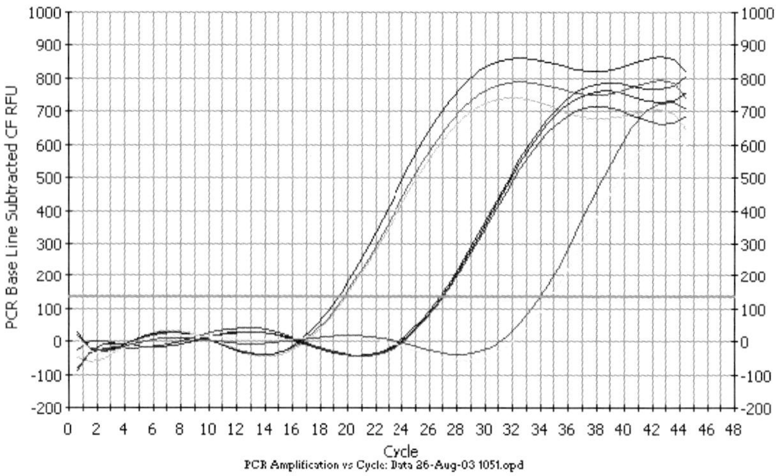

Fig. 2 Real time polymerase chain reaction (PCR) output of serially diluted input cDNA showing displacement of curves to the right with increasing dilution of RNA. PCR was generated using 1.0 μg of total RNA to make cDNA using oligo-dT primers, and 1:1000 and 1:10 000 dilution of cDNA and negative control with primers to amplify the GAPDH housekeeping transcript.

is a double stranded DNA intercalating dye that fluoresces upon laser light excitation), where an internal passive reference dye (ROX) is automatically detected by the machine to normalise for inconsistent pipetting. PCR product amplification is measured in real time by SYBR green I fluorescence emission upon binding to amplified PCR products after each PCR cycle. After 40 cycles, an end point or plateau is reached whereby no further amplification can take place owing to competitive PCR effects.[33–35] SYBR green assays provide a cheaper alternative to the more expensive 5′ endonuclease TAMRA probes (Taqman™), minor groove binder (MGB) probes, and Molecular Beacon detection chemistries.[17] These methods use primers where the fluorescent detection molecule is incorporated to the PCR primer and is quenched by the presence of either secondary structure or an additional quencher molecule.[16] MGB probe assays represent a new type of 5′ nuclease assay incorporating a different probe design and fluorescent detection to that of the more traditional TAMRA probe assays. A non-specific MGB is incorporated at the probe 3′ end leading to an increase in melting temperature (Tm)[36] allowing for shorter probes, typically 12–14 bp in length. A non-fluorescent quencher molecule (NFQ) is also situated at the MGB probe 3′ end in close proximity to the 6-FAM™ fluorophore. Q-PCR clearly is a robust method of verifying initial gene expression results from microarrays and of investigating novel genes that may be involved in neurological disease.[30,37,38]

While q-PCR is a common approach to validation of microarray data and is suited to high throughput, alternative methods of validation of data are also available. As many arrays are based on cloned cDNA sequences, the availability of these clones makes *in situ* hybridisation possible for determining not only levels of transcripts, but also cellular localisation, which is vital in neuroscience given the numerous cell types present in the CNS. Similarly, if antibodies are available to the particular gene, western or slot blotting for quantitation, and immunocytochemistry for cellular localisation are methods of analysis, the use of protein methods providing the ultimate validation that a gene change is accompanied by a protein change.

PROTEOMICS

Proteomics is the analysis of the protein complement of a given cell or tissue at a given point in time, and as such represents the natural extension of functional genomic analysis. While there are about 30 000 genes in the human genome (see http://www.ncbi.nlm.nih.gov/genome/guide/human/ for a list of human gene resources), the protein complement of a cell or tissue, the proteome, is much larger and also much more dynamic in nature. This is because most mammalian genes show alternative splicing of transcripts, leading to different isoforms of a given protein. This, coupled with post-translational modification such as glycosylation, myristoylation, and phosphorylation, leads to two or more effectively different proteins per gene. Proteomics can therefore generate much larger datasets requiring more resources to handle and analyse the data effectively.

It is doubtful whether one single method in a single pass will be able to identify accurately all the proteins that are expressed by a cell or tissue, owing to the huge numbers of protein isoforms that are present and their highly variable physical and chemical properties. The biggest problem is undoubtedly the large number of proteins that are expressed as this gives enormous analytical problems, and tissues or cells therefore need to be divided into manageable sized chunks. Proteomic approaches normally use fractionation of the sample into, for example, plasma membrane, nuclear, cytoplasmic, mitochondrial, lysosomal, and endoplasmic reticulum/Golgi fractions, to generate the maximum amount of protein information for a given cell or tissue. This can be achieved by various centrifugation methods, such as using increasing centrifugal forces or by centrifuging over media of different buoyant densities. Cocktails of protease and phosphatase/kinase inhibitors allow the various constituent proteins to be maintained in an intact state ready for analysis. Additionally, the use of various detergents and chaotropes can be applied to the samples to extract insoluble proteins (such as those proteins integrated into the cell membrane) for analysis. By this route, it is possible to generate a series of fractions that are sufficiently refined to allow analysis.

TWO DIMENSIONAL GEL ELECTROPHORESIS

Two dimensional polyacrylamide gel electrophoresis (2D-PAGE) is the most popular technique employed in proteomic studies as it can simultaneously resolve thousands of proteins with one format. Despite first being described in the mid-1970s,[39,40] the basic principles of separating proteins first according to charge (isoelectric point), and then according to mass remain unchanged. As isoelectric focusing can resolve 70 proteins while SDS-PAGE on a gradient gel is capable of separating out around 100 protein spots, 2D-PAGE should be able to resolve a "spot map" of around 7000 proteins, though the first use of 2D-PAGE only resolved 1100 *E coli* proteins on a single gel.[39] Recent advances have greatly increased this original number and it is now possible to distinguish up to 10 000 individual protein spots on a single large format gel.[41]

The technologies employed in 2D-PAGE have evolved rapidly. The first stage involves isoelectric focusing, where proteins are separated according to pI, using immobilised pH gradients (IPG) that result in ampholytes (charged carrier molecules) being incorporated into a thin gel strip. Proteins applied to these IPG strips migrate in an electric field and, owing to the pH gradient effect created by the ampholytes, stop migrating through the IPG strip when their charge is net neutral. These IPG strips are widely available in various formats, from wide range covering many pH units, to narrow pH ranges covering just one or two pH units. These narrow range strips provide much greater resolution and can be overlapped. This stretching of the protein pattern allows visualisation of proteins within the same pH range as a standard wide range gel yet at a much greater resolution.[40] This was demonstrated by the use of eight overlapping pH gradients to define the proteome of the *Mycoplasma genitalium*.[41] Studies of *Sacchromyces cerevisiae* proteins have compared overlapping narrow IPG strips between pH 4 and pH 9 with one pH 3–10 strip, producing patterns of 2286 and 755 distinct spots, respectively.[42] Despite these advances in first dimension technology, the method is unsuitable for highly acidic or basic proteins as separation below pH 3 and above pH 11 is poor because of lack of good ampholytes. This has often led to poor separation for membrane bound proteins (as they are often very acidic) owing to poor solubility in the absence of detergents. Additional preseparation methods are therefore needed to solubilise membrane proteins before separation.

After completion of first dimension separation by pI, proteins are separated by molecular weight in polyacrylamide gels containing sodium dodecyl sulphate (SDS-PAGE). SDS is an anionic detergent which denatures the proteins, converting them to a linear molecule by relaxing secondary structure, and because of its anionic nature it gives proteins a net negative charge. When combined with a reducing agent such as dithiothreitol (DTT), proteins are therefore separated exclusively by mass. Once the protein spots are separated they can be visualised using a variety of stains. Silver and Coomassie stains are relatively simple, require little specialist equipment, and are therefore the most frequently used methods. Silver staining is the most common method and has a higher sensitivity than traditional radiolabelling or Coomassie brilliant blue staining. Coomassie staining typically detects 8–10 ng of protein, while silver staining can be 100 times more sensitive.[43] Both methods, however, work by the stains reacting with functional groups on the proteins and therefore some polypeptides are not effectively stained. Recently developments in fluorescence technologies have led to the production of fluorescent protein dyes such as SYPRO and CyDyes with sensitivity similar to silver stains, though specialist equipment is required for their detection. These fluorescent dyes are also more compatible with peptide identification by mass spectrometry, silver and Coomassie stained gels often requiring destaining before mass spectrometry.[43–45]

Silver and Coomassie stains are routinely used in many laboratories, providing a simple method for building up proteome databases of various organisms or tissues (see http://ca.expasy.org/ch2d/). For example, two dimensional gels of protein samples from Alzheimer's disease and control brains have been described recently using silver staining to visualise the spot map of over 1500 proteins, and quadrupole time of flight tandem mass spectrometry (Q-TOF MS/MS) for protein identification (see the section on mass spectrometry below).[43] The proteome of human cerebrospinal fluid (CSF) has also been analysed, producing a list of over 480 proteins.[44,45] Such studies have identified variations in apolipoprotein E in the CSF between patients with sporadic or variant Creutzfeldt-Jakob disease,[46] and between patients with Alzheimer's disease and schizophrenia.[47] Nuclear proteins from human blood lymphocytes have been identified, showing that the method can be applied to any tissue,[48] including protein expression in the mouse cerebellum.[49]

A new development in two dimensional technology is fluorescence 2D difference gel electrophoresis (DIGE), where up to three different samples are separated on a single gel because they have each been labelled before 2D electrophoresis with a different fluorescent cyanine dye (Cy2, Cy3, Cy5). Each dye can then be visualised under a different wavelength and the images overlayed, giving a combined image that can be analysed using various software packages (Fig. 3).[50,51] This technique allows direct comparison between samples to show the presence of a particular protein — for instance, in a test sample compared with control — reducing the effects of inter-gel variation and improving reproducibility, which has traditionally been one of the major problems associated with 2D-PAGE.[52] For example, a control sample is labelled with Cy3 and disease sample with Cy5; a pool of all samples is created and labelled with a third label, Cy2, and this is included in all gels as an internal standard; the three protein samples are then combined and run on a single gel. The three different spot maps can be directly overlayed to allow comparison of the two protein samples, and reference to the internal standard will confirm actual protein changes or experimental artefacts.[52,53]

2D-PAGE is also a flexible technique in terms of the format of the gels, as in some cases mini 2D gels may be preferable to large format gels — for example, in experiments involving large numbers of samples. Thus 7 cm gels have been used to investigate protein oxidation in Alzheimer's disease brain tissue,[54] where proteins in the 2D gels were transferred to a PVDF membrane for western blotting and the spot map visualised by SYPRO

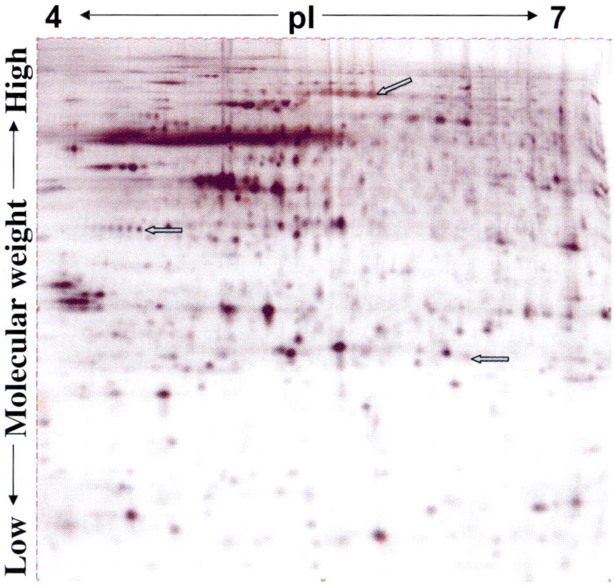

Fig. 3 Two dimensional difference gel electrophoresis (2D-DIGE), demonstrating the use of new "sat-dye" thiol group reactive labelling reagents on small samples (transgenic and control rat hippocampus). There are approximately 5000 different protein/peptide spots, and the identification of differences in protein expression is shown by the presence of differences in fluorescence of protein spots (a few very clear examples are arrowed).

ruby fluorescent stain, followed by probing with an antibody against oxidised protein groups. This method demonstrated over 100 proteins on these "oxyblots," with many showing significant changes in Alzheimer's disease compared with controls.[54,55]

ICAT — ISOTOPE CODED AFFINITY TAGS

Isotope coded affinity tagging (ICAT) is a relatively recent advance in proteomic analysis first reported in 1999 by Gygi and colleagues.[56] The technique, coupled with mass spectrometry, allows both identification and quantitation of proteins within complex mixtures and, as with DIGE, permits simultaneous analysis of two protein samples. In ICAT, one sample is labelled with a reagent which contains normal hydrogen while the other sample is labelled with a reagent containing deuterium, and both samples are then mixed together. Following separation, application of the samples to a mass spectrometer allows the differences of a few Daltons to be resolved, with direct quantitation of both mass peaks.

The two isotopic forms of the ICAT labelling reagent contain an isotopically light or heavy linker region, a protein reactive group, and a biotin affinity tag. The reactive group of the commercially available ICAT reagents and those first described by Gygi[57] are specific for cysteine residues through the third group. The linker region contains eight hydrogen atoms (d0) for the light (H) chain reagent, or eight deuterium (D) atoms (d8) for the heavy chain. In a standard ICAT experiment two protein samples, such as control and disease, are labelled with light or heavy reagents, respectively. The two are then combined and subjected to proteolytic digestion, typically by trypsin. The resulting peptide mixture is then fractionated by avidin affinity chromatography, which isolates only the cysteine containing peptides by binding to the biotin moiety on the ICAT reagent. This step results in 10-fold fewer peptides than in the original mixture, simplifying subsequent analysis. Identification and quantitation is then determined by liquid chromatography and tandem mass spectrometry.[56,58,59] However, the original ICAT reagents are relatively large and therefore identifying the peptide fragments is complicated by their presence, which causes substantial shifts in the peptide mass. Current reagents for ICAT have been improved by incorporating an acid cleavable linker, allowing removal of the biotin affinity tag before mass spectrometry, but leaving the peptide isotopically labelled (http://appliedbiosystems.com). This simplifies the analysis so that greater numbers of peptides can be identified and quantified.[60]

ICAT labelling has been employed to study protein changes induced in cultures of cortical neurones by the chemotherapeutic agent camptothecin. ICAT labelled peptides were purified on an avidin affinity column and analysed by liquid chromatography and mass spectrometry, with 129 proteins identified and their relative abundance quantified. This has demonstrated ICAT's usefulness in detecting low abundance proteins from many different subcellular compartments, including those involved in protein synthesis, transcription regulation, and signal transduction.[61] This technique is also suited to the study of relatively insoluble proteins such as membrane proteins which are often not compatible with 2D-PAGE, as these proteins can be extracted with strong ionic detergents and then labelled, the digestion step also creating peptides that are also more soluble than whole proteins.[59]

A problem with ICAT is that not only does it require proteins to contain cysteine residues, but these residues must be in the region of a peptide that is produced during proteolytic cleavage. This was highlighted recently in a study of a multi-subunit membrane protein of *E coli*,[62] which revealed that a high proportion (10–15%) of proteins lack cysteine residues. This problem may be overcome by the development of ICAT reagents reactive for different amino acid residues or by incorporating isotopic tags during the proteolytic step. For example, Glu-C proteolysis has been described using regular water ($H_2{}^{16}O$) and heavy water ($H_2{}^{18}O$). This results in two ^{16}O or two ^{18}O atoms being incorporated into the peptide fragments, giving a 4 Da difference. Likewise, different affinity matrices can be used to isolate different peptides, for example nickel immobilised metal affinity chromatography for histidine containing peptides, or a lectin affinity column for selectively isolating glycoproteins.[58] ICAT also fails to identify post-translational modifications such as phosphorylation and glycosylation unless these changes occur on the peptide containing the cysteine residue.

Recently a hybrid method for using ICAT in conjunction with two dimensional electrophoresis has been described.[63] Two protein samples were labelled with light and heavy ICAT reagents, pooled, and separated on the same 2D gel. The gel was then stained to visualise the protein spots, which were then excised from the gel and enzymatically digested, providing the peptide mixture for identification by mass spectrometry. The 8 Da difference, while not detectable on the 2D gel, was still identifiable on the peptide mass fingerprint obtained by mass spectrometry, providing quantitative data on the differences in protein expression. The potential for ICAT to become comparable with 2D-PAGE for identification of protein differences without the need for time consuming gel analysis is likely to be an emerging technique in the next few years which will allow the high throughput analysis of complex protein patterns provided by CNS samples.

PROTEIN MICROARRAYS

DNA and oligonucleotide based microarrays have been routinely used for many years now, and more recently their protein counterparts have been developed. Protein microarrays or "chips" are gaining in popularity as miniaturised ligand binding assays which can be used for complex protein samples, because they allow simultaneous detection and quantitation of biomolecules. In this instance, capture molecules such as antibodies are immobilised at a high density in a small area on a solid support such as a treated glass microscope slide. When exposed, each individual antibody captures its target protein from, for instance, a cell lysate or a serum sample. This technique allows large scale and high throughput analysis, using small sample volumes and relatively low protein concentrations. For these reasons microarrays are likely to find routine applications in basic research, disease diagnosis, and the identification of therapeutic targets.[64,65] Protein microarrays effectively allow quantitation of several hundred to several thousand analytes with one system (Fig. 4).

Antibody microarrays are the most accessible medium used in proteomics. In one of the first papers reporting the use of protein microarrays, 115 antibodies or antigens were immobilised using a robotic arrayer and probed with the corresponding ligands in mixtures of varying but known concentrations. Interactions were visualised by labelling the protein mixtures with Cy3 or Cy5 fluorescent dyes, and the relative intensities provide data on relative abundance.[64] For the production of antibody based microarrays there is a vast library of antibodies that are relatively stable and well characterised and are already routinely used in various techniques. The disadvantages with the use of antibodies include a large molecular

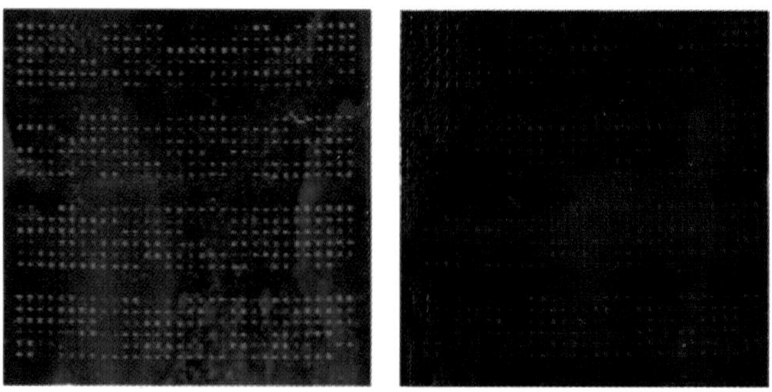

Fig. 4 Protein antibody microarray showing separate colour channels of the same array labelled with Cy3 (green; control sample) and Cy5 (red; test sample). By running two samples at once, differences in expression can be determined in only a few experiments.

size, and in the case of polyclonal antibodies, a possible lack of specificity and limited resource. Commercial companies are, however, developing methods for overcoming these problems using antibody fragments or phage technology, the latter using phage (bacterial viruses) which have been genetically modified to express immunoglobulin fragments on their surface.[66] Antibody function may also vary between assay types; therefore in a clinical setting it is important that antibody performance can be validated. A recent assessment of protein microarrays reports that, of over 100 commercially available antibodies tested, as few as 5% are suitable for use in microarray based analyses.[67] To overcome the problem it may be necessary to select antibodies specifically for a particular protocol, which in turn requires some knowledge of the system being studied.[68] This will, however, allow specific systems to be studied, such as apoptosis or inflammation, providing a better understating of certain pathways — for example, signal transduction. Alternatively it is possible to simplify complex protein samples such as cell lysates using chromatographic methods like ion exchange or affinity chromatography, liquid phase isoelectric focusing, or 1D-PAGE.[69] Synthetic alternatives to antibodies have been described which show high specificity and affinity and are stable; this may be a future route for protein microarrays.[70] Antibody based microarrays are, however, likely to have potential for clinical applications such as detection of diagnostic proteins in serum or CSF, as well as being of use in a research setting. The microarray format has been described for high throughput detection of clinical analytes, albeit in a low density, 6×6 (36 analytes), format.[71]

Variations on the antibody array format that have been described include tissue arrays, peptide arrays, and carbohydrate arrays where these molecules (or tissue) are arrayed and probed with single or multiple analytes to determine their binding partner on the array. For instance, tissue arrays have small sections of normal or pathological tissue from various organs gridded onto their surface, which are probed with antibodies to novel proteins to determine expression patterns. Protein–protein interactions in yeast have been described whereby 80% of the proteome was cloned and purified and then spotted onto a glass slide, forming a proteome array. Probing with protein extracts then provided information on protein–protein and protein–phospholipid interactions.[72] These protein arrays may be more likely to become the method of choice for determining protein binding partners to identify cellular pathways. In the future it should be possible to make cDNA expression libraries using the human genome database as a resource, and to spot all known proteins producing a proteome microarray. For example, if a particular neuronal protein is being studied, application of the recombinant or isolated protein to the array labelled with a suitable reporter would then allow potential interacting partners. Sequential analysis of these binding partners will permit such protein–protein pathways to be generated.

The use of microarrays in proteomic studies is still very much in its infancy but it is clear from the few publications there are on this subject that it will one day be an invaluable tool in the clinical setting, particularly where protein concentrations or sample quantity may be limited. A recent report identified five serum proteins which differed significantly between prostate cancer and control samples,[73] while a second study used small amounts of oral cancer material obtained using laser capture microdissection.[74] Here differences in protein expression could be quantified for a number of proteins during tumour progression.

MASS SPECTROMETRY

Mass spectrometry is the preferred method for the identification of proteins, forming an essential part of proteomic analysis. Mass spectrometry measures the mass to charge ratio (m/z) of gaseous ions produced by accelerating an ionised particle, in this case the protein or peptide, through a rarefied atmosphere to a detector. By providing structural information such as peptide mass and amino acid sequence, as well as information on protein modifications, the data obtained can then be used to identify a protein by searching various databases available.

Because with a large protein — for instance, albumin at (~65 kDa) — the mass charge ratio could be derived from multiple combinations of amino acids, and as the accuracy of mass spectrometers is reduced at higher molecular mass, methods are required that improve the accuracy of detection. The protein to be analysed is purified, often by 2D-PAGE although chromatographic methods can be used, and digested enzymatically (for example by trypsin or Lys-C) to cleave the protein at specific bonds, giving a reproducible pattern of digestion. Mass spectrometry is then used on the complex peptide mixture, giving peptide masses with high accuracy — a peptide mass fingerprint (PMF) (Fig. 5). With this information the likely amino acid

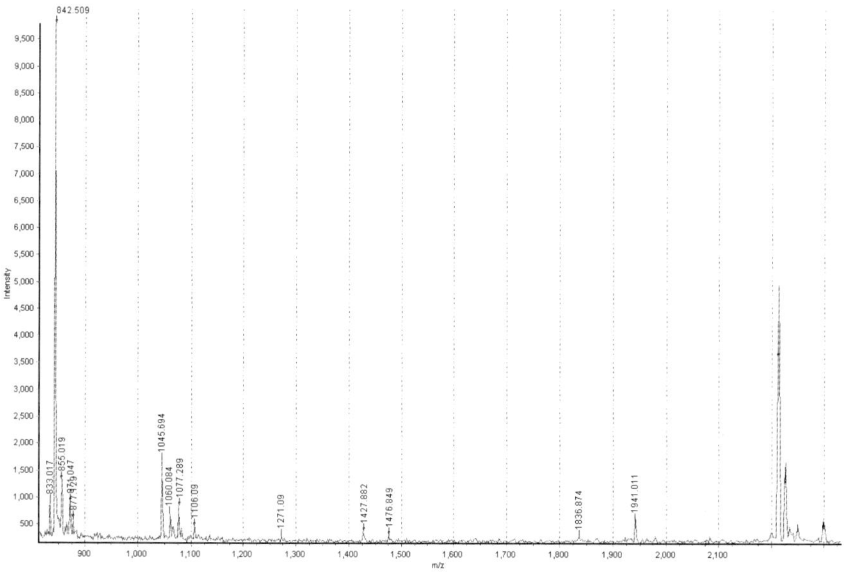

Fig. 5 Typical MALDI-TOF spectrum or "peptide mass fingerprint" for a protein excised from a preparative 2-D PAGE gel and trypsin digested, showing the individual peptide mass peaks that contribute to the whole protein and allow its identification by database searches.

composition of each peptide from the protein digest can be derived, which is then compared with databases containing theoretical protein cleavage data using sophisticated computer search engines, producing a list of the closest matching proteins. For this reason PMF is ideally suited to those species for which the genome has been completely sequenced, such as man and mouse, but is less useful in those cases where the genome sequence has not been fully determined. In such cases additional peptide sequence information can be obtained using, for instance, traditional Edman sequencing, where amino acids are sequentially removed from the peptide and analysed by high performance liquid chromatography (HPLC) with electrochemical detection or by mass spectrometry, or more directly by using tandem mass spectrometry (MS/MS), which combines two mass analysers — for example, a quadrupole with a TOF analyser.[75]

All mass spectrometers have three main components: the ionisation source, the mass detector, and the ion detector. Ionisation sources include matrix associated laser desorption/ionisation (MALDI) and electrospray ionisation (ESI). Both are ideal for detecting low protein concentrations and can be used with complex protein/peptide mixtures or with prefractionated samples. Both are "soft" ionisation techniques allowing ion formation without altering the native protein or peptide, thereby providing more accurate mass information.

MALDI requires picomoles or less of sample, is relatively insensitive to contaminants such as salts and non-ionic detergents, and samples can be in the solid, liquid, or gaseous phase. The analyte is co-crystalised with an ultraviolet absorbing matrix solution on a target plate. A laser beam is fired at the target which is absorbed by the matrix, transferring energy to the analyte and causing it to transfer into the gas phase.[76] Traditionally MALDI instruments are coupled to TOF mass analysers, which measure the time lag between the point at which ions are accelerated to the point at which they reach the ion detector, ions with a smaller mass reaching the detector before those with a greater mass. Other analysers include the commonly used quadrupole which consists of four parallel metal rods which can act as a filter to allow only the passage of ions with a certain m/z. By placing multiple quadrupoles in series the amino acid sequence of a peptide can be determined.[76]

ESI requires the analyte to be in solution and therefore is ideally coupled to liquid chromatographic separation methods. As the sample is injected into the mass spectrometer it is sprayed across a high potential difference, resulting in the formation of a fine mist of charged droplets. More recent advances in ESI include nanospray ionisation in which the microcapillary tube used for injection of the sample has a diameter as small as 1–2 μm, allowing flow rates as low as 5 nl/min,[77] which greatly reduces th e amount of sample needed for analysis.

As previously mentioned, MALDI-TOF is a commonly used combination for PMF, but the MALDI-Q-TOF hybrid allows the amino acid sequence to be determined for any peptide that is not identified by PMF.[77] These newer mass spectrometry instruments allow greater sensitivity and high mass accuracy, and offer high throughput analysis by being coupled to automated systems for either robotic sampling of 2D gels or direct capillary based separation of protein/peptide mixtures. They are also able to detect and characterise post-translational modifications and identify different isomers.[78]

There are numerous examples of the use of mass spectrometry in neuroscience — for example, increased platelet activating factor in the plasma and CSF in multiple sclerosis was identified using HPLC with tandem mass spectrometry,[79] and the processing of neuropeptide Y in the CSF of patients with depression has been monitored using MALDI-TOF.[80] The structural variants of β-amyloid in brain tissue from patients with Alzheimer's disease compared with healthy controls was determined by ESI-MS, and other neuropeptides such as substance P and dynorphin A have been analysed in plasma, CSF, and brain in various conditions, using mass spectrometry techniques (see Nilsson *et al.* for a review[81]).

SELDI

Recently, SELDI (surface enhanced laser desorption/ionisation) has increased in popularity. First detailed in 1993,[82] the principles behind SELDI are similar to MALDI in that it uses a laser beam to desorb analyte ions from a solid for analysis by mass spectrometry.[83] Sample preparation is simplified compared with MALDI as proteins are captured onto a solid phase chromatographic surface.[84] For example, the sample is applied to strong cationic support and washed with an appropriate buffer so that only proteins and peptides with affinity for the support are retained. Analysis with SELDI therefore produces a series of mass peaks for each affinity matrix, effectively a protein peptide signature for each tissue or cell type. By using different types of support, it is thus possible to analyse different subsets of proteins to build up a picture of the proteins present. The technique has been used in various studies and is particularly suited to analysis of diagnostic biomarkers in plasma or CSF. For example, cystatin C — a secreted cysteine protease inhibitor — has been identified in CSF as a marker of chronic pain,[85] and β-amyloid has been identified in the lens of Alzheimer's disease patients, suggesting that the pathological features of the disease overlap between brain and lens.[86] Recently SELDI has been used to analyse plasma samples from individuals with ovarian tumours compared with normal individuals and individuals with benign ovarian cysts.[87] By comparing the peptide profile, a series of peptide peaks was identified which together provided 99% sensitivity and 99% specificity in the diagnosis of ovarian tumour. The identification of prostate cancer associated biomarkers has also shown the value of SELDI for rapid discovery of potential clinical markers,[88] suggesting that its application to samples such as CSF may have considerable utility.

One drawback of SELDI is that it is most suited to the analysis of peptides and proteins of less than 20 kDa, as these are more likely to be retained by the affinity support and are more likely to fly when hit by the laser. This can be overcome by simple prefractionation by size before affinity separation, so that only proteins greater than 20 kDa are applied to the affinity support. Also, as SELDI uses a mild ionisation procedure, it has a limited capacity to identify the peptide unless it is less than about 10 kDa. A peptide identified in this way requires further purification regimens and analytical methods for precise identification. SELDI is therefore more suited to high throughput prescreening of large sample numbers, but nonetheless provides an effective tool for identifying the presence of differentially expressed proteins. Advances in MALDI-TOF to provide high throughput, along with different isolation and enrichment strategies, will possibly make the principles of SELDI-TOF much more amenable to protein identification.

FUTURE PROSPECTS

One of the major hurdles to the application of genomics in neuroscience is the complexity of the brain itself and the numerous different cell types, even within a relatively small area. As genomics and proteomics aim to define the complement of a given cell, this poses major problems in attempting to decipher which genes and proteins are associated with a particular cell type. To overcome this, techniques such as laser capture microdissection are available, where an individual cell type such as a pyramidal neurone in the hippocampus can be isolated from its neighbours by very focused laser light.[89] The use of cells isolated in this way will allow these technologies to be applied to individual cell types, and enable comparison of, for instance, the proteomes of pools of neurones affected by degenerative pathology compared with their unaffected neighbours. With advances in sensitivity in fluorescent detection and mass spectrometry, it may even be possible to analyse the expression of a single cell.[90] No one single technique will be capable of analysing the entire genome or proteome of a given cell or tissue, but with selective use of various methods it should be possible to determine the gene and protein expression patterns of key brain regions in health and disease. Ultimately, these global profiling technologies will help to unravel both the genetic and environmental factors that predispose to and precipitate complex neuropsychiatric disorders.

ACKNOWLEDGEMENTS

MMR and SB gratefully acknowledge the support of the National Alliance for Research on Schizophrenia and Depression, the BBSRC, the Stanley Medical Research Institute, and the Stanley Foundation. CM and KEW are supported by the MRC, the Alzheimer's Research Trust, Amersham Biosciences plc, and the Newcastle Hospitals Special Trustees.

REFERENCES

1. Bahn S, Augood SJ, Ryan M, *et al.* Gene expression profiling in the post-mortem human brain – no cause for dismay. *J Chem Neuroanat* 2001; **22**:79–94.
2. Castensson A, Emilsson L, Preece P, *et al.* High-resolution quantification of specific mRNA levels in human brain autopsies and biopsies. *Genome Res* 2000; **10**:1219–1229.
3. Harrison PJ, Heath PR, Eastwood SL, *et al.* The relative importance of premortem acidosis and postmortem interval for human brain gene expression studies: Selective mRNA vulnerability and comparison with their encoded proteins. *Neurosci Lett* 1995; **200**:151–154.
4. Chomczynski P, Sacchi N. Single-step method of RNA isolation by acid guanidinium thiocyanate-phenol-chloroform extraction. *Anal Biochem* 1987; **162**: 156–159.
5. Mikulowska-Mennis A, Taylor TB, Vishnu P, *et al.* High-quality RNA from cells isolated by laser capture microdissection. *Biotechniques* 2002; **33**:176–179.
6. Ellis M, Davis N, Coop A, *et al.* Development and validation of a method for using breast core needle biopsies for gene expression microarray analyses. *Clin Cancer Res* 2002; **8**:1155–1166.
7. Rapley R. Enhancing PCR amplification and sequencing using DNA-binding proteins. *Mol Biotechnol* 1994; **2**:295–298.
8. 8. Hara E, Kato T, Nakada S, *et al.* Subtractive cDNA cloning using oligo(dT)30-latex and PCR: Isolation of cDNA clones specific to undifferentiated human embryonal carcinoma cells. *Nucleic Acids Res* 1991; **19**:7097–7104.
9. Velculescu VE, Zhang L, Vogelstein B, *et al.* Serial analysis of gene expression. *Science* 1995; **270**:484–487.

10. Duguid JR, Dinauer MC. Library subtraction of *in vitro* cDNA libraries to identify differentially expressed genes in scrapie infection. *Nucleic Acids Res* 1990; **18**:2789–2792.
11. Liang P, Pardee AB. Differential display of eukaryotic messenger RNA by means of the polymerase chain reaction. *Science* 1992; **257**:967–971.
12. Mahadeva H, Starkey MP, Sheikh FN, *et al.* A simple and efficient method for the isolation of differentially expressed genes. *J Mol Biol* 1998; **284**:1391–1398.
13. Ryan M, Starkey M, Faull R, *et al.* Indexing-based differential display — studies on post-mortem Alzheimer's brains. *Brain Res Mol Brain Res* 2001; **88**:199–202.
14. Bahn S, Mimmack M, Ryan M, *et al.* Neuronal target genes of the neuron-restrictive silencer factor in neurospheres derived from fetuses with Down's syndrome: A gene expression study. *Lancet* 2002; **359**:310–315.
15. Kim SH, Shim KS, Lubec G. Human brain nascent polypeptide-associated complex alpha subunit is decreased in patients with Alzheimer' s disease and Down syndrome. *J Invest Med* 2002; **50**:293–301.
16. Bustin SA. Quantification of mRNA using real-time reverse transcription PCR (RT-PCR): Trends and problems. *J Mol Endocrinol* 2002; **29**:23–39.
17. Pasinetti GM. Use of cDNA microarray in the search for molecular markers involved in the onset of Alzheimer's disease dementia. *J Neurosci Res* 2001; **65**:471–476.
18. Mandel S, O Weinreb, Youdim MB. Using cDNA microarray to assess Parkinson's disease models and the effects of neuroprotective drugs. *Trends Pharmacol Sci* 2003; **24**:184–191.
19. Pongrac J, Middleton FA, Lewis DA, *et al.* Gene expression profiling with DNA microarrays: Advancing our understanding of psychiatric disorders. *Neurochem Res* 2002; **27**:1049–1063.
20. Vawter MP, Crook JM, Hyde TM, *et al.* Microarray analysis of gene expression in the prefrontal cortex in schizophrenia: A preliminary study. *Schizophr Res* 2002; **58**(1):11–20.
21. Mimmack ML, Ryan M, Baba H, *et al.* Gene expression analysis in schizophrenia: Reproducible up-regulation of several members of the apolipoprotein L family located in a high-susceptibility locus for schizophrenia on chromosome 22. *Proc Natl Acad Sci USA* 2002; **99**:4680–4685.
22. Watson A, Mazumder A, Stewart M, *et al.* Technology for microarray analysis of gene expression. *Curr Opin Biotechnol* 1998; **9**:609–614.
23. Mirnics K, Middleton FA, Marquez A, *et al.* Molecular characterization of schizophrenia viewed by microarray analysis of gene expression in prefrontal cortex. *Neuron* 2000; **28**:53–67.

24. Hakak Y, Walker JR, Li C, *et al.* Genome-wide expression analysis reveals dysregulation of myelination-related genes in chronic schizophrenia. *Proc Natl Acad Sci USA* 2001; **98**:4746–4751.
25. Colangelo V, Schurr J, Ball MJ, *et al.* Gene expression profiling of 12633 genes in Alzheimer hippocampal CA1: Transcription and neurotrophic factor down-regulation and up-regulation of apoptotic and pro-inflammatory signaling. *J Neurosci Res* 2002; **70**:462–473.
26. Napolitano M, Centonze D, Calce A, *et al.* Experimental parkinsonism modulates multiple genes involved in the transduction of dopaminergic signals in the striatum. *Neurobiol Dis* 2002; **10**:387–395.
27. Polacek DC, Passerini AG, Shi C, *et al.* Fidelity and enhanced sensitivity of differential transcription profiles following linear amplification of nanogram amounts of endothelial mRNA. *Physiol Genomics* 2003; **13**:147–156.
28. Anderson L, Seilhamer J. A comparison of selected mRNA and protein abundances in human liver. *Electrophoresis* 1997; **18**:533–537.
29. Weinreb O, Mandel S, Youdim MB. cDNA gene expression profile homology of antioxidants and their antiapoptotic and proapoptotic activities in human neuroblastoma cells. *FASEB J* 2003; **17**:935–937.
30. Bunney WE, *et al.* Microarray technology: A review of new strategies to discover candidate vulnerability genes in psychiatric disorders. *Am J Psychiatry* 2003; **160**(4):657–666.
31. Higuchi R, Fockler C, Dollinger G, *et al.* Kinetic PCR analysis: Real-time monitoring of DNA amplification reactions. *Biotechnology (NY)* 1993; **11**:1026–1030.
32. Bustin SA, Dorudi S. The value of microarray techniques for quantitative gene profiling in molecular diagnostics. *Trends Mol Med* 2002; **8**:269–272.
33. Giulietti A, Overbergh L, Valckx D, *et al.* An overview of real-time quantitative PCR: Applications to quantify cytokine gene expression. *Methods* 2001; **25**: 386–401.
34. Gibson UE, Heid CA, Williams PM. A novel method for real time quantitative RT-PCR. *Genome Res* 1996; **6**:995–1001.
35. Heid CA, Stevens J, Livak KJ, *et al.* Real time quantitative PCR. *Genome Res* 1996; **6**:986–994.
36. Afonina I, Zivarts M, Kutyavin I, *et al.* Efficient priming of PCR with short oligonucleotides conjugated to a minor groove binder. *Nucleic Acids Res* 1997; **25**:2657–2660.
37. Aldape K, Ginzinger DG, Godfrey TE. Real-time quantitative polymerase chain reaction: A potential tool for genetic analysis in neuropathology. *Brain Pathol* 2002; **12**:54–66.

38. Tkachev D, Mimmack ML, Ryan MM, *et al.* Oligodendrocyte dysfunction in schizophrenia and bipolar disorder. *Lancet* 2003; **362**:798–805.
39. O'Farrell PH. High resolution two-dimensional electrophoresis of proteins. *J Biol Chem* 1975; **250**:4007–4021.
40. Klose J. Protein mapping by combined isoelectric focusing and electrophoresis of mouse tissues. A novel approach to testing for induced point mutations in mammals. *Humangenetik* 1975; **26**:231–243.
41. Klose J, Kobalz U. Two-dimensional electrophoresis of proteins: An updated protocol and implications for a functional analysis of the genome. *Electrophoresis* 1995; **16**:1034–1059.
42. Wildgruber R, Harder A, Obermaier C, *et al.* Towards higher resolution: Two-dimensional electrophoresis of Saccharomyces cerevisiae proteins using overlapping narrow immobilized pH gradients. *Electrophoresis* 2000; **21**:2610–2616.
43. Tsuji T, Shiozaki A, Kohno R, *et al.* Proteomic profiling and neurodegeneration in Alzheimer's disease. *Neurochem Res* 2002; **27**:1245–1253.
44. Sickmann A, Dormeyer W, Wortelkamp S, *et al.* Towards a high resolution separation of human cerebrospinal fluid. *J Chromatogr B Analyt Technol Biomed Life Sci* 2002; **771**:167–196.
45. Davidsson P, Folkesson S, Christiansson M, *et al.* Identification of proteins in human cerebrospinal fluid using liquid-phase isoelectric focusing as a prefractionation step followed by two-dimensional gel electrophoresis and matrix-assisted laser desorption/ionisation mass spectrometry. *Rapid Commun Mass Spectrom* 2002; **16**:2083–2088.
46. Choe LH, Green A, Knight RS, *et al.* Apolipoprotein E and other cerebrospinal fluid proteins differentiate ante mortem variant Creutzfeldt-Jakob disease from ante mortem sporadic Creutzfeldt-Jakob disease. *Electrophoresis* 2002; **23**:2242–2246.
47. Johnson G, Brane D, Block W, *et al.* Cerebrospinal fluid protein variations in common to Alzheimer's disease and schizophrenia. *Appl Theor Electrophor* 1992; **3**:47–53.
48. Bergquist J, Gobom J, Blomberg A, *et al.* Identification of nuclei associated proteins by 2D-gel electrophoresis and mass spectrometry. *J Neurosci Methods* 2001; **109**:3–11.
49. Beranova-Giorgianni S, Pabst MJ, Russell TM, *et al.* Preliminary analysis of the mouse cerebellum proteome. *Mol Brain Res* 2002; **98**:135–140.
50. Leimgruber RM, Malone JP, Radabaugh MR, *et al.* Development of improved cell lysis, solubilization and imaging approaches for proteomic analyses. *Proteomics* 2002; **2**:135–144.

51. Witzman FA, Li J. Cutting-edge technology II. Proteomics: Core technologies and applications in physiology. *Am J Physiol Gastrointest Liver Physiol* 2002; **282**:G735–G741.
52. Tonge R, Shaw J, Middleton B, *et al.* Validation and development of fluorescence two-dimensional gel electrophoresis proteomics technology. *Proteomics* 2001; **1**:377–396.
53. Yan JX, Devenish AT, Wait R, *et al.* Fluorescence two-dimensional difference gel electrophoresis and mass spectrometry based proteomic analysis of Escherichia coli. *Proteomics* 2002; **2**:1682–1698.
54. Aksenov MY, Aksenova MV, Butterfield DA, *et al.* Protein oxidation in the brain in Alzheimer's disease. *Neuroscience* 2001; **103**:373–383.
55. Korolainen MA, Goldsteins G, Alafuzoff I, *et al.* Proteomic analysis of protein oxidation in Alzheimer's disease brain. *Electrophoresis* 2002; **23**:3428–3433.
56. Gygi SP, Rist B, Gerber SA, *et al.* Quantitative analysis of complex protein mixtures using isotope-coded affinity tags. *Nat Biotechnol* 1999; **17**:994–999.
57. Gygi SP, Rist B, Gerber SA, *et al.* Quantitative analysis of complex protein mixtures using isotope-coded affinity tags. *Nat Biotechnol* 1999; **17**:994–999.
58. Patton WF. Detection technologies in proteome analysis. *J Chromatogr B* 2002; **771**:3–31.
59. Moseley MA. Current trends in differential expression proteomics: Isotopically coded tags. *Trends Biotechnol* 2001; **19**(suppl 10):S10–S16.
60. Sechi S, Oda Y. Quantitative proteomics using mass spectrometry. *Curr Opin Chem Biol* 2003; **7**:70–77.
61. Yu L-R, Johnson MD, Conrads TP, *et al.* Proteome analysis of camptothecin-treated cortical neurons using isotope-coded affinity tags. *Electrophoresis* 2002; **23**:1591–1598.
62. Berggren KN, Chernokalskaya E, Lopez MF, *et al.* Comparison of three different fluorescent visualisation strategies for detecting Escherichia coli ATP synthase subunits after sodium dodecyl sulfate-polyacrylamide gel electrophoresis. *Proteomics* 2001; **1**:54–65.
63. Smolka M, Zhou H, Aebersold R. Quantitative protein profiling using two-dimensional gel electrophoresis, isotope-coded affinity tag labeling, and mass spectrometry. *Mol Cell Proteomics* 2002; **1**:19–29.
64. Haab BB, Dunham MJ, Brown PO. Protein microarrays for highly parallel detection and quantitation of specific proteins and antibodies in complex solutions. *Genome Biol* 2001; **2**:research0004.1–0004.13.
65. Templin MF, Stoll D, Schrenk M, *et al.* Protein microarray technology. *Drug Discovery Today* 2002; **7**:815–822.

66. Lopez MF, Pluskal MG. Protein micro- and macroarrays: Digitizing the proteome. *J Chromatogr B* 2003; **787**:19–27.
67. MacBeath G. Protein microarrays and proteomics. *Nat Genet Suppl* 2002; **32**: 526–532.
68. Wingren C, Ingvarsson J, Lindstedt M, *et al.* Recombinant antibody microarrays — a viable option? *Nat Biotechnol* 2003; **21**:223.
69. Madoz-Gurpide J, Wang H, Misek DE, *et al.* Protein based microarrays: A tool for probing the proteome of cancer cells and tissues. *Proteomics* 2001; **1**:1279–1387.
70. Lal SP, Christpherson RI, dos Remedios CG. Antibody arrays: An embryonic but rapidly growing technology. *Drug Discovery Today* 2002; **7**(suppl 18):S143–S149.
71. Rowe CA, Scruggs SB, Feldstein MJ, *et al.* An array immunosensor for simultaneous detection of clinical analytes. *Anal Chem* 1999; **71**:433–439.
72. Zhu H, Bilgin M, Bangham R, *et al.* Global analysis of protein activities using proteome chips. *Science* 2001; **293**:2101–2105.
73. Miller JC, Zhou H, Kwekel J, *et al.* Antibody microarray profiling of human prostate cancer sera: Antibody screening and identification of potential biomarkers. *Proteomics* 2003; **3**:56–63.
74. Knezevic V, Leethanakul C, Bichsel VE, *et al.* Proteomic profiling of the cancer microenvironment by antibody arrays. *Proteomics* 2001; **1**:1271–1278.
75. Dunn MJ. Studying heart disease using the proteomic approach. *Drug Discovery Today* 2000; **5**:76–84.
76. Jonsson AP. Mass spectrometry for protein and peptide characterisation. *Cell Mol Life Sci* 2001; **58**:868–884.
77. Graves PR, Haystead TAJ. Molecular biologist's guide to proteomics. *Microbiol Mol Biol Rev* 2002; **66**:39–63.
78. Mo W, Karger BL. Analytical aspects of mass spectrometry and proteomics. *Curr Opin Chem Biol* 2002; **6**:666–675.
79. Callea L, Arese M, Orlandini A, *et al.* Platelet activating factor is elevated in cerebral spinal fluid and plasma of patients with relapsing-remitting multiple sclerosis. *J Neuroimmunol* 1999; **94**:212–221.
80. Ekman R, Juhasz P, Heilig M, *et al.* Novel neuropeptide Y processing in human cerebrospinal fluid from depressed patients. *Peptides* 1996; **17**:1107–1111.
81. Nilsson CL, Karlsson G, Bergqvist J, *et al.* Mass spectrometry of peptides in neuroscience. *Peptides* 1998; **19**:781–789.
82. Hutchens TW, Yip TT. New desoprtion strategies for the mass-spectrometric analysis of macromolecules. *Rapid Commun Mass Spectrom* 1993; **7**:576–580.

83. Merchant M, Weinberger SJ. Recent developments in surface-enhanced laser desorption/ionisation-time of flight-mass spectrometry. *Electrophoresis* 2000; **21**:1164–1167.
84. Issaq HJ, Veenstra TD, Conrads TP, *et al.* The SELDI-TOF MS approach to proteomics: Protein profiling and biomarker identification. *Biochem Biophys Res Commun* 2002; **292**:587–592.
85. Mannes AJ, Martin BM, Yang HY, *et al.* Cystatin C as a cerebrospinal fluid biomarker for pain in humans. *Pain* 2003; **102**:251–256.
86. Goldstein LE, Muffat JA, Cherny RA, *et al.* Cytosolic beta-amyloid deposition and supranuclear cataracts in lenses from people with Alzheimer's disease. *Lancet* 2003; **361**:1258–1265.
87. Petricoin EF, Ardekani AM, Hitt BA, *et al.* Use of proteomic patterns in serum to identify ovarian cancer. *Lancet* 2002; **359**:572–577.
88. Wright GL, *et al.* Proteinchip surface enhanced laser desorption/ionisation (SELDI) mass spectrometry: A novel protein biochip technology for detection of prostate cancer biomarkers in complex protein mixtures. *Prostate Cancer and Prostatic Diseases* 1999; **2**.
89. Emmert-Buck MR, Bonner RF, Smith PD, *et al.* Laser capture microdissection. *Science* 1996; **274**:998–1001.
90. Xiang CC, Chen M, Ma L, *et al.* A new strategy to amplify degraded RNA from small tissue samples for microarray studies. *Nucleic Acids Res* 2003; **31**:e53.

4

Mitochondria

Patrick F Chinnery and Eric A Schon*

HISTORICAL INTRODUCTION

It is over forty years since the first human mitochondrial disease was described in a patient with non-thyroidal hypermetabolism (Luft disease).[1] Although this disorder is exceptionally rare (in fact, only two cases have been described), the clinical description and biochemical studies paved the way for three decades of clinical and pathological research on patients with suspected mitochondrial disease. Patients were classified into groups based upon the pattern of clinical involvement, histological and ultrastructural abnormalities of mitochondria, and biochemical assays of mitochondrial function. It was clear that there were clinical similarities among some patients, allowing for the definition of syndromes such as the Kearns-Sayre syndrome (KSS) or chronic progressive external ophthalmoplegia (CPEO), but it was recognised that there was considerable phenotypic diversity and that many patients did not fit neatly into a specific diagnostic group. The inheritance pattern also varied. Some patients appeared to be sporadic cases, whereas others were clearly familial. It was known for some time that

*Neurology, The Medical School, Newcastle upon Tyne, NE2 4HH; Tel: 0191 222 8334; Fax: 0191 222 8553; P.F.Chinnery@ncl.ac.uk

mitochondrial DNA (mtDNA) was maternally inherited, and whilst some families displayed a clear maternal inheritance pattern, others did not. There were attempts to classify based upon the number and size of mitochondria in skeletal muscle, leading to terms such as pleoconial or megaconial myopathies,[2] and also on the pattern of respiratory chain involvement. There were those who wanted to sub-divide suspected mitochondrial disease into discrete categories (the "splitters"[3]) and those who thought of all mitochondrial disease as a single, if wide, spectrum of disorders (the "lumpers"[4]). At this early stage it was apparent that mitochondrial disorders were a heterogeneous group: clinically, histologically, biochemically, and probably genetically.

Following the discovery in the early 1960's that mitochondria contain their own DNA (mtDNA),[5] there were two major advances, both in the 1980s: the human mtDNA sequence was published in 1981,[6] and in 1988 the first pathogenic mtDNA mutations were identified.[7,8] The floodgates were opened, and the 1990s became the decade of the mitochondrial genome. Over 150 different pathogenic point mutations and a larger number of different rearrangements (i.e. partial deletions and duplications) of mtDNA were associated with disease,[9] and there were major advances in our understanding of the molecular pathophysiology.[10,11] There has been a change of emphasis in the first few years of the new millennium, away from the "magic circle" of mtDNA and back to the nuclear genome.[12] A number of nuclear genes have been identified that are fundamentally important for mitochondrial homeostasis, and when these genes are disrupted, they cause autosomally inherited mitochondrial disease.[13] Moreover, mitochondrial dysfunction plays an important role in the pathophysiology of a number of well-established nuclear genetic disorders, such as dominant optic atrophy (mutations in *OPA1*),[14] Friedreich's ataxia (*FRDA*),[15] hereditary spastic paraplegia (*SPG7*)[16] and Wilson's disease (*ATP7B*).[17] The next major challenge is to define the more subtle interactions between nuclear and mitochondrial genes in health and disease. It is likely that these mechanisms will have broader relevance for our understanding of many inherited and sporadic neurological disorders.

In this article we will review the basic scientific principles that underpin our understanding of mitochondrial pathology. Rather than giving a comprehensive description of mitochondrial biology, we will focus on the bare essential facts that will help the practicing general neurologist to understand, identify, investigate, and manage patients with primary mitochondrial disease (by which we mean disorders that result directly from mutations either in mtDNA or in nuclear genes affecting mtDNA homeostasis). Mitochondrial abnormalities have been identified in more common sporadic neurological disorders, including Alzheimer's disease and Parkinson's disease, and they also occur as part of normal aging.[18] The role of these secondary mitochondrial abnormalities is uncertain, and they will be discussed in other articles in this series.

WHAT ARE MITOCHONDRIA AND WHAT DO THEY DO?

Mitochondria are a sub-compartment of the cell bound by a double membrane. Although some mitochondria probably do look like the traditional cigar-shaped structures that appear in standard textbooks, it is more accurate to think of mitochondria as a budding and fusing network similar to the endoplasmic reticulum (Fig. 1). Mitochondria are intimately involved in cellular homeostasis. They play a part in intracellular signalling and apoptosis, intermediate metabolism, and in the metabolism of amino acids,

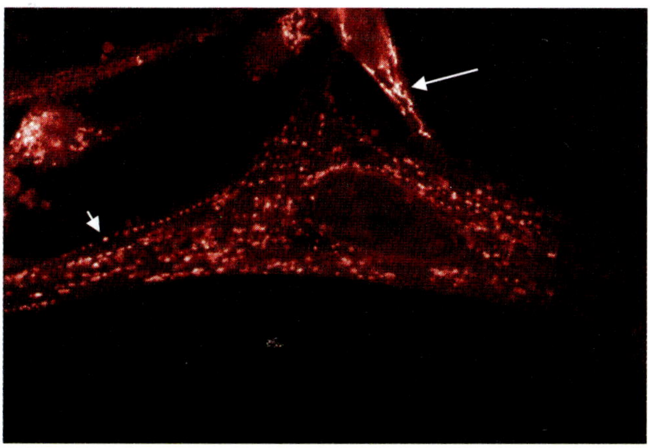

Fig. 1 Human mitochondria. Scanning fluorescence confocal micrograph of a cultured human myoblasts stained with the Mitotracker, which is a fluorescent potentiometric dye taken up specifically by living mitochondria because of their membrane potential. Some mitochondria form discrete organelles (short arrow), but others form a reticulate network (long arrow).

lipids, cholesterol, steroids, and nucleotides, among other functions. Apoptosis has been discussed in other chapters in this book and will not be considered here. Perhaps most importantly, mitochondria have a fundamental role in cellular energy metabolism. This includes fatty acid β oxidation, the urea cycle and the final common pathway for ATP production — the respiratory chain.

The mitochondrial respiratory chain is a group of five enzyme complexes situated on the inner mitochondrial membrane (Fig. 2). Each

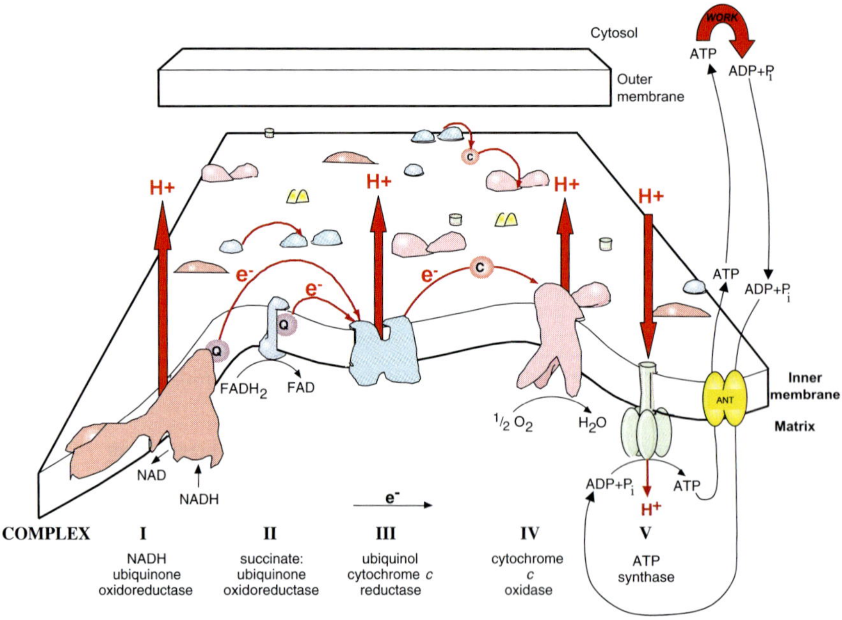

Fig. 2 The respiratory chain. Schematic diagram of the respiratory chain. Reduced cofactors (NADH and $FADH_2$) are produced from the intermediary metabolism of carbohydrates, proteins, and fats. These co-factors donate electrons to (e^-) to complex I (NADH-ubiquinone oxidoreductase) and complex II (succinate-ubiquinone oxidoreductase). These electrons flow between the complexes down an electrochemical gradient (black arrow), shuttled by ubiquinone (Q) and cytochrome *c* (C), involving complex III (ubiquinol-cytochrome *c* oxidase reductase) and complex IV (cytochrome *c* oxidase, or COX). Complex IV donates an electron to oxygen which results in the formation of water. Protons (H^+) are pumped from the mitochondrial matrix into the inter-membrane space (red arrows). This proton gradient generates the mitochondrial membrane potential which is harnessed by complex V to synthesise adenosine triphosphate (ATP) from adenosine diphosphate (ADP) and inorganic phosphate (Pi). ANT = adenine nucleotide translocator which exchanges ADP for ATP across the mitochondrial membrane.

complex is composed of multiple subunits, the largest being complex I with over 40 polypeptide components. Reduced co-factors (NADH and $FADH_2$) generated from the intermediary metabolism of carbohydrates, proteins, and fats donate electrons to complex I and complex II. These electrons flow between the complexes down an electrochemical gradient, shuttled by complexes III and IV and by two mobile electron carriers, ubiquinone (ubiquinol, co-enzyme Q10) and cytochrome *c*. The electron-transfer function of complexes I-IV is accomplished via subunits harbouring prosthetic groups (e.g. iron-sulfur groups in complexes I, II, and III, and heme iron in cytochrome *c* and in complex IV). The liberated energy is used by complexes I, III, and IV to pump protons (H^+) out of the mitochondrial matrix into the inter-membrane space. This proton gradient, which generates the bulk of the mitochondrial membrane potential (the asymmetric distribution of ions, such as Na^+, K^+, and Ca^{++}, across the inner membrane makes up the "chemical" portion of the gradient), is harnessed by complex V to synthesise adenosine triphosphate (ATP) from adenosine diphosphate (ADP) and inorganic phosphate. The overall process is called oxidative phosphorylation (OXPHOS). ATP is the high energy source used for essentially all active metabolic processes within the cell, and it must be released from the mitochondrion in exchange for cytosolic ADP. This is carried out by the adenine nucleotide translocator (ANT), which has a number of tissue specific isoforms.

Thus, the respiratory chain is an elaborate system that must respond to the energy requirements of the cell. While these requirements may be constant (e.g. in hepatocytes), the energy requirements may also change dramatically over short periods of time (e.g. in skeletal muscle). We are only just beginning to understand the mechanisms that maintain and regulate a healthy respiratory chain, and it is likely that many additional unknown genetic and environmental factors will be involved.

THE GENETIC BASIS OF MITOCHONDRIAL BIOGENESIS

Two distinct genetic systems encode mitochondrial proteins: mitochondrial DNA (mtDNA) and nuclear DNA (nDNA). MtDNA is a small 16.6-kb circle of double stranded DNA that specifies 13 respiratory chain polypeptides and 24 nucleic acids (2 ribosomal RNAs [rRNAs] and 22 transfer

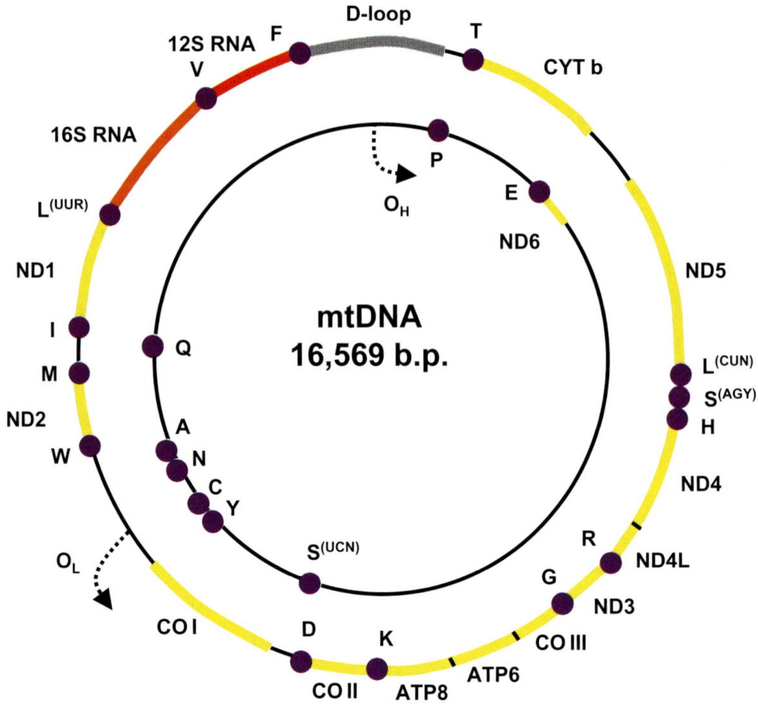

Fig. 3 The human mitochondrial genome. The human mitochondrial genome (mtDNA) is a small 16,569 kb molecule of double stranded DNA. MtDNA encodes for 13 essential components of the respiratory chain. ND1-ND6, and ND4L encode 7 subunits of complex I (NADH-ubiquinone oxidoreductase). Cyt b is the only mtDNA encoded complex III subunit (ubiquinol-cytochrome *c* oxidase reductase). COX I to III encode for three of the complex IV (cytochrome *c* oxidase, or COX) subunits, and the ATP 6 and ATP 8 genes encode for two subunits of complex V (ATP synthase). Two ribosomal RNA genes (12S and 16S rRNA), and 22 transfer RNA genes are interspaced between the protein-encoding genes. These provide the necessary RNA components for intra-mitochondrial protein synthesis. D-loop = the 1.1 kb non-coding region which is involved in the regulation of transcription and replication of the molecule, and is the only region not directly involved in the synthesis of respiratory chain polypeptides. O_H and O_L are the origins of heavy and light strand mtDNA replication.

RNAS [tRNAs]) that are needed for intra-mitochondrial protein synthesis (Fig. 3).[6] Nuclear genes code for the majority of mitochondrial respiratory chain polypeptides.[13] These polypeptides are synthesised in the cytoplasm with a mitochondrial targeting sequence that directs them through the translocation machinery spanning the outer and inner membranes. The targeting sequence is then cleaved before the subunit is assembled with its counterparts on the inner mitochondrial membrane. The components of the import machinery ("TIM" and "TOM" proteins), the importation processing enzymes, and the respiratory chain assembly proteins are all the products of nuclear genes.

Nuclear genes are also important for maintaining the mitochondrial genome, including those encoding the mitochondrial DNA polymerase γ (*POLG1*)[19] and products that maintain an appropriate balance of free nucleotides within the mitochondrion (*TP, TK, DGK* and *ANT1*).[20–23] A recently described gene, *C10orf2*, codes for a helicase-like protein called Twinkle that also appears to be important for mtDNA maintenance.[24] Nuclear DNA also codes for essential factors needed for intramitochondrial transcription and translation, including *TFAM, TFBM1 and TFBM2*.[25,26] A disruption of both nuclear and mitochondrial genes can therefore cause mitochondrial dysfunction and human disease (Table 1).

MITOCHONDRIAL DISEASE: GENOTYPE AND PHENOTYPE

A neurologist who has seen a few patients with mitochondrial disease will be puzzled by a number of questions. If mitochondria are so important, why don't mitochondrial diseases affect every tissue in the body, and if primary mitochondrial disorders are all genetically determined, why do they have such a varied clinical phenotype, even within the same family? Recent advances in our understanding of the molecular pathology of mitochondrial disease have provided us with some explanations and also raised new questions.

The clinical features of mitochondrial disease have been discussed widely elsewhere,[27–29] and they will not be considered in detail here (see Table 2 and Fig. 4 for an overview). In general terms, tissues and organs that are heavily dependent upon oxidative phosphorylation bear the brunt of the pathology. This means that neurological features are common, but cardiac, endocrine and ophthalmological features are often prominent. Other tissues are less dependent upon sustained oxidative phosphorylation, so they are less likely to be involved in mitochondrial disease, but there are some notable exceptions.

Table 1 Genetic classification of human mitochondrial disorders.

Mitochondrial genetic disorders (mtDNA nucleotide positions refer to the L-chain, and are taken from the "standard Cambridge" sequence)	*Inheritance pattern*
Rearrangements (large-scale partial deletions and duplications)	
Chronic progressive external ophthalmoplegia (CPEO)	S
Kearn's Sayre syndrome	S
Diabetes and deafness	S
Pearson marrow-pancreas syndrome	S
Sporadic tubulopathy	S
Point mutations	
Protein-encoding genes	
LHON (G11778A, T14484C, G3460A)	M
NARP/Leigh syndrome (T8993G/C)	M
tRNA genes	
MELAS (A3243G, T3271C, A3251G)	M
MERRF (A8344G, T8356C)	M
CPEO (A3243G, T4274C)	M
Myopathy (T14709C, A12320G)	M
Cardiomyopathy (A3243G, A4269G, A4300G)	M
Diabetes and deafness (A3243G, C12258A)	M
Encephalomyopathy (G1606A, T10010C)	M
rRNA genes	
Non-syndromic sensorineural deafness (A7445G)	M
Aminoglycoside induced non-syndromic deafness (A1555G)	M
Nuclear genetic disorders	*Inheritance pattern*
Disorders of mtDNA maintenance	
Autosomal dominant progressive external ophthalmoplegia (with 2° multiple mtDNA deletions)	
Mutations in adenine nucleotide translocator (*ANT1*)	AD
Mutations in DNA polymerase γ (*POLG*)	AD or AR
Mutations in Twinkle helicase (*C10orf2*)	AD
Mitochondrial neuro-gastrointestinal encephalomyopathy (with 2° multiple mtDNA deletions)	
Mutations in thymidine phosphorylase (*TP*)	AR
Myopathy with mtDNA depletion	
Mutations in thymidine kinase (*TK2*)	AR
Encephalopathy with liver failure	
Mutations in deoxyguanosine kinase (*DGK*)	AR

Table 1 (*Continued*)

Primary disorders of the respiratory chain	
Leigh syndrome	
Complex I deficiency — mutations in complex I subunits (*NDUFS2,4,7,8* and *NDUFV1*)	AR
Complex II deficiency — mutations in complex II flavoprotein subunit (*SDHA*)	AR
Leukodystrophy and myoclonic epilepsy	
Complex I deficiency — mutations in complex I subunit (*NDUFV1*)	AR
Cardioencephalomyopathy	
Complex I deficiency — mutations in complex I subunit (*NDUFS2*)	AR
Optic atrophy and ataxia	
Complex II deficiency — mutations in complex II flavoprotein subunit (*SDHA*)	AD
Disorders of mitochondrial protein import	
Dystonia-deafness	
Mutations in deafness-dystonia protein DDP1 (*TIMM8A*)	XLR
Disorders of assembly of the respiratory chain	
Leigh syndrome	
Complex IV deficiency — mutations in COX assembly protein (*SURFI*)	AR
Complex IV deficiency — mutations in COX assembly protein (*COX10*)	AR
Cardioencephalomyopathy	
Complex IV deficiency — mutations in COX assembly protein (*SCO2*)	AR
Hepatic failure and encephalopathy	
Complex IV deficiency — mutations in COX assembly protein (*SCO1*)	AR
Complex IV deficiency — mutations in protein affecting COX mRNA stability (*LRPPRC*)	AR
Tubulopathy, encephalopathy and liver failure	
Complex III deficiency — mutations in complex III assembly (*BCS1L*)	AR

(AD = autosomal dominant, AR = autosomal recessive, M = maternal, S = sporadic, XLR = X linked recessive)

Whilst this general rule is helpful, it can only be part of the explanation. Mitochondrial disease is ultimately due to a defect of oxidative phosphorylation within a cell, but the pattern of cellular involvement will determine the clinical features of the disease. On the one hand, there must be common mechanisms explaining why patients with mutations in nDNA and

Table 2 Clinical syndromes associated with mitochondrial disease.

	Primary features	Additional features
Chronic progressive external ophthalmoplegia (CPEO)	External ophthalmoplegia and bilateral ptosis	Mild proximal myopathy
Infantile myopathy and lactic acidosis (fatal and non-fatal forms)	Hypotonia in the first year of life. Feeding and respiratory difficulties	Fatal form may be associated with a cardiomyopathy and/or the Toni-Fanconi-Debre syndrome
Kearns-Sayre syndrome (KSS)	PEO onset before age 20 with pigmentary retinopathy Plus one of the following: CSF protein greater than 1 g/l, cerebellar ataxia, heart block.	Bilateral deafness Myopathy Dysphagia Diabetes mellitus and hypoparathyroidism Dementia
Leber hereditary optic neuropathy (LHON)	Subacute painless bilateral visual failure Males:females approx. 4:1 Median age of onset 24 years	Dystonia Cardiac pre-excitation syndromes
Leigh syndrome (LS)	Subacute relapsing encephalopathy with cerebellar and brain-stem signs presenting during infancy	Basal ganglia lucencies
Mitochondrial encephalomyopathy with lactic acidosis and stroke-like episodes (MELAS)	Stroke-like episodes before age 40 years Seizures and/or dementia Ragged-red fibres and/or lactic acidosis	Diabetes mellitus Cardiomyopathy (hypertrophic leading to dilated) Bilateral deafness Pigmentary retinopathy Cerebellar ataxia

Table 2 *(Continued)*

	Primary features	**Additional features**
Myoclonic epilepsy with ragged-red fibers (MERRF)	Myoclonus Seizures Cerebellar ataxia Myopathy	Dementia, optic atrophy Bilateral deafness Peripheral neuropathy Spasticity Multiple lipomata
Neurogenic weakness with ataxia and retinitis pigmentosa (NARP)	Late childhood or adult onset peripheral neuropathy with associated ataxia and pigmentary retinopathy	Basal ganglia lucencies Abnormal electroretinogram Sensori-motor neuropathy
Pearson Syndome	Sideroblastic anemia of childhood Pancytopenia Exocrine pancreatic failure	Renal tubular defects

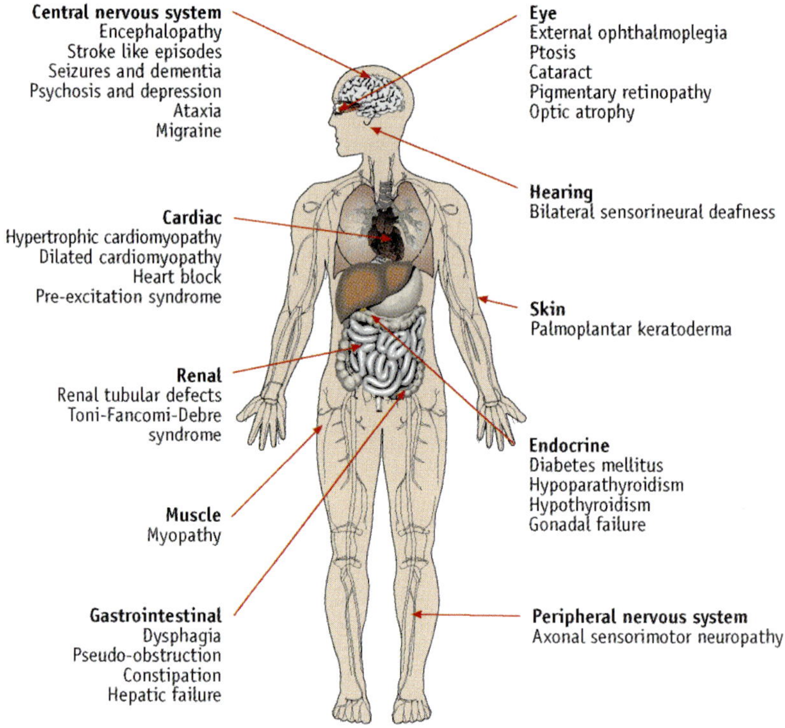

Fig. 4 Clinical features of mitochondrial disease Mitochondrial disease may present with single organ involvement (sensorineural deafness, diabetes, visual failure, myopathy or cardiomyopathy), or multi-system involvement. It may be possible to recognize a specific clinical syndrome (Table 2), but often patients do not fit neatly into one particular category. The combination of neurological disease and extra-neurological involvement should raise the suspicion of a mitochondrial disorder.

mtDNA can have a similar clinical phenotype (for example, clinically indistinguishable Leigh syndrome can be due to mutations both in the nuclear COX assembly gene *SURF1* and the mtDNA ATPase 6 gene[30,31]). On the other hand, there must be different mechanisms to explain why the same genetic defect can cause very different clinical phenotypes (for example, the A3243G mtDNA mutation in the $tRNA^{Leu(UUR)}$ gene can present with classical *m*itochondrial *e*ncephalomyopathy with *l*actic *a*cidosis and *s*troke like episodes (MELAS), with CPEO, or with diabetes and deafness[32–34]). It is also puzzling that many mitochondrial disorders affect multiple organ systems, whereas others have a highly stereotyped and organ-specific phenotype (such as Leber hereditary optic neuropathy and aminoglycoside induced deafness).[35,36] Nuclear genetic mechanisms have been discussed in the preceding articles in this series, so this article will concentrate on mitochondrial genetic factors.

MITOCHONDRIAL GENETIC FACTORS

MtDNA has unusual properties that are important for our understanding of mitochondrial disease due to mtDNA mutations.

Heteroplasmy and the threshold effect. Whilst most human cells contain two copies of nuclear DNA, they contain many more copies of mtDNA (from 1,000 to 100,000, depending on the cell type). These are all identical in a healthy individual at birth (*homoplasmy*). By contrast, patients harbouring pathogenic mtDNA defects often have a mixture of mutated and wild-type mtDNA (*heteroplasmy*).[30] The percentage of mutated mtDNA can vary widely among different patients, and also from organ to organ, and even between cells within the same individual. *In vitro* studies using "transmitochondrial cytoplasmic hybrid (cybrid)" cells[37] containing different amounts of mutated mtDNA have shown that most mtDNA mutations are highly recessive. In other words, the cells were able to tolerate high percentage levels of mutated mtDNA (typically 70–90%) before they developed a biochemical respiratory chain defect. The precise threshold for biochemical expression varies from mutation to mutation, and from tissue to tissue. Large retrospective studies have shown that the percentage level of mutated mtDNA in clinically relevant tissues does correlate with the severity of disease.[38,39]

Maternal inheritance and the genetic bottleneck. Although it has been known for some time that mtDNA is transmitted from mother to offspring,[40] the mechanisms are only just becoming clear. Sperm contain ~100 mtDNAs which enter the zygote on fertilization before being actively degraded.[41] There has been a recent report of a pathogenic mtDNA microdeletion in a patient with a sporadic muscle-specific mitochondrial disorder.[42] The mutated mtDNA arose on a mitochondrial genome that was paternal in origin, bringing in to question the traditional dogma of strict maternal inheritance. However, many families with mtDNA disease have been studied in detail over the last decade, and there are no other reports of paternal mtDNA transmission. Based upon the available evidence, paternal leakage is unlikely to be clinically significant.

One of the most remarkable features of mitochondrial disease due to mtDNA defects is the clinical variability amongst siblings. This is thought to be due to the mitochondrial "genetic bottleneck".[43] Our understanding of the transmission of mtDNA heteroplasmy has been greatly advanced by detailed studies of heteroplasmic mice generated by karyoplast transfer.[44,45] These mice transmit heteroplasmic mtDNA polymorphisms (see Table 3). By measuring the variation in heteroplasmy between the offspring of a single female, and comparing this to the variation between oocytes at different stages of development, it was shown that the transmitted percentage level of heteroplasmy is determined at an early stage during oogenesis in a heteroplasmic female developing *in utero*.[44] It is likely that there is a restriction in the number of mitochondrial genomes during early oogenesis, creating a functional "genetic bottleneck". This creates a sampling effect, akin to taking a small handful of marbles from a bag containing a large number of well-mixed black and white marbles whilst wearing a blindfold. Each independent sample will contain different proportions of the two types, corresponding to mutated and wild-type mtDNA in the offspring. Recent work suggests that the same random mechanism operates during the transmission of pathogenic mtDNA mutations in humans.[46,47] Whilst this generates variability in the transmitted mutation load to the offspring, this occurs within a given confidence interval, explaining why retrospective family studies have shown a relationship between the level of mutated mtDNA in the mother and the outcome of pregnancy.[39,48]

Although differences in the transmitted mutation load provide some explanation for the difference in severity between different family members, it does not explain why one sibling might present with neurological disease, whilst another might develop heart failure. Clearly additional factors must come in to play.

The percentage level of mutated mtDNA in individual tissues may also change during development and throughout adult life, potentially influencing the phenotype within an individual. Two mechanisms contribute to this process: relaxed replication and mitotic segregation.

Relaxed replication. Unlike nuclear DNA which replicates only once during each cell cycle, mtDNA is continuously recycled, even in non-dividing

Table 3 Animal models of mitochondrial disease. Adapted from.[97]

Type	Mouse model	Gene(s)[ref]	Biochemical abnormality	Phenotype
Nuclear gene knockouts	Adenosine nucleotide translocase	*ANT1*[98]	Defect of coupled respiration	Myopathy and cardiomyopathy
	Mitochondrial superoxide dismutase	*SOD2*[99]	Mitochondrial superoxide deficiency	Myopathy and cardiomyopathy
	Mitochondrial transcription factor A			
	Germ line	*TFAM*, germ line[25]	Respiratory chain defect	Embryonic lethal. Abnormal development with absence of heart and optic disc.
	Heart specific	*TFAM*, tissue specific[100]	Respiratory chain defect	Dilated cardiomyopathy and cardiac conduction block
	Pancreatic β-cell specific	*TFAM*, tissue specific[101]	Respiratory chain defect	Diabetes
	COX assembly protein SURF1	*SURF1*[102]	Respiratory chain defect	Embryonic lethality
	Thymidine phosphorylase	*TP*[103]	Reduction in liver TP activity	None
Mitochondrial DNA	BALB/NZB heteroplasmic	NZB/BALB mtDNA[44]	None	Random drift during transmission. Tissue specific selection of different genotypes.
	Chloramphenicol resistance	CAP-R T2443C mtDNA[104]	None	Myopathy, cardiomyopathy, perinatal death
	ΔmtDNA	4.7 kb mtDNA deletion[96]	Respiratory chain defect	Growth delay, nephropathy, myopathy
Spontaneous mutants	Defect of nuclear-mitochondrial communication	Not known[105]	None	Deafness

tissues such as skeletal muscle and brain.[49,50] MtDNA replication is therefore independent of the cell cycle (i.e. it is relaxed). In a heteroplasmic cell, it is possible that mutated and wild-type mtDNA replicate at subtly different rates — either because one type was selected for destruction or replication by chance, or because of a subtle selective effect in favour of one particular type. In theory, this mechanism could lead to changes in the proportion of mutated mtDNA that have been described in patients with mtDNA disease, providing an explanation for the late onset and progression of some mtDNA disorders.[51]

Mitotic (vegetative) segregation. When a heteroplasmic cell divides, subtle differences in the proportion of mutated mtDNA may be passed on to the daughter cells, leading to changes in the level of mutated mtDNA within a dividing tissue.[50,52] The unequal partitioning may be a purely random process, independent of any selection due to an effect of the mutation on mitochondrial function. On the other hand, presumed shifts due to functional selection may explain why the level of some pathogenic mtDNA mutations decreases in blood during life (e.g. 0.5 to 1% per annum for A3243G[53]).

MtDNA "background," nuclear genes, and the environment

Whilst there are a great many different heteroplasmic mtDNA mutations, in epidemiological terms most patients harbouring a pathogenic mtDNA defect harbour only mutated mtDNA (i.e. they are homoplasmic mutated).[54] The most common example is *L*eber *h*ereditary *o*ptic *n*europathy (LHON).[54] LHON is a mitochondrial genetic disorder that is primarily due to mutations in mtDNA complex I (ND) genes and is characterised by sub-acute bilateral visual failure presenting in early adult life.[55] LHON is intriguing because it is essentially an organ-specific disease that principally affects the retinal ganglion cells and the optic nerve.[56] LHON also has a markedly reduced penetrance with a clear gender bias, with only ~50% of males and ~10% of females developing visual failure.[35,57,58] Most patients with LHON are homoplasmic mutated for one of three mtDNA ND gene mutations (Fig. 4),[59] so heteroplasmy cannot explain the varied disease penetrance, and a number of unknown additional factors appear to be important.

Wild-type (normal) mtDNA can be subdivided into different genetic groups (haplogroups) based upon a characteristic pattern of polymorphisms that occur within the normal population.[60] Two of the three principal LHON mtDNA mutations (T14484C in the ND6 gene and G11778A in the ND4 gene) are preferentially associated with haplogroup J which is found in ~15% of northern Europeans.[61] The reason for this association is not known, but it seems likely that haplogroup J increases the penetrance of the T14484C and G11778A mutations.[62] It therefore appears that the mitochondrial genetic background can influence disease expression — but this cannot explain the gender bias in LHON.

The segregation pattern of disease in some LHON families suggests that there may be a nuclear genetic modifier locus modulating the clinical expression of the LHON mtDNA mutations. A recessive visual loss susceptibility locus on the X-chromosome would explain the gender bias in LHON,[63] but attempts to identify the locus have not been successful.[64] Environmental factors may also play a part in LHON. There are many anecdotal reports of visual failure following alcohol intoxication, starvation, heavy smoking, and head trauma,[58] but large studies have yielded conflicting results.[65,66]

In many ways LHON is best considered as a complex trait, where the disease phenotype arises through multiple genetic factors (bothmitochondrial and nuclear) interacting with the environment. A similar mechanism might explain the variable penetrance of other homoplasmic mtDNA mutations that cause organ-specific disease — such as the A1555G mtDNA mutation in the 12S rRNA gene that causes maternally inherited susceptibility to aminoglycoside-induced deafness, and possibly the A4300G mtDNA mutation in $tRNA^{Ile}$ that causes maternally-inherited cardiomyopathy (see Table 1). Similar nuclear-mitochondrial interactions are also likely to contribute to the varied phenotype seen in other mitochondrial disorders — be they due to primary nDNA or primary mtDNA defects.

Nuclear genes and mtDNA heteroplasmy

After heteroplasmic mice were generated from laboratory strains with two different mtDNA genotypes in the mid-'90's (see Table 3),[44] it became clear that a particular mitochondrial genome was favoured in some tissues, and the other mitochondrial genome was favoured in others.[67] Detailed experiments showed that this selective effect was not due to detectable differences in respiratory chain activity or rates of mtDNA replication, and that the selection appeared to be controlled at the level of the mtDNA molecule itself.[68] Recent work has identified three specific nuclear genetic loci that influence this process.[69] This has important implications for our understanding of mtDNA diseases because the equivalent genes in humans might influence the level of heteroplasmy in different tissues and organs, and therefore modulate the clinical phenotype.

Conclusion

The last five years has seen major advances in our understanding of mitochondrial genetics and how mtDNA mutations cause disease. Clinical expression is influenced by heteroplasmy, mtDNA background, nuclear genes and their interaction with the environment. Evolutionary studies are casting light on this complex relationship. For example, in the char (a fish), different environments, and particularly the water temperature, have selected in favour of a particular mitochondrial genotype.[70] Recent work on humans suggests that the same phenomenon may have occurred during population migrations throughout the world.[71] Understanding these processes is of fundamental importance for the clinical management of patients — from genetic counselling to developing new treatments.

CONFIRMING SUSPECTED MITOCHONDRIAL DISEASE

Our understanding of mitochondrial biochemistry and genetics has important implication for the investigation of suspected mitochondrial disease. In patients with a clearly defined clinical syndrome it may be possible to confirm the diagnosis with a simple molecular genetic test carried out on DNA extracted from blood. A good example of this is Leber hereditary optic neuropathy, where over 97% of cases are due to one three defined mtDNA point mutations that are usually homoplasmic in blood.[72] A similar approach may also be possible for nuclear genetic mitochondrial disorders (see Table 1, although most of these genetic tests are still within the realms of research and are not part of a routine diagnostic service). Investigating the remaining patients is more complex, partly because many disorders may mimic mitochondrial disease and also because there is no one single test that will prove or disprove that a patient has a mitochondrial disorder. Many different genetic defects in both mitochondrial and nuclear DNA can cause similar neurological disorders, so rather than carry out a series of random genetic tests, it is better to approach the problem systematically to identify and characterise the underlying metabolic defect.

Heteroplasmy is the main problem when investigating mtDNA disorders. Pathogenic mtDNA mutations may not be detectable in blood using conventional techniques, and, almost counter-intuitively, direct sequencing of mtDNA is the least robust technique of all. This means that a negative blood test result does not exclude a particular genetic diagnosis (for example see[73]). If mitochondrial disease is suspected, and the blood DNA tests are negative, the patient should have a muscle biopsy (usually the first choice in adults) or a skin biopsy (usually the first choice in children). Urine sediment, and to a lesser degree hair follicles, are excellent sources for non-invasive mtDNA testing.[74]

Fresh muscle can be analysed histologically and histochemically for evidence of mitochondrial disease. Characteristic features include ragged-red fibres which can be seen with the Gomori-trichrome stain.[75] or with succinate dehydrogenase histochemistry.[76] The ragged-red appearance is due to the sub-sarcolemmal accumulation of mitochondria and is thought to be a response to metabolic stress within a diseased muscle cell.[77] There may also be a reduction in cytochrome *c* oxidase activity (COX, complex IV) either within some of the fibres (a mosaic defect, suggestive of a mtDNA disorder[78] but see[79]) or affecting all the fibres within the entire biopsy (suggestive of a nuclear genetic defect).

Specialist centres carry out measurements of the individual respiratory chain complexes, which may also provide a clue to the underlying genetic defect. These can be carried out on fresh muscle or cultured fibroblasts grown from a skin biopsy. If a single complex is deficient, this points to a genetic defect in the relevant coding region of mtDNA or nuclear DNA, or a gene involved in the assembly of that particular complex. If there are multiple complex defects, that suggests a generalised defect of protein synthesis, and an underlying mtDNA defect involving a tRNA gene (including deletions that remove tRNA genes), or perhaps a nuclear gene defect with secondary effects on mtDNA. It is worth remembering that mitochondrial biochemical tests carried out on muscle and fibroblasts in the laboratory measure mitochondrial function under optimal conditions. It is therefore possible that there is a *functional* defect of mitochondrial metabolism that is not detectable in the laboratory (mutations in the ATPase 6 gene causing NARP are a good example of this pitfall). Evidence of impaired mitochondrial function may only be apparent on clinical testing using techniques such as exercise testing with lactate measurements,[80] magnetic resonance spectroscopy, [81] or infrared spectroscopy.[82]

A structured approach to investigation allows targeted genetic analysis, which often means a Southern blot of muscle mtDNA looking for a mtDNA rearrangement, a series of allele-specific assays looking for common point mutations of mtDNA or nuclear DNA, and direct sequencing of the relevant genes. A significant proportion of adults have rare or unique mtDNA defects (i.e. "private" mutations). These are identified by mtDNA sequencing which should also be carried out on DNA extracted from muscle (but with the caveat noted above).

Proving a mtDNA mutation is pathogenic

MtDNA is highly polymorphic, with any two individuals differing at up to 60 base pairs (see the mtDNA sequence databases in the websites box). The variation is so great that it is not unusual to find unique base changes in control individuals. This presents a particular problem when investigating patients with suspected mtDNA disorders — when is the base change a neutral polymorphism and when is it pathogenic?

Five "canonical" criteria suggest that a novel base change is pathogenic[29]:

1. The mutation must not be a known polymorphism (as described on one of the established sequence data bases, see the websites box).
2. The base change must affect a site that has been conserved during evolution. If the site is conserved across species then it implies that it is functionally important, and a mutation at this site is likely to be deleterious. The mutation must also be in a region that is functionally important. This essentially means anywhere in the tRNA genes, certain regions of the rRNA genes, or causing an amino acid change in the protein encoding genes.

BOX 1 AREAS FOR FUTURE DEVELOPMENT

- GENETIC COUNSELLING — There are currently no statistically-derived genetic counselling guidelines for mtDNA disease. A multi-national consortium is currently collecting data, and robust guidelines should become available in the next few years when a large cohort has been assembled.
- NUCLEAR-MITOCHONDRIAL INTERACTIONS — By studying relatively rare mitochondrial disorders we will increase our understanding of the way that mitochondria interact with cellular metabolism and particularly the cell nucleus. It is likely that this will have a broader relevance for other neurological diseases.
- ANIMAL MODELS — We currently do not have a good model for heteroplasmic mtDNA point mutation disorders (such as "MELAS" or "MERRF" — see Table 2), but there is considerable effort world wide to generate the model. Many of the unanswered questions about genotype and phenotype can be addressed when the model becomes available, and it may be used to test new treatments.
- NOVEL TREATMENTS — There are currently no treatments for mitochondrial disease, but a number of avenues show promise, and are likely to enter clinical practice within the next decade (see text).
- ENVIRONMENTAL/EXTERNAL FACTORS — Although it is generally accepted that exogenous factors influence mitochondrial function in humans, identifying them is proving difficult. Population based studies and the further investigation of animal models are likely to provide some insight. If these exogenous factors can be changed (by dietary or pharmacological manipulation), this approach may open up avenues for new treatments.

3. Deleterious mutations are usually (but not exclusively) heteroplasmic. This implies that the mutation occurred recently and it has not had time to "fix" in the female line, or that there has been selection against fixation acting at the level of the organism.
4. The mutation segregates with the disease clinically. For heteroplasmic mutations this means that severely affected individuals have a high percentage level of mutated mtDNA, and unaffected individuals have a lower percentage level of mutated mtDNA.
5. The mutation segregates with the disease biochemically. This is usually achieved by single cell mtDNA analysis.[83] Individual muscle fibres are micro-dissected from thick cross sections of muscle and the percentage level of mutated mtDNA is measured in histochemically normal and abnormal muscle fibres (either because they are ragged-red or COX deficient). For a pathogenic mutation, the percentage level of mutated mtDNA will be higher in the pathologically abnormal fibres.

BOX 2 MITOCHONDRIA ON THE WORLD WIDE WEB

Information	Name	URL
Information for patients	United Mitochondrial Diseases Foundation (USA)	www.umdf.org
	Mitolinks (UK)	http://www.communigate.co.uk/ne/mitolinks/index.phtml
	Leber hereditary optic neuropathy trust (UK)	www.leberstrust.btinternet.co.uk
Clinical and molecular information for clinicians and scientists	On-line Mendelian Inheritance in Man	www.ncbi.nlm.nih.gov
	Mitomap	www.mitomap.org
Clinical and genetic information for clinicians and patients	Geneclinics	www.geneclinics.org
Molecular data for scientists	Uppsala	http://www.genpat.uu.se/mtDB/index.html
Biochemical and molecular data for scientists	Mitop	mips.gsf.de/proj/medgen/mitop
	Mitodat	http://www-lecb.ncifcrf.gov/mitoDat/

Managing Mitochondrial Disease — the Future

Making a specific genetic diagnosis is helpful in a number of ways. It allows a comparison of that individual with other patients described in the literature, providing some guide to prognosis and highlighting complications that may evolve over time (Table 2). It also has implications for genetic counselling (Table 1).[84] Nuclear defects may be autosomal recessive, autosomal dominant, or sex-linked. MtDNA defects may be sporadic or maternally transmitted. There are no statistically-based robust counselling guidelines for mtDNA disease,[85] but data collection is underway, and they should become available over the next five years. Retrospective studies suggest that measuring the percentage level of mutated mtDNA in the mother will provide some guidance.[39,48]

At present the management of mitochondrial disease is largely supportive and aimed at identifying, preventing and treating complications wherever possible. A number of pharmacological treatments have been used with varying degrees of success (recently reviewed in[86]). Limited clinical trials have been carried out, but no consistent clinical improvements have been demonstrated. A multi-centre trial is currently underway for dichloracetate to reduce lactic acidosis in MELAS patients.

Our understanding of the basic biology of mitochondrial disease provides a basis for developing new treatments. A number of strategies have been employed to try and correct the underlying genetic defect. The overall aim is to reduce the proportion of mutated mtDNA to sub-threshold levels. This could be achieved by adding more wild-type mtDNA, or by removing mutated mtDNA.

1. *Adding wild-type mtDNA*. Despite initial promise,[87] attempts to deliver synthetic wild-type mtDNA into cells have not been successful. A more attractive strategy is to move wild-type mitochondrial genomes from one compartment to another — an approach called "gene shifting".[88–90] Healthy skeletal muscle contains small precursors called satellite cells. Satellite cells proliferate and fuse with the juxtapositionary mature skeletal fibres in response to stress and exercise. In some patients with mtDNA myopathy the percentage level of mutated mtDNA in satellite cells is lower that the level in affected skeletal muscle. It is possible to induce satellite cell proliferation by injecting a toxin into muscle (such as bupivacaine)[88,89] or by exercising the muscle.[90] Both techniques have been shown to deliver wild-type mtDNA from the satellite cell compartment into mature muscle fibres, to reduce the proportion of mutated mtDNA within affected tissues, and to correct the biochemical defect. Exercise also improves the strength and stamina of patients with mtDNA myopathy[91] — but there are concerns that it may also increase the amount of mutated mtDNA in the muscle, leading to short term improvements that may be detrimental in the longer term.[92]
2. *Removing mutated mtDNA*. Two strategies have been employed to remove mutated mtDNA. Both are at the experimental stage, and both

require considerable development before they can be used on patients. One approach has been to develop synthetic molecules that bind to mutated mtDNA molecules and prevent them from replicating, but allowing wild-type mtDNA replication to continue unimpeded.[93] Whilst this strategy works *in vitro*, and it appears that the "antigenomic" molecules can be delivered into mitochondria,[94] so far it has not been possible to influence the level of heteroplasmy in living cells. An alternative approach is to use drugs that select against mutated mtDNA in dividing cells, allowing wild-type mtDNA levels to increase.[95]

All of these approaches have the same drawback — even if they are effective, how can the treatments be delivered to the nervous system and alter the mtDNA levels in non-dividing cells? For this reason perhaps the best strategy is to remove all mutated mtDNA at an early stage in development — by nuclear transfer. By removing the nucleus from an affected zygote with a mtDNA mutation, and inserting it into a healthy enucleated donor with normal mtDNA, it should be possible to form healthy offspring that do not harbour the mtDNA defect, thereby preventing the disease in that individual, and also preventing further transmission of the disease. This approach is currently at an experimental stage, but provides some hope for the future.

ANIMAL MODELS

A number of mouse models for mitochondrial disease have been developed over the last five years. It is hoped that these models will advance our understanding of the pathophysiology and will also be useful for developing new treatments (Table 3). Most of the models were produced by nuclear gene manipulation, and some bear clinical or pathological resemblances to human mitochondrial diseases. Developing a good model for human mtDNA disorders is proving difficult because it is currently not possible to transfect mammalian mitochondria with exogenous DNA. To get around this problem, Hayashi and colleagues fused synaptic nerve terminals (synaptosomes) from aged mice harbouring low levels of mtDNA deletions with cybrid cells.[96] They screened the cybrid clones for detectable levels of mtDNA deletions and fused one with a mouse zygote that was implanted into a foster mother. The offspring contained a mixture of wild-type mtDNA and mtDNA with a 4.7-kb deletion. These mice share some clinical features with human mtDNA deletion disorders, but they were remarkable in two respects. First, the mice developed a nephropathy (a feature not typically found in KSS), and second, the female offspring also harboured deleted mtDNA (also not typical of KSS). Thus, like many other mouse models, there is not a complete correspondence between the human and the murine phenotype. To some extent this is inevitable — human mtDNA disorders are late-onset diseases, and it is difficult to mimic the effects of aging in other shorter-lived mammals. Attempts to generate mice transmitting mtDNA point mutations similar to those found in humans have not yet been forthcoming.

CONCLUSION

Mitochondria have an essential role in maintaining cellular homeostasis, and their many functions integrate closely with the cellular metabolic network. It is therefore naïve to think of mitochondria in isolation, and although we recognise primary disorders of mitochondrial function, the cellular and clinical consequences will depend on many other genetic and environmental factors. Precisely how the jigsaw fits together will vary from person to person. This presents a unique challenge to neurologists wanting to identify, diagnose and manage patients and families with mitochondrial disease.

These are not rare disorders. Recent epidemiological studies have shown that as a group primary mitochondrial disorders affect at least one in 8,000 of the general population,[86] and that number is likely to increase as the genotypic and phenotypic spectrum expands. It is also becoming clear that mitochondrial dysfunction occurs in many common sporadic neurological disorders, and there may be common mitochondrial mechanisms associated with certain neurological phenotypes (a good example being the various forms of hereditary spastic paraplegia). Our grasp of primary mitochondrial disorders will therefore have a much broader relevance, helping us to understand many other diseases and hopefully leading to novel generic treatments for neurological diseases.

Glossary

Term	Definition
ANT	Adenine nucleotide translocator. A transporter protein that exchanges ADP for ATP across the mitochondrial inner membrane. There are a number of tissue specific isoforms of ANT. Mutations in the nuclear gene *ANT1* cause autosomal dominant chronic progressive external ophthalmoplegia with secondary mtDNA deletion formation (see Table 1). There is a mouse knockout model for the cardiac isoform of ANT (see Table 3).
Apoptosis	Programmed cell death. A critical process essential for normal development and cell turnover. The release of cytochrome *c* from the inner mitochondrial membrane is one mechanism that can trigger apoptosis. This is discussed in detail in other articles in this series.
ATP	Adenosine triphosphate. A high energy phosphate molecule produced by the respiratory chain from ADP. ATP is required for all active cellular processes.
Cybrid cell	Cultured cell generated by fusing the cytoplasm of one cell (cytoplast) with that of another. This technique can be used to generate cell lines with different levels of heteroplasmy on a specific nuclear genetic background.
Cytoplasmic transfer	Technique used to generate heteroplasmic mice. Cytoplasm containing mitochondria from one inbred mouse strain (e.g. NZB) are transferred and fused with an early embryonic cell from another cell inbred mouse strain (e.g. BALB).
Genetic bottleneck (mitochondrial)	A restriction in the number of mitochondrial genomes during early development of the female germ line causes a "sampling effect" which results in great variation in the level of heteroplasmy amongst the offspring of a heteroplasmic mother.

Term	Definition
Haplogroup (mtDNA)	MtDNA is highly polymorphic within the general population. There are certain groups of polymorphisms that reflect the maternal ancestry of a particular individual. These large groups are called haplogroups, and they contain individual haplotypes.
Haplotype (mtDNA)	A specific mitochondrial genotype defined by a characteristic collection of mtDNA polymorphisms
Heteroplasmy	Mammalian cells contain many copies of mtDNA. Patients with mtDNA disease often harbour a mixture of mutated and wild-type mtDNA — a situation known as heteroplasmy. The proportion of mutated mtDNA can vary between 0 and 100%.
Homoplasmy	Mammalian cells contain many copies of mtDNA. Usually all of these copies are identical — a situation known as homoplasmy.
Knockout	Type of mouse model where a specific nuclear gene is removed.
Mitotic (vegetative) segregation	When a heteroplasmic cell divides, the daughter cells may receive different amounts of mutated mtDNA by chance
mtDNA	Mitochondrial DNA. The 16,569 base pair circular mitochondrial genome. Each cell contains thousands of copies.
Oxidative phosphorylation (OXPHOS)	Electrochemical process carried out by the respiratory chain. Electrons received from reduced co-factors are passed between the different respiratory chain complexes, protons are pumped from the mitochondrial matrix into the inter-membrane space generating the mitochondrial membrane potential. The influx of protons through complex V is used to generate ATP from ADP and inorganic phosphate.

Term	Definition
POLG	Polymerase γ — the only mitochondrial DNA polymerase. Mutations in the nuclear gene *POLG1* are a common cause of autosomal dominant and recessive chronic progressive external ophthalmoplegia with secondary mtDNA deletions (see Table 1).
Polymorphism	A phenotypically neutral natural variation in DNA code at a particular site. In strict terms a particular sequence change must be present in 1% of the population to be a polymorphism, but the term is often used to describe any phenotypically neutral sequence variant.
Relaxed replication	Unlike nuclear DNA which is only copied during cell division, mtDNA is continuously recycled within the cell (whilst the overall amount is maintained at roughly constant levels).
Respiratory chain	A group of four enzyme complexes (I–IV) situated on the inner mitochondrial membrane. Together with ATP synthase (complex V), the respiratory chain is the final common pathway for aerobic energy metabolism and the production of ATP
Satellite cell	Muscle cell precursor situated adjacent to the mature muscle fibre. Satellite cells proliferate in response to muscle stress and fuse with the mature muscle fibre.
Threshold effect	Most mtDNA mutations only cause a biochemical defect of the respiratory chain when the proportion exceeds a critical threshold level. This threshold varies from tissue to tissue, and from mutation to mutation.
Wild-type	Normal DNA found in the outbred population

ACKNOWLEDGEMENTS

PFC is funded by the Wellcome Trust and Ataxia (UK). EAS is funded by the U.S. National Institutes of Health, the Muscular Dystrophy Association, and by the National Organization for Rare Diseases. We are grateful to Dr Zofia Chrzanowska-Lightowlers for Fig. 2.

REFERENCES

1. Luft R, Ikkos D, Palmieri G, Ernster L, Afzelius B. A case of severe hypermetabolism of nonthyroid origin with a defect in the maintenance of mitochondrial respiratory control: a correlated clinical, biochemical and morphological study. *J Clin Invest* 1962; **41**:1776–1804.
2. Shy GM, Gonatas NK, Perez M. Two childhood myopathies with abnormal mitochondria. I. Megaconial myopathy. II. Pleioconial myopathy. *Brain* 1966; **89**:133–158.
3. Rowland LP, Hays AP, Di Mauro S, *et al.* Diverse clinical disorders associated with abnomalities of mitochondria. In: Scarlato G, Cerri C, eds. Mitochondrial Pathology in Muscle Diseases. Padova: Piccin, 1983; 141–158.
4. Petty RK, Harding AE, Morgan-Hughes JA. The clinical features of mitochondrial myopathy. *Brain* 1986; **109**:915–938.
5. Nass S, Nass MMK. Intramitochondrial fibres with DNA characteristics. *J Cell Biol* 1963; **19**:593–629.
6. Anderson S, Bankier AT, Barrell BG, *et al.* Sequence and organization of the human mitochondrial genome. *Nature* 1981; **290**:457–465.
7. Wallace DC, Singh G, Lott MT, *et al.* Mitochondrial DNA mutation associated with Leber's hereditary optic neuropathy. *Science* 1988; **242**:1427–1430.
8. Holt I, Harding AE, Morgan-Hughes JA. Deletion of muscle mitochondrial DNA in patients with mitochondrial myopathies. *Nature* 1988; **331**:717–719.
9. Servidei S. Mitochondrial encephalomyopathies: gene mutation. *Neuromuscul Disord* 2003; **13**:109–114.
10. Schon EA, Bonilla E, DiMauro S. Mitochondrial DNA mutations and pathogenesis. *J Bioenerget Biomemb* 1997; **29**:131–149.
11. Wallace DC. Mitochondrial diseases in mouse and man. *Science* 1999; **283**: 1482–1488.
12. DiMauro S. Mitochondrial encephalomyopathies: back to Mendelian genetics. *Ann Neurol* 1999; **45**:693–694.
13. Shoubridge EA. Nuclear genetic defects of oxidative phosphorylation. *Hum Mol Genet* 2001; **10**:2277–2284.

14. Alexander C, Votruba M, Pesch UE, *et al.* OPA1, encoding a dynamin-related GTPase, is mutated in autosomal dominant optic atrophy linked to chromosome 3q28. *Nat Genet* 2000; **26**:211–215.
15. Rotig A, de Lonlay P, Chretien D, *et al.* Aconitase and mitochondrial iron-sulphur protein deficiency in Friedreich ataxia. *Nat Genet* 1997; **17**:215–217.
16. Casari G, De Fusco M, Ciarmatori S, *et al.* Spastic paraplegia and OXPHOS impairment caused by mutations in paraplegin, a nuclear-encoded mitochondrial metalloprotease. *Cell* 1998; **93**:973–983.
17. Lutsenko S, Cooper MJ. Localization of the Wilson's disease protein product to mitochondria. *Proc Natl Acad Sci USA* 1998; **95**:6004–6009.
18. Leonard JV, Schapira AVH. Mitochondrial respiratory chain disorders II: neurodegenerative disorders and nuclear gene defects. *Lancet* 2000; **355**:389–394.
19. Van Goethem G, Dermaut B, Lofgren A, Martin J-J, Van Broeckhoven C. Mutation of *POLG* is associated with progressive external ophthalmoplegia characterized by mtDNA deletions. *Nat Genet* 2001; **28**:211–212.
20. Nishino I, Spinazzola A, Hirano M. Thymidine phosphorylase gene mutations in MNGIE, a human mitochondrial disorder. *Science* 1999; **283**:689–692.
21. Saada A, Shaag A, Mandel H, Nevo Y, Eriksson S, Elpeleg O. Mutant mitochondrial thymidine kinase in mitochondrial DNA depletion myopathy. *Nat Genet* 2001; **29**:342–344.
22. Mandel H, Szargel R, Labay V, *et al.* The deoyguanosine kinase gene is mutated in individuals with depleted hepatocerebral mitochondrial DNA. *Nat Genet* 2001; **29**:337–341.
23. Kaukonen J, Juselius JK, Tiranti V, *et al.* Role of adenine nucleotide translocator 1 in mtDNA maintenance. *Science* 2000; **289**:782–785.
24. Spelbrink JN, Li FY, Tiranti V, *et al.* Human mitochondrial DNA deletions associated with mutations in the gene encoding Twinkle, a phage T7 gene 4-like protein localised in mitochondria. *Nat Genet* 2001; **28**:223–231.
25. Larsson NG, Wang J, Wilhelmsson H, *et al.* Mitochondrial transcription factor A is necessary for mtDNA maintenance and embryogenesis in mice. *Nat Genet* 1998; **18**:231–236.
26. Falkenberg M, Gaspari M, Rantanen A, Trifunovic A, Larsson NG, Gustafsson CM. Mitochondrial transcription factors B1 and B2 activate transcription of human mtDNA. *Nat Genet* 2002; **31**:289–294.
27. Chinnery PF, Howell N, Andrews RA, Turnbull DM. Clinical mitochondrial genetics. *J Med Genet* 1999; **36**:425–436.
28. Leonard JV, Schapira AH. Mitochondrial respiratory chain disorders I: mitochondrial DNA defects. *Lancet* 2000; **355**:299–304.

29. DiMauro S, Schon EA. Mitochondrial DNA mutations in human disease. *Am J Med Genet* 2001; **106**:18–26.
30. Holt IJ, Harding AE, Petty RK, Morgan-Hughes JA. A new mitochondrial disease associated with mitochondrial DNA heteroplasmy. *Am J Hum Genet* 1990; **46**:428–433.
31. Tiranti V, Hoertnagel K, Carrozzo R, *et al.* Mutations of SURF-1 in Leigh disease associated with cytochrome c oxidase deficiency. *Am J Hum Genet* 1998; **63**:1609–1621.
32. Ciafaloni E, Ricci E, Shanske S, *et al.* MELAS: clinical features, biochemistry, and molecular genetics. *Ann Neurol* 1992; **31**:391–398.
33. Moraes CT, Ciacci F, Silvestri G, *et al.* Atypical clinical presentations associated with the MELAS mutation at position 3243 of human mitochondrial DNA. *Neuromuscul Disord* 1993; **3**:43–50.
34. Reardon W, Ross RJ, Sweeney MG, *et al.* Diabetes mellitus associated with a pathogenic point mutation in mitochondrial DNA. *Lancet* 1992; **340**:1376–1379.
35. Nikoskelainen EK. Clinical picture of LHON. *Clin Neurosci* 1994; **2**:115–120.
36. Prezant TR, Agapian JV, Bohlman MC, *et al.* Mitochondrial ribosomal RNA mutations associated with both antibiotic-induced and non-sydromic deafness. *Nat Genet* 1993; **4**:289–294.
37. King MP, Attardi G. Injection of mitochondria into human cells leads to a rapid replacement of the endogenous mitochondrial DNA. *Cell* 1988; **52**:811–819.
38. Chinnery PF, Howell N, Lightowlers RN, Turnbull DM. Molecular pathology of MELAS and MERRF. The relationship between mutation load and clinical phenotypes. *Brain* 1997; **120**:1713–1721.
39. White SL, Collins VA, Woolfe R, *et al.* Genetic counseling and prenatal diagnosis for the mitochondrial DNA mutations at nucleotide 8993. *Am J Hum Genet* 1999; **65**:474–482.
40. Giles RE, Blanc H, Cann HM, Wallace DC. Maternal inheritance of human mitochondrial DNA. *Proc Natl Acad Sci USA* 1980; **77**:6715–6719.
41. Sutovsky P, Moreno RD, Ramalho-Santos J, Domiko T, Simerly C, Schatten G. Ubiquitin tag for sperm mitochondria. *Nature* 1999; **403**:371–372.
42. Schwartz M, Vissing J. Paternal inheritance of mitochondrial DNA. *N Engl J Med* 2002; **347**:576–580.
43. Poulton J, Macaulay V, Marchington DR. Mitochondrial genetics '98: Is the bottleneck cracked? *Am J Hum Genet* 1998; **62**:752–757.
44. Jenuth J, Peterson AC, Fu K, Shoubridge EA. Random genetic drift in the female germ line explains the rapid segregation of mammalian mitochondrial DNA. *Nat Genet* 1996; **14**:146–151.

45. Meirelles F, Smith LC. Mitochondrial genotype segregation in a mouse heteroplasmic lineage produced by embyonic karyoplast transplantation. *Genetics* 1997; **145**:445–451.
46. Chinnery PF, Thorburn DR, Samuels DC, *et al.* The inheritance of mitochondrial DNA heteroplasmy: random drift, selection or both? *Trends Genet* 2000; **16**: 500–505.
47. Brown DT, Samuels DC, Michael EM, Turnbull DM, Chinnery PF. Random genetic drift determines the level of mutant mitochondrial DNA in human primary oocytes. *Am J Hum Genet* 2001; **68**:553–536.
48. Chinnery PF, Howell N, Lightowlers RN, Turnbull DM. MELAS and MERRF: the relationship between maternal mutation load and the frequency of clinically affected offspring. *Brain* 1998; **121**:1889–1894.
49. Bogenhagen D, Clayton DA. Mouse L cell mitochondrial DNA molecules are selected randomly for replication throughout the cell cycle. *Cell* 1977; **11**: 719–727.
50. Birky CW, Jr. The inheritance of genes in mitochondria and chloroplasts: laws, mechanisms, and models. *Annu Rev Genet* 2001; **35**:125–148.
51. Chinnery PF, Samuels DC. Relaxed replication of mtDNA: a model with implications for the expression of disease. *Am J Hum Genet* 1999; **64**:1158–1165.
52. Birky CW. Relaxed and stringent genomes: why cytoplasmic genes don't obey Mendel's laws. *J Heredity* 1994; **85**:355–365.
53. 't Hart LM, Jansen JJ, Lemkes HHPJ, de Knijff P, Maassen JA. Heteroplasmy levels of a mitochondrial gene mutation associated with diabetes mellitus decrease in leucocyte DNA upon aging. *Hum Mutat* 1996; **7**:193–197.
54. Chinnery PF, Johnson MA, Wardell TM, *et al.* Epidemiology of pathogenic mitochondrial DNA mutations. *Ann Neurol* 2000; **48**:188–193.
55. Howell N. Leber hereditary optic neuropathy: mitochondrial mutations and degeneration of the optic nerve. *Vision Res* 1997; **37**:3495–3507.
56. Saadati HG, Hsu HY, Heller KB, Sadun AA. A histopathologic and morphometric differentiation of nerves in optic nerve hypoplasia and Leber hereditary optic neuropathy. *Arch Ophthalmol* 1998; **116**:911–916.
57. Newman NJ, Lott MT, Wallace DC. The clinical characteristics of pedigrees of Leber's hereditary optic neuropathy with the 11 778 mutation. *Am J Ophthalmol* 1991; **111**:750–762.
58. Riordan-Eva P, Sanders MD, Govan GG, Sweeney MG, Da Costa J, Harding AE. The clinical features of Leber's hereditary optic neuropathy defined by the presence of a pathogenic mitochondrial DNA mutation. *Brain* 1995; **118**: 319–337.

59. Man PY, Griffiths PG, Brown DT, Howell N, Turnbull DM, Chinnery PF. The epidemiology of leber hereditary optic neuropathy in the north East of England. *Am J Hum Genet* 2003; **72**:333–339.
60. Torroni A, Huoponen K, Francalacci P, *et al.* Classification of European mtDNAs from an analysis of three European populations. *Genetics* 1996; **144**: 1835–1850.
61. Torroni A, Petrozzi M, D'Urbano L, *et al.* Haplotype and phylogenetic analyses suggest that one European-specific mtDNA background plays a role in the expression of Leber hereditary optic neuropathy by increasing the penetrance of the primary mutations 11778 and 14484. *Am J Hum Genet* 1997; **60**: 1107–1121.
62. Howell N, Kubacka I, Halvorson S, Howell B, McCullough DA, Mackey D. Phylogenetic analysis of the mitochondrial genomes from Leber hereditary optic neuropathy pedigrees. *Genetics* 1995; **140**:285–302.
63. Bu X, Rotter JI. X chromosomal-linked and mitochondrial gene control of Leber hereditary optic neuropathy: Evidence from segregation analysis for dependence on X-chromosome inactivation. *Proc Natl Acad Sci USA* 1991; **88**: 8198–8202.
64. Chalmers RM, Davis MB, Sweeney MG, Wood NW, Harding AE. Evidence against an X-linked visual loss susceptibility locus in Leber hereditary optic neuropathy. *Am J Hum Genet* 1996; **59**:103–108.
65. Kerrison JB, Miller NR, Hsu F, *et al.* A case-control study of tobacco and alcohol consumption in Leber hereditary optic neuropathy. *Am J Ophthalmol* 2000; **130**:803–812.
66. Tsao K, Aitken PA, Johns DR. Smoking as an aetiological factor in a pedigree with Leber's hereditary optic neuropathy. *Br J Ophthalmol* 1999; **83**:577–581.
67. Jenuth JP, Peterson AC, Shoubridge EA. Tissue-specific selection for different mtDNA genotypes in heteroplasmic mice. *Nat Genet* 1997; **16**:93–95.
68. Battersby BJ, Shoubridge E. Selection of a mtDNA sequence variant in hepatocytes of heteroplasmic mice is not due to difference in respiratory chain function or efficiency of replication. *Hum Mol Genet* 2001; **10**:2469–2479.
69. Battersby BJ, Loredo-Osti JC, Shoubridge EA. Nuclear genetic control of mitochondrial DNA segregation. *Nat Genet* 2003; **33**:183–186.
70. Blier PU, Dufresne F, Burton RS. Natural selection and the evolution of mtDNA-encoded peptides: evidence for intergenomic co-adaptation. *Trends Genet* 2001; **17**:400–406.
71. Mishmar D, Ruiz-Pesini E, Golik P, *et al.* Natural selection shaped regional mtDNA variation in humans. *Proc Natl Acad Sci USA* 2003; **100**:171–176.

72. Mackey DA, Oostra R-J, Rosenberg T, *et al.* Primary pathogenic mtDNA mutations in multigeneration pedigrees with Leber hereditary optic neuropathy. *Am J Hum Genet* 1996; **59**:481–485.
73. Chinnery PF, Reading PJ, Walls TJ, Turnbull DM. Recurrent strokes in a 34 year old man. *Lancet* 1997; **350**:560.
74. Dubeau F, De Stefano N, Zifkin BG, Arnold DL, Shoubridge EA. Oxidative phosphorylation defect in the brains of carriers of the $tRNA^{leu(UUR)}$ A3243G mutation in a MELAS pedigree. *Ann Neurol* 2000; **47**:179–185.
75. Engel WK, Cunningham CG. Rapid examination of muscle tissue: an improved trichrome stain for fresh-frozen biopsy sections. *Neurology* 1963; **13**:919–923.
76. Old SL, Johnson MA. Methods of microphotometric assay of succinate dehydrogenase and cytochrome c oxidase activities for use on human skeletal muscle. *Histochem J* 1989; **21**:545–555.
77. Hayashi J, Ohta S, Kikuchi A, Takemitsu M, Goto Y, Nonaka I. Introduction of disease-related mitochondrial DNA deletions into HeLa cells lacking mitochondrial DNA results in mitochondrial dysfunction. *Proc Natl Acad Sci USA* 1991; **88**:10614–10618.
78. Johnson MA, Bindoff LA, Turnbull DM. Cytochrome*c* oxidase activity in single muscle fibres: assay techniques and diagnostic applications. *Ann Neurol* 1993; **33**:28–35.
79. Sasarman F, Karpati G, Shoubridge EA. Nuclear genetic control of mitochondrial translation in skeletal muscle revealed in patients with mitochondrial myopathy. *Hum Mol Genet* 2002; **11**:1669–1681.
80. Taivassalo T, Abbott A, Wyrick P, Haller RG. Venous oxygen levels during aerobic forearm exercise: An index of impaired oxidative metabolism in mitochondrial myopathy. *Ann Neurol* 2002; **51**:38–44.
81. Argov Z, Bank W. Phosphorus magnetic resonance spectroscopy (31-P MRS) in Neuromuscular disorders. *Ann Neurol* 1991; **30**:90–97.
82. van Beekvelt MC, van Engelen BG, Wevers RA, Colier WN. Quantitative near-infrared spectroscopy discriminates between mitochondrial myopathies and normal muscle. *Ann Neurol* 1999; **46**:667–670.
83. Moraes CT, Schon EA. Detection and analysis of mitochondrial DNA and RNA in muscle by in situ hybridization and single-fiber PCR. *Methods Enzymol* 1996; **264**:522–540.
84. Thorburn DR, Dahl HHM. Mitochondrial disorders: genetics, counseling, prenatal diagnosis and reproductive options. *Am J Med Genet* 2001; **106**:102–114.
85. Poulton J, Turnbull DM. 74th ENMC International workshop: Mitochondrial diseases. *Neuromuscul Disord* 2000; **10**:460–462.

86. Chinnery PF, Turnbull DM. The epidemiology and treatment of mitochondrial disease. *Am J Med Genet* 2001; **106**:94–101.
87. Seibel P, Trappe J, Villani G, Klopstock T, Papa S, Reichmann H. Transfection of mitochondria: strategy towards a gene therapy of mitochondrial DNA diseases. *Nucleic Acids Res* 1995; **23**:10–17.
88. Fu K, Hartien R, Johns T, Genge A, Karpati G, Shoubridge EA. A novel heteroplasmic $tRNA^{leu(UUR)}$ mtDNA point mutation in a sporadic patient with mitochondrial encephalomyopathy segregates rapidly in muscle and suggests an approach to therapy. *Hum Mol Genet* 1996; **5**:1835–1840.
89. Clark K, Bindoff LA, Lightowlers RN, *et al.* Correction of a mitochondrial DNA defect in human skeletal muscle. *Nat Genet* 1997; **16**:222–224.
90. Taivassalo T, Fu K, Johns T, Arnold D, Karpati G, Shoubridge EA. Gene shifting: a novel therapy for mitochondrial myopathy. *Hum Mol Genet* 1999; **8**:1047–1052.
91. Taivassalo T, Matthews PM, De Stefano N, *et al.* Combined aerobic training and dichloracetate improve exercise capacity and indices on aerobic metabolism in muscle cytochrome oxidase deficiency. *Neurology* 1996; **47**:529–534.
92. Taivassalo T, Shoubridge EA, Chen J, *et al.* Aerobic conditioning in patients with mitochondrial myopathies: physiological, biochemical, and genetic effects. *Ann Neurol* 2001; **50**:133–141.
93. Taylor RW, Chinnery PF, Turnbull DM, Lightowlers RN. Selective inhibition of mutant human mitochondrial DNA replication in vitro by peptide nucleic acids. *Nat Genet* 1997; **15**:212–215.
94. Chinnery PF, Taylor RW, Diekert K, Lill R, Turnbull DM, Lightowlers RN. Peptide nucleic acid delivery into human mitochondria. *Gene Therapy* 1999; **6**: 1919–1928.
95. Manfredi G, Gupta N, Vazquez-Memije ME, *et al.* Oligomycin induces a decrease in the cellular content of a pathogenic mutation in the human mitochondrial ATPase6 gene. *J Biol Chem* 1999; **274**:9386–9381.
96. Inoue K, Nakada K, Ogura A, *et al.* Generation of mice with mitochondrial dysfunction by introducing mouse mtDNA carrying a deletion into zygotes [In Process Citation]. *Nat Genet* 2000; **26**:176–181.
97. Larsson NG, Rustin P. Animal models for respiratory chain disease. *Trends Mol Med* 2001; **7**:578–581.
98. Graham BH, Waymire KG, Cottrell B, Trounce IA, MacGregor GR, Wallace DC. A mouse model for mitochondrial myopathy and cardiomyopathy resulting from a deficiency in the heart/muscle isoform of the adenine nucleotide translocator. *Nat Genet* 1997; **16**:226–234.

99. Li Y, Huang TT, Carlson EJ, *et al.* Dilated cardiomyopathy and neonatal lethality in mutant mice lacking manganese superoxide dismutase. *Nat Genet* 1995; **11**:376–381.
100. Wang J, Wilhelmsson H, Graff C, *et al.* Dilated cardiomyopathy and atrioventricular conduction blocks induced by heart-specific inactivation of mitochondrial DNA expression. *Nat Genet* 1999; **21**:133–137.
101. Silva JP, Kohler M, Graff C, *et al.* Impaired insulin secretion and beta-cell loss in tissue-specific knockout mice with mitochondrial diabetes. *Nat Genet* 2000; **26**:336–340.
102. Agostino A, Invernizzi F, Tiveron C, *et al.* Constitutive knockout of Surf1 is associated with high embryonic lethality, mitochondrial disease and cytochrome c oxidase deficiency in mice. *Hum Mol Genet* 2003; **12**:399–413.
103. Haraguchi M, Tsujimoto H, Fukushima M, *et al.* Targeted deletion of both thymidine phosphorylase and uridine phosphorylase and consequent disorders in mice. *Mol Cell Biol* 2002; **22**:5212–5221.
104. Sligh JE, Levy SE, Waymire KG, *et al.* Maternal germ-line transmission of mutant mtDNAs from embryonic stem cell-derived chimeric mice. *Proc Natl Acad Sci USA* 2000; **97**:14461–14466.
105. Johnson KR, Zheng QY, Bykhovskaya Y, Spirina O, Fischel-Ghodsian N. A nuclear-mitochondrial DNA interaction affecting hearing impairment in mice. *Nat Genet* 2001; **27**:191–194.

5

Nanotechnology for Neuronal Ion Channels

F Lehmann-Horn and K Jurkat-Rott*

Ion channels provide the basis for the regulation of electrical excitability in the central and peripheral nervous systems. This review deals with the techniques that make the study of structure and function of single channel molecules in living cells possible. These are the patch clamp technique, which was derived from the conventional voltage clamp method and is currently being developed for automated and high throughput measurements; and fluorescence and nano-techniques, which were originally applied to non-biological surfaces and are only recently being used to study cell membranes and their proteins, especially in combination with the patch clamp technique. The characterisation of the membrane channels by techniques that resolve their morphological and physical properties and dynamics in space and time in the nano range is termed nanoscopy.

*Professor Frank Lehmann-Horn, Abteilung Angewandte Physiologie, Universität Ulm, D-89081 Ulm, Germany; frank.lehmann-horn@medizin.uni-ulm.de

CHANNELS — ESSENTIAL FOR NEURONAL EXCITABILITY

Ion channels are proteins that are equipped with a membrane spanning ion conducting pore. Most channels possess gates and voltage or ligand sensors. They activate — that is, open — in response to ligands, especially neurotransmitters, or voltage changes, and desensitise or inactivate — that is, close — by a normally intrinsic inactivation process. They are the molecular basis for intracellular signal transduction, maintenance of the resting potential, and the generation of excitatory and inhibitory potentials, particularly the action potential, the basic element of information in the brain.

Ion channels are integral proteins in the outer and inner membranes of excitable and non-excitable cells. Their essential property is a membrane spanning pore that acts as a conducting pathway for the flow of ions between the intracellular and extracellular spaces (Fig. 1). As a special feature, ion channels are opened and closed by gates. Channel opening (activation) may be effected by ligands, transmitters, a force directly acting on the channel, or changes of the transmembrane voltage. The ensuing closure (inactivation) is usually an intrinsic process. Special voltage sensing transmembrane segments of the protein control the gates in those channels that are responsible for the generation of excitatory and inhibitory subsynaptic potentials and the action potentials. They provide the basis for the regulation of excitability in the central and peripheral nervous system and the skeletal muscle (Fig. 2).

Structures of importance like those determining selectivity filters, voltage sensors, ligand binding sites, and gates have been highly conserved for more than 600 million years. Thus by creating closely related channel siblings, evolution has developed on the basis of a few mechanisms a way to generate a host of physiological functions, including possibilities for compensating for functional defects.

HISTORY OF THE VOLTAGE CLAMP TECHNIQUE

Modern research into the properties of ionic channels was initiated by the pioneering work of Hodgkin and Huxley,[3] who employed the voltage clamp technique using intracellular microelectrodes to provide the first detailed description of the ionic basis of the action potential in nerve axons.

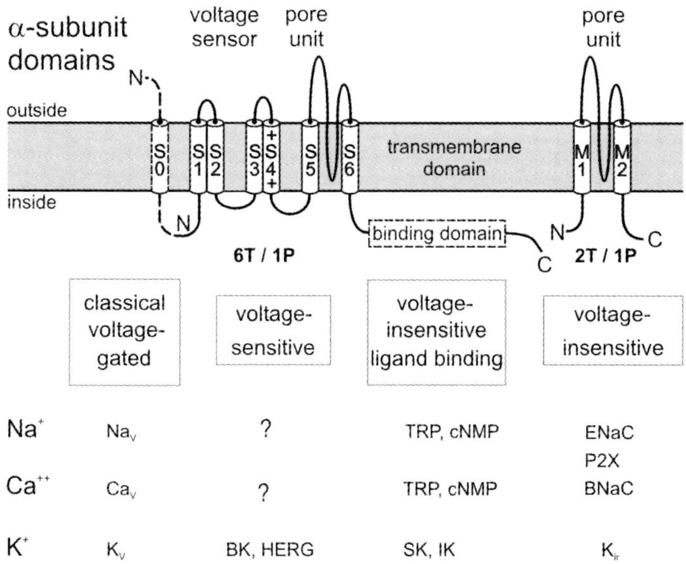

Fig. 1 Diversity of domains forming cation channel α-subunits. The most simple domain, typically used for ingoing rectifier potassium channel α-subunits, is a pore unit (2T/1P) that consists of two transmembrane segments M1 and M2, an extracellular loop dipping into the membrane and lining the pore, and intracellular N- and C-terminals. The transmembrane segments are thought to be α-helices. All voltage gated α-subunit domains are 6T/1P domains as they contain a four transmembrane segment unit, S1 to S4, acting as voltage sensor, and the two transmembrane pore unit. S4 is the particular voltage sensing segment that contains positive charges at each third amino acid residue. Ligand gated cation channel α-subunit domains usually possess a C-terminal binding site in addition to the 6T/1P domain. Although some ligand gated channels — for example, the calcium activated SK potassium channel — contain a positively charged S4 segment, they are not voltage sensitive at all, maybe because of uncoupling of sensor and activation gate. BK (big conductance K^+) potassium channels possess an additional S0 segment. The following channels complete the classification of the α-subunit domains: HERG, a potassium channel encoded by the human ether-a-go-go related gene that is similar to the *drosophila* ether-a-go-go gene (eag); IK, a calcium activated potassium channel with intermediate conductance; cNMP or CNG, sodium or calcium channels that are cyclo-nucleotide monophosphates gated such as cGMP; and voltage insensitive sodium channels of epithelial cells (ENaC) and in free nerve terminals of the brain (BNaC). BNaC was later found to conduct calcium. Not shown is another group of α-subunit domains, 4T/2P, which contain four transmembrane segments and two pore units.

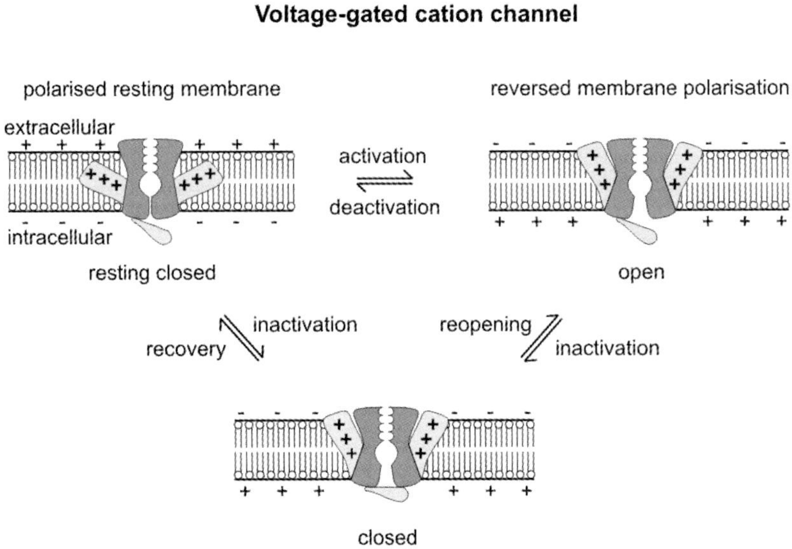

Fig. 2 Scheme of three potential states of voltage gated cation channels such as those for sodium, calcium, and potassium ions. The resting closed channel state (upper left panel) is activated by membrane depolarisation which causes a fast transition to the open state (upper right). Owing to an intrinsic inactivation as present in sodium and calcium channels and also in some potassium channels, the channel closes (lower panel) and reopens very rarely. Repolarisation of the membrane leads to recovery from the inactivated (refractory) state back to the resting state (upper left) from activation is again possible. There are probably more than one open and at least two inactivated (fast and slow) states (not shown). Note that transition from the resting to the inactivated state is also possible without channel opening, particularly during slow depolarisation (so called accommodation). Note also that the amphipathic voltage sensor helices move within the lipid bilayer when the membrane polarity changes. The voltage sensors are formed by the S3/S4 segments (see Fig. 6) and act as a "paddle" in the membrane. Modified after Jiang *et al.*, 2003.[1,2]

Their work provided our first look at some of the functional properties of voltage gated sodium and potassium membrane *particles* before the first proof of the existence of channel proteins. For the following 50 years, the voltage clamp became the principal tool for the study of channels. Three more recent developments have revolutionised this field. The first is the patch clamp technique, developed by Neher and colleagues,[4] which is a specific application of voltage clamping developed to record the current through a membrane patch conducted by a single channel molecule. The second is the use of molecular cloning techniques to isolate channel genes, thereby determining the primary structure of the channel. The third is fluorescence microscopy, which makes functional analysis of channel ensembles and single channel molecules possible.

The patch clamp technique allows direct electrical measurement of ion channel currents while simultaneously controlling the cell's membrane potential. It relies on the use of a fine tipped glass capillary to make contact with a patch of a cell membrane in order to form a giga-ohm seal. This high resistance seal was originally applied to skeletal muscle fibres by enzymatic treatment that removed the basal membrane, glycocalix, and connective tissue. The treated preparation allowed Neher und Sakmann to attach the glass pipette to the plasma membrane with a leak resistance of 10 to 50 megohm. Suction resulted in a giga-ohm seal and enabled the measurements of currents in the 50 pA range with little noise.

Current variants of this technique make possible the application of solution on the exterior and interior of whole cells and on membrane patches torn from the cell (outside-out or inside-out) — every conceivable configuration of solution and ion channel orientation that the heart of ion channel researcher craves (Fig. 3). Usually, primary cultured cells or cell lines are preferred as they reveal a relatively clean surface membrane[5] and require no enzymatic treatment likely to damage the plasma membrane. The patch clamp technique is now the gold standard measurement for characterising and studying ion channels, and one of the most important methods applied to neurophysiology.

Single channel recordings have shown that many channels — for example, the voltage gated sodium channel — possess only two conductance levels: zero when the channel is closed, and a constant conductance when the channel is open (Fig. 4A, 4B). Following depolarisation, there is a brief delay before channels open. The intervals are not identical during each depolarisation; in fact, the opening and closing of a given single channel is a random process, even though the open probability depends on the voltage and is more sensitive to the voltage than an electronic device such as a transistor. After a subsequent short interval — the open time — the current jumps back to zero as the channels close.

The stochastic nature can be understood by certain energy barriers that must be overcome before a channel can flip from one conformation (for example, open) to another (closed). The energy needed for this purpose comes from the random thermal energy of the system. One can imagine

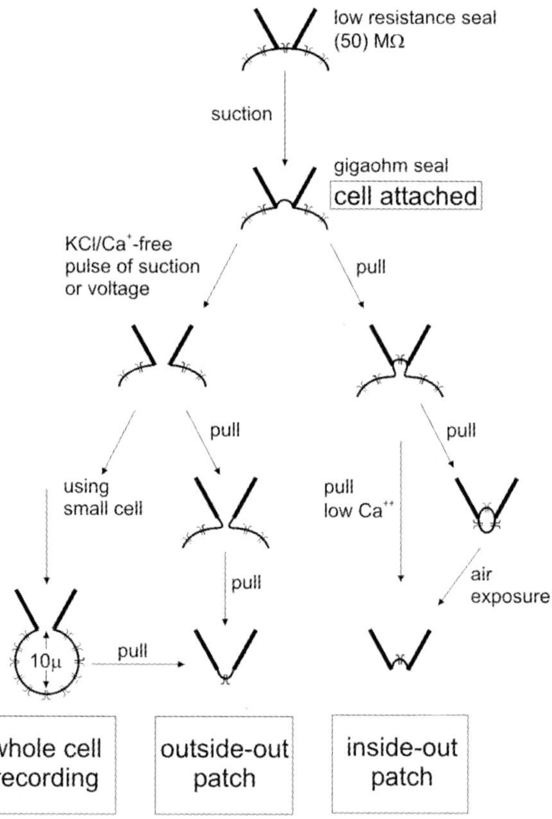

Fig. 3 Schematic representation of the procedures for various recording configurations. A fine tipped (about 0.5 to 5 μm in tip diameter) glass patch electrode is used as a current monitor and the voltage in the pipette is held at a desired level. The first step in applying the technique is the formation of a high resistance seal between the patch electrode and the surface of the cell. Once the seal is established, several recording configurations are available to the investigator, and these fall into two broad categories. On the one hand, current flow through the patch of membrane under the electrode tip can be monitored, in which case single channel currents are usually recorded. Alternatively, for whole cell recording, the patch of membrane can be disrupted so that the electrode monitors current flow through the entire cell surface. The symbol]⌈ is inserted for easier recognition of the orientation of the membrane and the channels. Modified after Sakmann and Neher, 1985.[6]

that each time the channel molecule vibrates, bends, or stretches, it has a chance to surmount the energy barrier. Each motion is like a binomial trial with a certain probability of success. As the protein movements are on a picosecond time scale, but the channel stays open for milliseconds, clearly

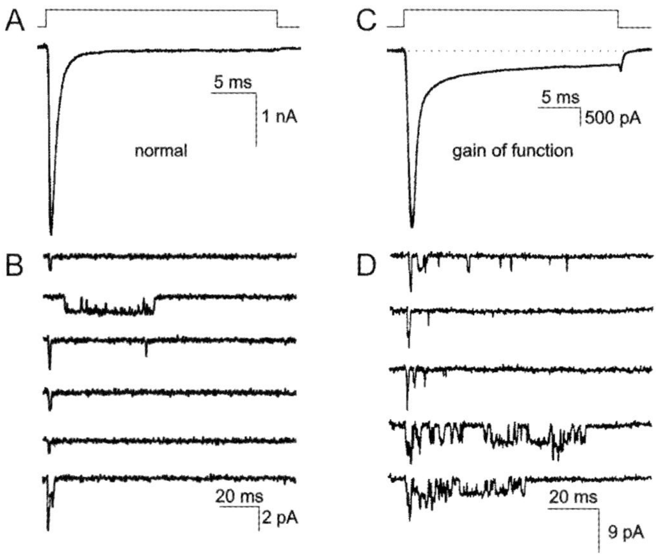

Fig. 4 Single channel recordings of sodium currents on two membrane patches, one from an adult muscle fibre from a healthy control (left), the other from a fibre from a patient with hyperkalaemic periodic paralysis, a sodium channelopathy. Both patches contained four sodium channels in the bleb attached mode. (A), (C): Superposition of 500 depolarisation steps from -100 mV to -30 mV, as indicated by the upper rectangular pulse, leads to a "macroscopic" sodium current peak. (B), (D): Six of the 500 original current traces were selected to show the downward deflections corresponding to channel openings. Note that channel bursts and late reopenings in the original traces occur much more often in the mutant channels than in the normal channels and lead to a persistent current in the averaged curve as long as the membrane is depolarised by the artificial voltage step, or — for example, *in vivo* — by increased extracellular potassium (hence the name *myotonia*, namely potassium aggravated myotonia).[7]

the chance of success at each trial must be small, and a large number of trials will be needed before the channel shuts.

Usually, a normal sodium channel does not reopen even though the depolarisation may be maintained by the voltage clamp step for a certain time (Fig. 4B). Remarkably, the average behaviour of a single channel is identical in time course to the macroscopic, whole cell sodium current (not shown).

By combining the patch clamp with molecular cloning techniques, the function and significance of potentially important amino acid residues are rapidly being elucidated. A common approach in these studies is as follows.

The gene of interest is cloned and then a designed or naturally occurring mutation is inserted into the clone by mutagenesis in the hope that the mutation will produce measurable changes in channel function. Then a heterologous expression system — for example, a cell line of a different tissue which does not endogenously express the gene of interest — is used to express the gene. This expression can either be transient, as in RNA injected *Xenopus* oocytes, or stable, as in virus infected or transfected cells. Finally, the patch clamp technique is used to characterise the function of a channel ensemble or a single protein molecule.

By combining this functional expression with fluorescence microscopy or other microscope techniques such as atomic force microscopy, structure–function correlations can be determined (see below).

APPLICATION TO CLINICAL NEUROLOGY

Molecular Biology and Patch Clamping Combined: The Key to Identifying Pathogenesis

In the last 10 years, the combination of electrophysiological and molecular genetic investigations has led to the exploration of the growing family of diseases caused by mutations in genes encoding voltage and ligand gated ion channels, the so called channelopathies. Most channelopathies require a non-normal situation, a so called trigger, before they present with recognisable symptoms, and only a few lead to permanent disability — for example, episodic ataxia type 2 or hypokalaemic periodic paralysis. Typically the symptoms occur as episodic attacks lasting from minutes to days, which show spontaneous and complete remission, onset in the first or second decade of life, and — for an unknown reason — amelioration after the age of 40 or 50. Channelopathies are found in very different tissues such as brain, muscle, and secretory tissue.[9] Skeletal muscle was the first tissue in which such diseases — namely the myotonias and periodic paralyses — were recognised as ion channelopathies. Examples of neuronal channelopathies are episodic ataxia, familial hemiplegic migraine, and various types of epilepsy (see Table 1). Channelopathies are also important models for more frequent disorders of non-monogenic aetiology.

For an understanding of the pathology of a given hereditary disorder, knowledge of the mutated gene is only a first step. More revealing is the

Table 1 Hereditary Channelopathies of the Central and Peripheral Nervous System.

Table width = E

Gene	Locus	Channel protein	Disease	Inheritance	Change
Cardiac muscle					
KCNQ1	11p15.5	Potassium channel α subunit, KCNQ1,	Long QT syndrome 1 Jervell and Lange-Nielsen	Dominant Recessive	Loss Loss
HERG	7q35-36	Potassium channel α subunit, HERG, eag related, Ikr	Long QT syndrome 2	Dominant	Loss
SCN5A	3p21	Sodium channel α subunit	Long QT syndrome 3 Brugada syndrome	Dominant Dominant	Gain Loss
KCNE1	21q22	β subunit of KVLQT1, MinK	Long QT syndrome 5	Dominant	Loss
KCNE2		β subunit of HERG, MiRP1	Long QT syndrome inducible	Dominant	Loss
RYR2	1q42-43	Ryanodine receptor 1, calcium release channel type 2	Catecholaminergic ventricular fibrillation	Dominant	Gain
CASQ2	1p13-11	Calsequestrin type 2, fast twitch	Catecholaminergic ventricular fibrillation	Recessive	Gain

Table 1 (*Continued*)

Gene	Locus	Channel protein	Disease	Inheritance	Change
Skeletal muscle					
SCN4A	17q23.1-25.3	Voltage gated sodium channel Nav1.4 α subunit	Hyperkalaemic periodic paralysis	Dominant	Gain
			Paramyotonia congenita	Dominant	Gain
			Potassium aggravated myotonia	Dominant	Gain
			Hypokalaemic periodic paralysis 2	Dominant	Loss
CACNA1S	1q31-32	Voltage gated L-type calcium channel α 1 subunit, dihydropyridine (DHP) receptor	Hypokalaemic periodic paralysis type 1	Dominant	Unclear
			Malignant hyperthermia type 5	Dominant	Unclear
RYR1	19q13.1	Ryanodine receptor 1, calcium release channel	Malignant hyperthermia type 1	Dominant	Gain
			Central core disease	Dominant	Gain
KCNJ2	17q23-24	Voltage insensitive potassium channel α subunit Kir2.1	Andersen's syndrome, long QT syndrome 7	Dominant	Loss
CLCN1	7q32-qter	Voltage gated chloride channel ClC1	Thomsen myotonia	Dominant	Loss
			Becker myotonia	Recessive	Loss
		Altered splicing	Myotonic dystrophy type 1	Dominant	Loss
			Myotonic dystrophy type 2	Dominant	Loss

Table 1 *(Continued)*

Gene	Locus	Channel protein	Disease	Inheritance	Change
ATP2A1	16p12	SERCA 1a, Ca-ATPase of sarcoplasmic reticulum, fast twitch 1	Brody's syndrome	Dominant, recessive	Loss
Endplate					
CHRNA1	2q24-32	nAChR α1 subunit	Congenital myasthenic syndrome (nAChR=nicotinic acetylcholine receptor)	Dominant and recessive	Gain and loss of function
CHRNB1	17p12-11	nAChR β1 subunit			
CHRND	2q33-34	nAChR δ subunit			
CHRNE	17	nAChR ε1 subunit			
RAPSN	11p11	Rapsyn, AChR associated protein	Congenital myasthenic syndrome	Recessive	Loss
Central nervous system					
SCN1A	2q24	Nav1.1 sodium channel α subunit 1	Generalised epilepsy with febrile seizures plus (GEFS+2)	Dominant	?
SCN1B	19q13.1	β1 subunit			
SCN2A	2q23	Nav1.2 sodium channel α subunit 2			
SCN2A	2q23	Nav1.2 sodium channel α subunit 2	Benign familial neonatal/infantile convulsions	Dominant	?
KCNA1	12p13	Potassium ch α1 subunit, A-type, Kv1.1	Episodic ataxia type 1, partial epilepsy (?)	Dominant	Loss

Table 1 *(Continued)*

Gene	Locus	Channel protein	Disease	Inheritance	Change
KCNQ2	20q13.3	Voltage gated potassium channel α subunit Kv7.2	Benign familial neonatal convulsions	Dominant	Loss
			Neuromyotonia	Dominant	Loss
KCNQ3	8q??	Voltage gated potassium channel α subunit Kv7.3	Benign familial neonatal convulsions	Dominant	Loss
CACNA1A	19p13.1	Calcium channel P/Q-type $\alpha 1$ subunit	Episodic ataxia type 2	Dominant	Loss
			Familial hemiplegic migraine 1	Dominant	Loss
			Spinocerebellar ataxia type 6	Dominant	Loss
			Absence epilepsy	Dominant	Loss
CACNA1F	Xp11.23	Calcium channel retinal L-type $\alpha 1$ subunit	Congenital stationary night blindness (CSNB2)	Recessive	Loss
CLCN2	3q26	Voltage gated chloride channel ClC-2	Idiopathic generalised epilepsies	Dominant	Loss or gain
CACNB4	2q22-23	Calcium channel L-type $\beta 4$ subunit	Generalised epilepsy	Dominant	Gain
			Episodic ataxia 3	Dominant	Gain
CHRNA4	20q13.3	Nicotinic acetylcholine receptor $\alpha 4$ subunit	Nocturnal frontal lobe, ADNFLE	Dominant	Loss
GLRA1	5q31.2	Glycine receptor $\alpha 1$ subunit	Hyperekplexia=startle disease=stiff baby syndrome (STHE)	Dominant	Loss
				Recessive	Loss

detection of the functional defect that is brought about by the disease causing mutation. Much progress on the road to this aim was achieved by the combination of molecular biology and electrophysiological patch clamp techniques.

The channel pathology of two diseases is given as an example, one caused by gain-of-function mutations in a sodium channel, the other by loss-of-function mutations.

Myotonia

An example of a gain-of-function mutation is myotonia. The term "myotonia" designates involuntary muscle contractions that are caused by membrane hyperexcitability of skeletal muscle and which lead clinically to slowed muscle relaxation following voluntary muscle contractions. Myotonia is felt as uncontrolled temporary muscle stiffness by the patient.

The voltage gated sodium channel is essential for the generation of an action potential. Its upstroke is mediated by opening the channels that passively conduct a fast sodium inward current in a feed forward mechanism along both an electrical and a concentration gradient. Owing to the resulting high conductance of the membrane for sodium ions, the membrane suddenly depolarises from the resting value of −84 mV to approximately +25 mV. Immediate repolarisation of the membrane to the highly negative resting value is made possible by fast inactivation of the sodium channels, resulting in the usual predominance of the membrane conductance for potassium ions that may be further supported by opening of additional voltage gated potassium channels (see second example). Gain-of-function sodium channel mutations — for example, those situated in the inactivation gate that is shown in Fig. 2 at the intracellular side of the channel protein — lead to an incomplete and destabilised channel inactivation that results in frequent channel reopenings and channel bursts, and hence in a pathologically increased sodium current (Fig. 4C, 4D). This persistent sodium ion influx generates repetitive muscle action potentials and thus myotonia.

If the sodium ion influx through mutant channels is large, an associated sustained membrane depolarisation may lead to a secondary loss of function of the 50% of sodium channels that are genetically normal. This loss of function can result in episodic weakness potentially occurring in myotonic conditions such as hyperkalaemic periodic paralysis and paramyotonia congenita. The patients are heterozygous for mutant and wild-type channels, but by dictating a change in cell excitability, the mutation results in a dominantly inherited disease.

Benign neonatal familial convulsions

Benign neonatal familial convulsions (BNFC) is the second example. BFNC is a rare dominantly inherited epileptic syndrome characterised by frequent brief seizures within the first days of life that typically disappear spontaneously after weeks to months. The disease is caused by various mutations in KCNQ2 or KCNQ3 — voltage gated potassium channels that form a heterotetrameric channel complex which is slowly activated by membrane depolarisation.[10,11] Once opened, both channels conduct potassium ions from inside to outside along the concentration gradient and against the electric field. This outward current is the so-called "M-current" known to play an important role in the regulation of the firing rate of neurones.[12] It stabilises the highly negative membrane potential and reduces the neuronal excitability.

BFNC-causing mutations in KCNQ channels are clustered in two regions of the protein: in the P loop between segments S5 and S6 constituting the pore region, and in the C-terminus that is unusually long in all KCNQ channels (see Fig. 1). All mutations drastically reduce or abolish the M current, the pore mutations probably by affecting the ionic conductance, and the C-terminus mutations by reduced assembly to heteromeric channels. Although it is now clear that the mutations exert loss-of-function at the channel level (see Fig. 5A, 5B) and thereby increase the excitability of the neurones that express the mutant genes, the question remains as to why the mutations result in seizures preferentially during the neonatal period. One possibility is that the premature brain is generally more likely to develop seizures than later in life.[13] Another explanation might be a differential expression of potassium channels during maturation in the hippocampus, leading to a dominant role of KCNQ channels in central neurones within the first days to weeks of life.[14]

As a result of the discovery of the pathogenesis of BNFC, a novel approach to the treatment of epilepsies emerged from identifying retigabine as an activator of M currents. Retigabine shifts the voltage dependence of steady state activation of these channels by about 20 mV in the negative

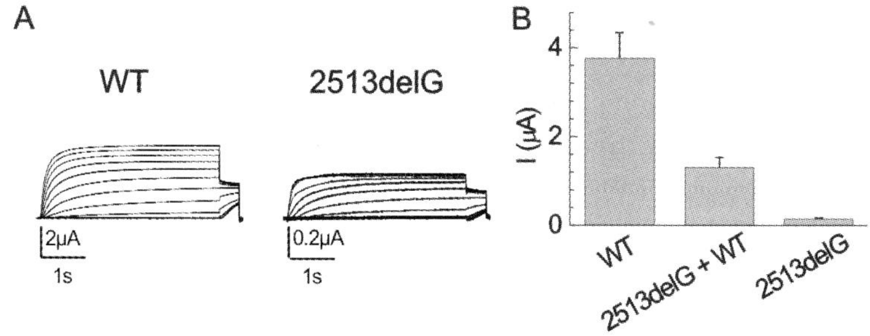

Fig. 5 Functional consequences of a mutation in KCNQ2, the Kv7.2 voltage gated potassium channel α subunit domain formed by 6T/1P. Several mutations in this domain cause benign neonatal familial convulsions (BNFC). The mutation, 2513delG, is a single base pair deletion that is located just seven codons before the regular stop codon and induces a frame shift, thereby resulting in a change in the last seven amino acids of the normal protein and in a prolongation of the mutant protein by another 56 amino acids before the new stop codon. The mutant KCNQ2 cDNA was functionally expressed in *Xenopus* oocytes. The mutation reduces the potassium current to less than 10%, as seen in the raw current traces in (A) (note the 10-fold higher current resolution of the mutant channel compared with the wild type channel, WT) and for the average of many such experiments in (B). Coexpression of both normal (WT) and mutant channels did not result in significantly less than 50% of WT current, suggesting there was no dominant negative effect of the mutation.[11]

direction, so that they are open at the resting membrane potential. This high channel-open probability "clamps" the neurones to resting potentials near to the potassium reversal potential.[15]

Knowledge of the functional defect that is brought about by the disease causing mutation is essential for an understanding of the aetiology and diagnosis of a given hereditary disorder. A unique feature is the availability of technical possibilities to study precisely the pathology of channel proteins by combining modern molecular biology with the patch clamp technique.

PRINCIPLES BEHIND THE TECHNIQUES

Channel Specific Techniques

Classical voltage clamp is done with two intracellular "sharp" microelectrodes (tip diameter <0.1 μm, resistance ~10 megohm when filled with 3 M KCl), one for clamping the membrane potential to values according to various pulse protocols for activation, inactivation, and so on; the other measuring the current conducted during the voltage steps. The advantage of this technique, which is still used, particularly for oocytes, is its simplicity. The disadvantage is that the membrane of larger cells and oocytes may not be homogeneously clamped to a certain value, which is especially problematic if rapidly activating ionic currents are studied.

Whole cell recordings are done with a glass pipette pulled as a "patch clamp microelectrode" (Fig. 6 left; tip diameter of $\sim 1\,\mu m$, resistance of ~ 1 megohm when filled with 100 mM KCl). Once the seal is established, a brief suction is applied to the interior of the electrode in order to rupture the patch of membrane under the electrode tip. After this break in, there will be a low resistance pathway for current and diffusional flow between the electrode and the cell interior. The cell membrane is voltage clamped at the pipette potential by virtue of this low resistance pathway, and the electrode monitors the current flowing across the entire cell surface. The ionic composition of the cell cytoplasm rapidly equilibrates with the pipette contents,

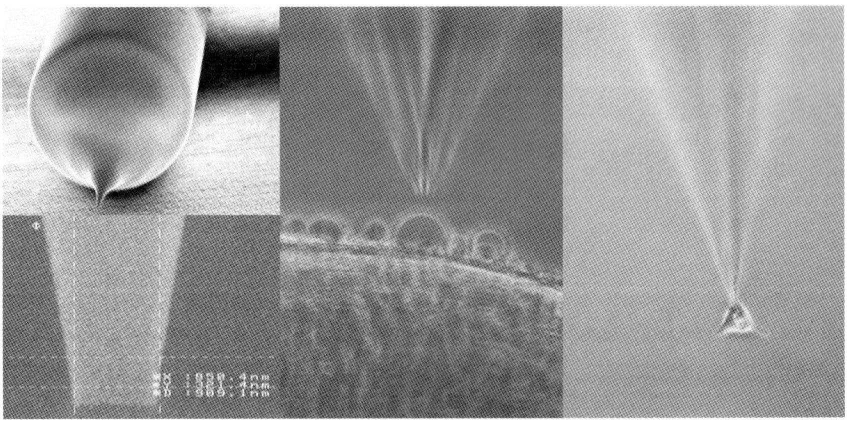

Fig. 6 Typical microelectrode suited for patch clamping (left panel, courtesy of Dr Ebert) and its attachment to either a membrane bleb formed from a native human skeletal muscle fibre (middle panel) or an HEK cell (right panel). The upper left panel shows the glass microelectrode, the lower part the tip, with a diameter of about 700 nm. The plasmalemmal bleb in the middle panel was formed by careful stretching of native muscle fibres at $\geq 10^{-5}$ M [Ca^{2+}] in the bathing solution. The membrane of native HEK and other cultured cells is so clean that giga-ohm seals can easily be produced without additional measures.

offering a pathway for the control of cellular constituents. Although whole cell recording can be highly accurate, two important limitations are associated with the technique. The first is connected with the resistance in series with the membrane (R_S), which in whole cell recording is the access resistance between the interior of the pipette and the cell cytoplasm.[16] Another limitation of the whole cell configuration is that important intracellular regulatory molecules — such as cAMP, Ca^{2+}, or GTP — can diffuse out of the cell through the patch electrode; thus the physiological regulation of these important second messenger substances is disrupted during whole cell recording. The *perforated patch* technique[17] provides a solution to this problem by making it possible to record macroscopic currents with a cell-intact recording configuration. This configuration is obtained by including a pore forming antibiotic, such as nystatin or amphotericin B, in the pipette solution. After a seal is formed on the cell, the antibiotic channels insert in the patch of membrane under the electrode tip, thereby providing electrical continuity between the pipette and the cell interior.

In conclusion, the whole cell mode measures the current through the total cell membrane superimposed with noise. This "macroscopic" current corresponds to the average of many simultaneously conducting channels and therefore resembles the intracellular recordings. Because of the simple and fast analysis, it is the configuration that is most frequently used.

Cell attached, *inside-out*, and *outside-out* modes allow the measurement of a small number of channel proteins or even a single channel in the electrically isolated membrane patch, depending on how many channels are embedded in the patch. The various configurations are available to the investigator once the seal is established (see Fig. 3). The *cell attached* mode is already made as soon as a giga-ohm seal between the patch electrode and the cell membrane is established (Fig. 6, left: same electrode as for whole cell recordings, but filled with an "extracellular" solution). The background noise can be sufficiently attenuated so that the current flowing through a single ionic channel can be resolved (single channel recordings).[4] A special form is the bleb attached mode that enables the measurement of adult native cells without enzymatic treatment (Fig. 6, middle). Blebs (or blisters) can be formed — for example, by stretching skeletal muscle fibres in a high calcium bath solution.[18,19] The advantage is that the plasma membrane is absolutely clean so that a giga-ohm seal can be achieved relatively easily.

If the recording electrode is withdrawn from the cell after a cell-attached patch is formed, the patch of membrane can be excised from the cell with the inside surface of the membrane facing the bath solution; this is called the *inside-out configuration*. An *outside-out* patch can be formed by removing the electrode after entering the whole cell configuration. One advantage of these cell-free patch recordings is that they provide more accurate control of the membrane potential. In the cell attached mode, the membrane potential of the patch is equal to the resting potential of the cell minus the pipette potential. As the resting potential may not be known with certainty or might vary during an experiment, there will be uncertainty in the value of the patch membrane potential. However, in the isolated patch configurations, the magnitude of the patch potential is equal to the pipette potential and is therefore known with precision. A second advantage is that the solution bathing one surface of the patch can rapidly be changed simply by changing the bath solution.

Once a recording mode is established the recordings can be made with a typical patch clamp circuit which uses the electrode both to control the potential and to measure the currents conducted by the channels (Fig. 7, top).

Additional Cloning Techniques

Transient and permanent expression in oocytes and other heterologous cells

Oocytes from *Xenopus laevis* have become a widely used preparation for the expression of cloned ion channel genes. A series of recent papers has reviewed many of the technical aspects of the use of this expression system.[20–25] Isolated *Xenopus* oocytes are capable of translating injected mRNA from a variety of sources and of leading to a high channel density in the membrane and large ion currents. The mRNA can be extracted from the tissue of interest[23] or it can be synthesised from an isolated clone (cDNA) which codes for the channel protein.[24] The isolated mRNA is pressure injected into the oocyte or its nucleus through a relatively large diameter (10 μm) microelectrode. Within a few days, functional channels will be present in the surface membrane of the oocyte and can be studied with

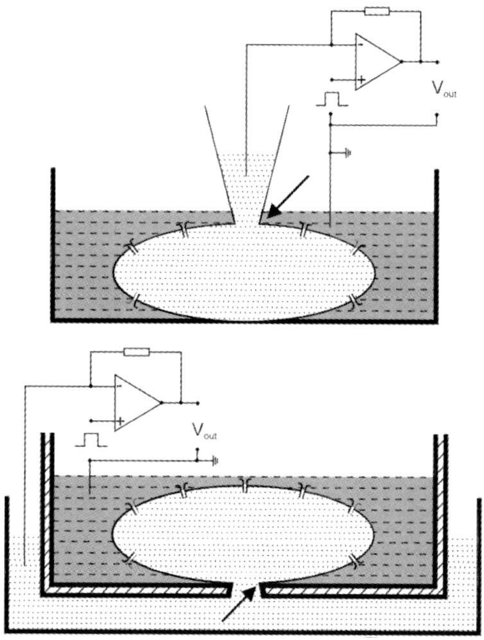

Fig. 7 Patch clamp arrangements. Upper panel: classical whole cell configuration and circuit using a glass microelectrode. Lower panel: patch clamp chip and circuit. This arrangement can be multiplied and automated for high throughput screening. In addition, the arrangement allows the scientist to install a second technique — for example, fluorescence or force microscopy — from the upside of the setup.

various voltage clamp techniques, including a conventional two microelectrode voltage clamp or single channel patch clamp recording.[25] Several types of ionic channel have now been expressed in oocytes using these techniques.[26,27]

In contrast to transient expression systems, stable expression involves the introduction of exogenous DNA into the genome of a cell so that it is transcribed with the cell's DNA and inherited by offspring cells during cell division. One of the obvious advantages of this type of system is that once the cell line is established, studies can be performed on a uniform cell population without re-establishing expression for each experiment (for an example see a patched HEK cell in Fig. 6, right panel). The foreign DNA can be introduced by microinjection, electroporation, viral expression vectors, or transfection using various techniques.[28]

POTENTIAL APPLICATION OF THE TECHNOLOGIES IN THE FUTURE

High throughput Screening for Neuroscience and Pharmaceutical Industry

Although very powerful, the patch clamp technique is extremely labour intensive and thereby limited to a throughput of 10–20 individual cell measurements a day. Thus several firms are trying to develop technologies for high throughput screening. Achieving this goal requires more than just the automation of existing patch clamp techniques; it requires the development of an entirely new paradigm for making electrophysiological measurements. All developments are based on positioning a cell on a small pore separating two isolated fluid chambers in a manner that requires no manual intervention or micromanipulation (Fig. 7, bottom). In order to perform whole cell electrophysiological measurements within this geometry, two criteria must be met. First, a high resistance seal must form between the cell membrane and peripheral region of the substrate pore. As in the case of patch clamp electrophysiology, this ensures that the current measured between the two electrodes passes through the cell membrane. Second, in order to be able to control the cell's membrane potential, a low resistance electrical pathway must form through the cell wall that covers the pore. This latter requirement in effect places the associated electrode at the interior of the cell and allows one to clamp the membrane potential over the rest of the cell membrane. Once these criteria have been met, and assuming no manual intervention, it is then possible to conceive a parallel format where many wells can be measured simultaneously.

Functional Testing of Tissue Specific Channel Splice Variants

Even though the human genome has been sequenced, it is not completely clear which genes are expressed in neurones and how these are tissue specifically spliced. In particular, the functional alterations of the splice products are almost unknown up to now. It may still take several years before more precise knowledge of the exact interplay between these proteins and other regulatory or signalling pathways is available.

Pharmaceuticals and Patch Clamping Combined

In vitro testing of drugs for channelopathies

The electrophysiological study of mutant channels expressed in cell systems allows one to characterise the functional alterations and to develop new strategies for the treatment of ion channelopathies — for example, by testing drugs that are already on the market for other indications; or by testing drugs that could be designed specifically either to block mutant channels that reveal a gain-of-function, or to activate non-mutant channels that could compensate for channels functionally lost by a mutation.

In vitro testing of drugs for neuroprotection

Part of the energy demand of neurones is required for active ionic pumps that compensate for the passive flux of ions through specific and non-specific ion channels along the diffusion gradients. A reduction of the transmembrane ionic flow diminishes the energy demand of neurones considerably. The pharmacological modification of the state of voltage or ligand activated ion channels thus provides potentially powerful strategies for neuroprotection.

TIRF Microscopy and Patch Clamping Combined: Improving Structure–Function Relations

Total internal reflection fluorescence (TIRF) microscopy is a powerful technique for visualising fluorescently labelled membrane proteins. The evanescent wave can selectively excite fluorescent molecules in or near to the cell membrane. TIRF microscopy is particularly suited for the combined study of voltage sensor movements and the resulting transmembranous gating and ion currents (Fig. 8A). Structural changes of channel ensembles can be studied in real time using a photomultiplier. Conformational changes of several single channels can be detected simultaneously by spot-like signals in the cell membrane changing their fluorescence intensity. The temporal and spatial resolutions of the changes are restricted by the characteristics of the detecting ccd camera.

In studies undertaken by Sonnleitner *et al.*,[29] a structural change of the gating subunit of an ion channel results in a change in the fluorescence

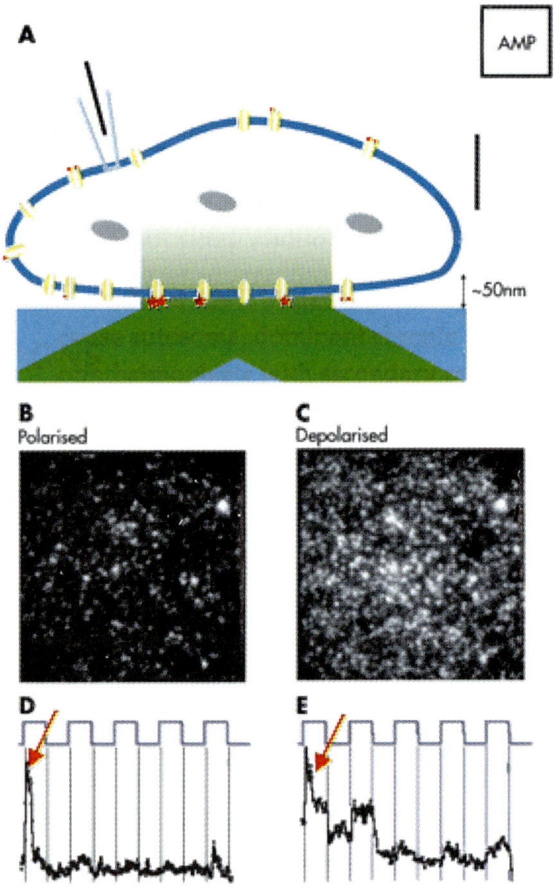

Fig. 8 Imaging of single gating charge movements. (A) Principle of TIRFM on channel proteins. (B), (C) CCD images off fluorescence signal from a $20 \times 20\,\mu m$ membrane patch from a *Xenopus* oocyte transfected with a shaker potassium channel labelled with a tetra-methyl-rhodamine-maleimide (TMRM) at the voltage sensing S4 domain. Individual spots correspond to single TMRM molecules attached to S4. Upon membrane depolarisation, an increase in fluorescence can be observed. (D), (E) Typical time traces of individual spots showing the response of a single voltage sensing S4 domain to depolarisation and subsequent photobleaching (red arrow) during the first depolarisation (an indication for a single flurophore attached to a single S4) (D). (E) Example of another spot, where bleaching of one TMR occurs during the first depolarisation, while the second flurophore bleaches only after the second depolarisation. Measurements of single gating charges are not possible by electrical means because of the low signal. Therefore the attachment of TMRM to the voltage sensing S4 domain presents the only practical way to detect single gating charge movements.[29]

intensity of the attached fluorophore (Fig. 8B *v* 8C). Hence the attached fluorophore acts as a sensor for structural changes in the ion channel that can be directly related to ion channel function through electrical measurements (Fig. 8D, 8E). TIRF microscopy can also be used in conjunction with FRET or other methods to detect relative movements between channel parts. This innovative combination of techniques allows molecular resolution of small motions underlying ion channel activation — for example, by ligands — and will probably find widespread use in the study of membrane associated molecules.[30] TIRF can also be combined with FRAP, a method that allows one to study the mobility of transmembrane proteins.

Atomic Force Microscopy and Patch Clamping: Characterisation of Mechanosensitive Channels

Formerly, atomic force microscopy (AFM[31]) and other scanning probe microscopies were applied exclusively to hard materials where the surface needed to be analysed — for example, for topographical, electrical, magnetic, thermal, and elastic properties and friction. About 10 years ago, these modern nanotechnologies also began to be used for the characterisation of soft and moist surfaces, such as the surface of living cells and biomembranes.[32–34] AFM is increasingly used to study the elastic and plastic properties of the cytoskeleton of cultured cells,[35–38] to image neurones and glia directly,[39] and even to study the binding forces between individual molecules and their respective subunits.[40,41] Studies of docking and fusion of synaptic secretory vesicles at the neurone membrane allow us to determine the proteins that make a principal contribution to the interactive binding force between the two membranes.[42] By using atomic force microscopy, laser confocal microscopy, electrical recording, and biochemical assays, the molecular conformations of reconstituted globular amyloid β protein (AbetaP) as well as their real time and acute effects on neuritic degeneration have been studied. AFM of AbetaP1-42 reconstituted in a planar lipid bilayer revealed multimeric channel-like structures, and electrophysiological recordings demonstrated the presence of multiple single channel currents of different sizes.[43]

AFM has also been combined directly with the patch clamp technique for the characterisation of biological mechanoelectrical transduction channels in living inner and outer hair cells of the cochlea.[44–46] Using an AFM

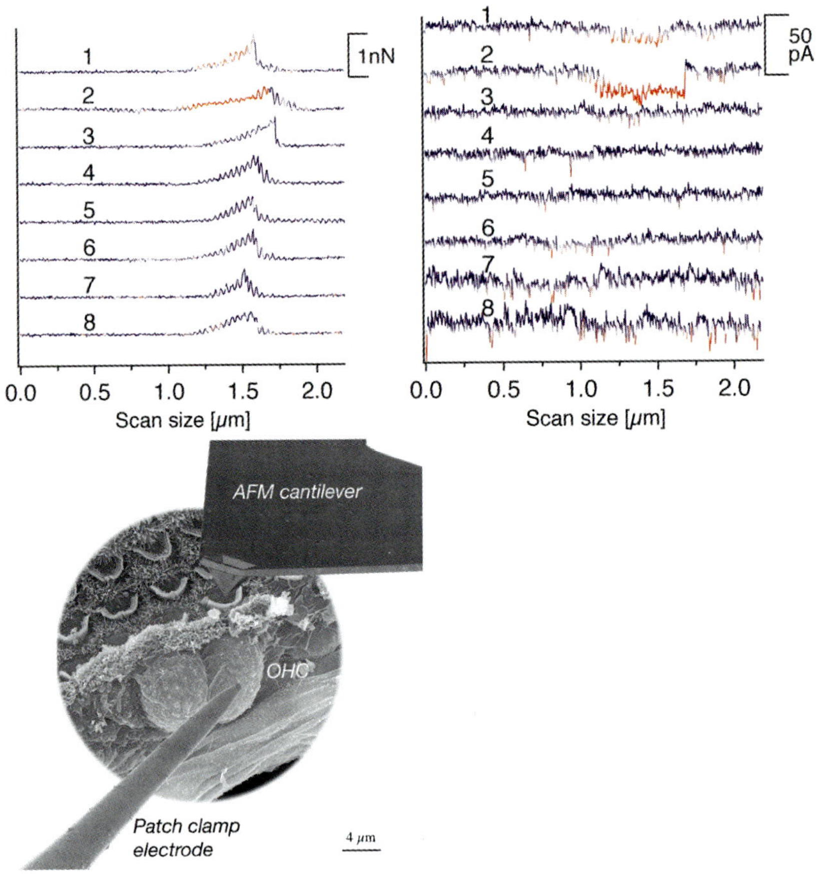

Fig. 9 Force and current responses to displacement of single stereocilia of postnatal rats. Upper left panel: eight adjacent stereocilia of a cochlear hair bundle are individually displaced, scanning an atomic force microscopy (AFM) tip in excitatory direction. Interaction between each stereocilium of the tallest row of stereocilia with the AFM tip results in an increasing force bending the AFM cantilever upward. After losing contact, the AFM force sensor moves down to its initial position. An additional sinusoidal modulation (f = 357 Hz) is added to the lateral movement of the AFM force sensor. Upper right panel: an in-phase current response (opening and closing) of the mechanosensitive transduction channel was observed for stereocilia Nos 1 and 2 until the transduction current through the channel pore reaches saturation where the channel is in the maximum open state (see traces 1 and 2). Although the maximum forces for the remaining six stereocilia are of the same magnitude, no clear response was observed. The last trace (No 8) shows an increased noise level, which is caused by an increased leak between patch clamp pipette and cell membrane. Lower panel: experimental setup showing the anatomy of the AFM cantilever, a partition of an organotypic culture of the organ of Corti with the three rows of outer hair cells (OHC), and the patch clamp electrode.[44,46]

stylus with a tip diameter of only a few nanometres, it was possible to displace individual stereocilia of cochlear hair cells, resulting in opening of single transduction channels. In contrast to the outside-out and the inside-out patch clamp configuration, this technique allows investigation of single mechanosensitive ion channels in entire cells (Fig. 9).

CONCLUSIONS

Methods that help to determine the electrical properties and the structure–function relations of neuronal channels are becoming increasingly adapted to study single proteins in their native environment. This is the basis for understanding the pathological mechanisms of the channelopathies and for developing new treatment strategies in the future.

ACKNOWLEDGEMENTS

We thank Drs M Langer, O Marti, and A Sonnleitner for fruitful discussions and U Richter for drawing the illustrations. This work was supported by the German Research Foundation (DFG) (JU 470/1), the Research Programme "Functional nanoscopy of pathologically altered biomembranes" of the Ministry for Science, Research and the Arts in the State of Baden-Württemberg, and the network on *Excitation-contraction coupling and calcium signalling in health and disease* of the IHP Programme funded by the European Community.

APPENDIX 1

GLOSSARY OF TERMS

Terms Related to Channel

- *Channelomics* is the short term for proteomics of voltage and ligand gated ion channel proteins of a cell and includes all aspects of an ion channel protein such as transcription, expression, and imprinting of the channel encoding genes; splicing and editing of the RNA; translational and post-translational effects on proteins and subunit composition; selectivity and gating of the channel pores; and modulatory effects of mutations, drugs, and toxins on the structure and function of the channels.

- *Channelopathies* are defined as disorders caused by pathology of ion channel *function*. These may be due to mutations in the corresponding genes or to binding of antibodies and toxins to the channel protein.

Terms Related to Patch Clamping

- *Cell attached patch*: recording microelectrode sealed onto the cell, current flowing through the channels (usually one to 10) situated in the membrane patch can be resolved.
- *Whole cell patch*: after a cell attached patch is formed, a brief suction is applied to the interior of the electrode in order to rupture the patch of membrane under the electrode tip; after this break in, there is a low resistance pathway for current and diffusional flow between the electrode and the cell interior; the cell membrane is voltage clamped at the pipette potential by virtue of this low resistance pathway, and the electrode monitors the current flowing across the entire cell surface; the ionic composition of the cell cytoplasm rapidly equilibrates with the pipette contents, offering a pathway for the control of cellular constituents.
- *Inside-out patch*: electrode is withdrawn from the cell after a cell attached patch is formed; the patch can be excised from the cell with the inside surface of the membrane facing the bath solution.
- *Outside-out patch*: electrode removed after entering whole cell configuration.

Terms Related to Force Microscopy

- *AFM* (atomic force microscopy) uses a mechanical probe to magnify surface features up to 100 000 000 times, and produces three dimensional images of the surface of rigid and soft materials. A fine microfabricated tip (<30 nm diameter) of a cantilever is brought into contact with the sample and scanned. The forces between the tip and the sample are used to image the surface. The interaction forces can be as low as the force needed to unwind a single coiled molecule. The dynamic modes allow visualisation of processes with a time resolution in the millisecond range.

Terms Related to Fluorescence Microscopy

- *Emission*: All fluorescent dyes emit light of one wave length (for example, green) after they have absorbed light of another wave length (for example, blue). However, if a very high intensity blue light is delivered to the dye, the dye will "photobleach," meaning that the high intensity light has rendered the dye unable to fluoresce.

- *FRET* (fluorescent resonance energy transfer): non-radiative transfer of photon energy from an excited fluorophore (the donor) to another fluorophore (the acceptor) when both are located within close proximity (1–10 nm); the technique can resolve the relative proximity of molecules beyond the optical limit of a light microscope.
- *FRAP* (fluorescence recovery after photobleaching): after high intensity light had rendered the dye unable to fluoresce, surrounding molecules that have not been photobleached migrate into this blackened area and can be made visible. Used to determine whether a protein is able to move within a membrane or is tethered to structural components of the cell.
- *TIRF* (total internal reflection fluorescence): microscopy allows visualisation of fluorescently labelled membrane proteins. The excitation light beam penetrates only a short distance when totally internally reflected at the interface plane. Depending on the angle of the excitation beam and the refractive index ratio, the penetration depth may vary between 50 and 300 nm. This surface electromagnetic field, called the "evanescent wave," can selectively excite fluorescent molecules in or near to the cell membrane. Background fluorescence from fluorophores either in the extracellular solution or inside the cells is suppressed. Also, because TIRF minimises the exposure of the cell interior to light, the survival of the cell during imaging procedures is much enhanced relative to standard illumination, which penetrates the cell (epi- or transillumination).

APPENDIX 2

ASSOCIATED WEBSITES OF INTEREST

http://www.essen-instruments.com

http://www.zeiss.de

http://www.uar.at

http://www.bio.davidson.edu

http://www.channelopathies.org/

http://beam.to/calcium_quark

REFERENCES

1. Jiang Y, Lee A, Chen J, *et al.* X-ray structure of a voltage-dependent K(+) channel. *Nature* 2003; **423**:33–41.
2. Jiang Y, Ruta V, Chen J, *et al.* The principle of gating charge movement in a voltage-dependent K(+) channel. *Nature* 2003; **423**:42–48.
3. Hodgkin AL, Huxley AF. A quantitative description of membrane current and its application to conduction and excitation in nerve. *J Physiol (Lond)* 1952; **117**:500–00.
4. Hamill OP, Marty A, Neher E, *et al.* Improved patch-clamp techniques for high-resolution current recording from cells and cell-free membrane patches. *Pflügers Arch* 1981; **391**:85–100.
5. Le Grimellec C, Lesniewska E, Giocondi MC, *et al.* Imaging of the surface of living cells by low-force contact-mode atomic force microscopy. *Biophys J* 1998; **75**:695–703.
6. Sakmann B, Neher. *Single-Channel Recording*. New York: Plenum Press, 1985.
7. Lerche H, Mitrovic N, Dubowitz V, *et al.* Paramyotonia congenita: the R1448P Na^+ channel mutation in adult human skeletal muscle. *Ann Neurol* 1996; **39**: 599–608.
8. Mitrovic N, George AL, Heine R, *et al.* Potassium-aggravated myotonia: the V1589M mutation destabilizes the inactivated state of the human muscle sodium channel. *J Physiol (Lond)* 1994; **478**:395–402.
9. Lehmann-Horn F, Jurkat-Rott K. Voltage-gated ion channels and hereditary disease. *Physiol Rev* 1999; **79**:1317–1371.
10. Biervert C, Schroeder BC, Kubisch C, *et al.* A potassium channel mutation in neonatal human epilepsy. *Science* 1998; **279**:403–406.
11. Lerche H, Biervert C, Alekov AK, *et al.* A reduced K^+ current due to a novel mutation in KCNQ2 causes neonatal convulsions. *Ann Neurol* 1999; **46**:305–312.
12. Wang HS, Pan Z, Shi W, *et al.* KCNQ2 and KCNQ3 potassium channel subunits: molecular correlates of the M-channel. *Science* 1998; **282**:1890–1893.
13. Swann JW, Smith KL, Brady RJ, *et al.* Neurophysiological studies of alterations in seizure susceptibility during brain development. In: Schwartzkroin PA, ed. *Epilepsy: Models, Mechanisms and Concepts*. Cambridge: Cambridge University Press, 1993; 209–243.
14. Lerche H, Jurkat-Rott K, Lehmann-Horn F. Ion channels and epilepsy. *Am J Med Genet* 2001; **106**:146–159.
15. Rundtfeld C, Netzer R. The novel anticonvulsant retigabine activates M-currents in Chinese hamster ovary-cells transfected with human KCNQ2/3 subunits. *Neurosci Lett* 2000; **282**:73–76.

16. Armstrong CM, Gilly WF. Access resistance and space clamp problems associated with whole-cell patch clamping. *Methods Enzymol* 1992; **207**:100–122.
17. Horn R, Korn SJ. Prevention of rundown in electrophysiological recording. *Methods Enzymol* 1992; **207**:149–155.
18. Quasthoff S, Franke C, Hatt H, *et al.* Two different types of potassium channels in human skeletal muscle activated by potassium channel openers. *Neurosci Lett* 1990; **119**:191–194.
19. Lerche H, Fahlke C, Iaizzo PA, *et al.* Characterization of the high-conductance Ca^{2+}-activated K^+ channel in adult human skeletal muscle. *Pflügers Arch* 1995; **429**:738–747.
20. Soreq H, Seidman S. Xenopus oocyte microinjection: from gene to protein. *Methods Enzymol* 1992; **207**:225–265.
21. Goldin AL. Maintenance of Xenopus laevis and oocyte injection. *Methods Enzymol* 1992; **207**:266–279.
22. Goldin AL, Sumikawa K. Preparation of RNA for injection into Xenopus oocytes. *Methods Enzymol* 1992; **207**:279–297.
23. Snutch TP, Mandel G. Tissue RNA as source of ion channels and receptors. *Methods Enzymol* 1992; **207**:297–309.
24. Swanson R, Folander K. In vitro synthesis of RNA for expression of ion channels in Xenopus oocytes. *Methods Enzymol* 1992; **207**:310–319.
25. Stühmer W. Electrophysiological recording from Xenopus oocytes. *Methods Enzymol* 1992; **207**:319–339.
26. Noda M, Ikeda T, Suzuki H, *et al.* Expression of functional sodium channels from cloned cDNA. *Nature* 1986; **322**:826–828.
27. Timpe LC, Schwarz TL, Tempel BL, *et al.* Expression of functional potassium channels from Shaker cDNA in Xenopus oocytes. *Nature* 1988; **331**:143–145.
28. Claudio T. Stable expression of heterologous multisubunit protein complexes established by calcium phosphate- or lipid-mediated cotransfection. *Methods Enzymol* 1992; **207**:391–408.
29. Sonnleitner A, Mannuzzu LM, Terakawa S, *et al.* Structural rearrangements in single ion channels detected optically in living cells. *Proc Natl Acad Sci USA* 2002; **99**:12759–12764.
30. Evans J, Yue DT. New Turf for CFP/YFP FRET imaging of membrane signaling molecules. *Neuron* 2003; **38**:145–147.
31. Binnig G, Gerber C, Quate CF. Atomic force microscope. *Phys Rev Lett* 1986; **56**:930–933.
32. Gould S, Marti O, Drake B, *et al.* Molecular resolution images of amino acid crystals with the atomic force microscope. *Nature* 1988; **332**:332–334.

33. Hörber JKH, Häberle W, Ohnesorge F, *et al.* Investigation of living cells in the nanometer regime with the scanning force microscope. *Scanning Micros* 1992; **6**:919–930.

34. Haydon PG, Henderson E, Stanley EF. Localization of individual calcium channels at the release face of a presynaptic nerve terminal. *Neuron* 1994; **13**: 1275–1280.

35. Radmacher M, Tillmann RW, Gaub HE. Imaging viscoelasticity by force modulation with the atomic force microscope. *Biophys J* 1993; **64**:735–742.

36. Hoh JH, Schoenenberger CA. Surface morphology and mechanical properties of MDCK monolayers by scanning force microscopy. *J Cell Sci* 1994; **107**: 1105–1114.

37. Shroff SG, Saner DR, Lal R. Dynamic micromechanical properties of cultured rat atrial myocytes measured by atomic force microscopy. *Am J Physiol* 1995; **269**:286–292.

38. Alonso JL, Goldmann WH. Feeling the forces: atomic force microscopy in cell biology. *Life Sci* 2003; **72**:2553–2560.

39. Parpura V, Haydon PG, Henderson E. Three-dimensional imaging of living neurons and glia with the atomic force microscope. *J Cell Sci* 1993; **104**:427–432.

40. Florin EL, Moy VT, Gaub HE. Adhesion forces between individual ligand-receptor pairs. *Science* 1994; **264**:415–417.

41. Seelert H, Poetsch A, Dencher NA, *et al.* Structural biology. Proton-powered turbine of a plant motor. *Nature* 2000; **405**:418–419.

42. Sritharan KC, Quinn AS, Taatjes DJ, *et al.* Binding contribution between synaptic vesicle membrane and plasma membrane proteins in neurons: an AFM study. *Cell Biol Int* 1998; **22**:649–655.

43. Lin H, Bhatia R, Lal R. Amyloid beta protein forms ion channels: implications for Alzheimer's disease pathophysiology. *FASEB J* 2001; **15**:2433–2444.

44. Langer MG, Koitschev A, Haase H, *et al.* Mechanical stimulation of individual stereocilia of living cochlear hair cells by atomic force microscopy. *Ultramicroscopy* 2000; **82**:269–278.

45. Langer MG, Fink S, Koitschev A, *et al.* Lateral mechanical coupling of stereocilia in cochlear hair bundles. *Biophys J* 2001; **80**:2608–2621.

46. Langer MG, Fink S, Löffler K, *et al.* Investigation of the mechanoelectrical transduction at single stereocilia by AFM. In: *The Biophysics of the Cochlea: Molecules to Models*. World Scientific, 2003.

6

Molecular and Cellular Pathways of Neurodegeneration in Motor Neurone Disease

*P J Shaw**

The process of neuronal degeneration in motor neurone disease is complex. Several genetic alterations may be involved in motor neurone injury in familial amyotrophic lateral sclerosis, less is known about the genetic and environmental factors involved in the commoner sporadic form of the disease. Most in known about the mechanisms of motor neurone degeneration in the subtype of disease caused by SOD1 mutations, but even here there appears to be a complex interplay between multiple pathogenic processes including oxidative stress, protein aggregation, mitochondrial dysfunction excitotoxicity, and impaired axonal transport. There is new evidence that non-neuronal cells in the vicinity of motor neurones may contribute to neuronal injury. The final demise of motor neurones is likely to involve a programmed cell death pathway resembling apoptosis.

*Professor of Neurology, Academic Neurology Unit, University of Sheffield.
Professor Pamela J Shaw, Academic Neurology Unit, E Floor, School of Medicine and Biomedical Sciences, Beech Hill Road, Sheffield S10 2RX; Pamela.Shaw@sheffield.ac.uk

Motor neurone disease (MND), known in many countries as amyotrophic lateral sclerosis (ALS), is the commonest adult onset disorder of motor neurones, and among the most common of adult onset neurodegenerative diseases. The incidence of the disease is 1–2 per 100 000 and is fairly uniform throughout the world, with the exception of several high incidence foci, for example on the island of Guam in the Western Pacific and on the Kii peninsula of Japan. The lifetime risk of developing ALS/MND is approximately 1 in 2000. At any one time there are approximately 5000 individuals suffering from MND in the United Kingdom. Some epidemiological studies indicate that the incidence of MND may be increasing, though the contribution to this of population aging and better developed neurological diagnostic services is currently unknown.[1] The disease is predominantly a condition of middle age and elderly life, with an average age of onset between 50 and 60 years, though rare juvenile onset forms of the condition also exist. For reasons that are not currently understood, MND affects men more commonly than women, with a male to female ratio of approximately 1.6/1.[2]

The pathology accompanying the clinical phenotype was first described by Jean-Martin Charcot in 1869.[3] The salient pathological feature is progressive injury and cell death of lower motor neurone groups in the spinal cord and brain stem and of upper motor neurones in the motor cortex. Characteristic cytopathological features in motor neurones not yet eliminated during the disease process are the presence of ubiquinitated proteinaceous inclusions within motor neurone cell bodies, and neurofilament accumulations within motor neurone axons.[4]

The person afflicted by MND typically develops a combination of upper and lower motor neurone signs, with progressive muscle weakness and wasting, usually accompanied by pathologically brisk reflexes, eventually involving the limb and bulbar muscles. Clinical variants of the disease may initially affect only the spinal lower motor neurones (progressive muscular atrophy variant); only the upper motor neurones (primary lateral sclerosis variant); or only the bulbar musculature (progressive bulbar palsy variant). With disease progression, the majority of patients will develop features of ALS. Certain motor neurone groups are less vulnerable to the pathological process, including those in upper brain stem nuclei controlling eye movements, and those in Onuf's nucleus within the sacral spinal cord controlling the pelvic floor musculature.[5] The rate of disease progression varies between individuals, but the average survival is only of the order of three years from symptom onset.[6] However, approximately 10% of patients will have a slower disease course with survival beyond 10 years.[7] Death in most patients results from neuromuscular respiratory failure. The World Federation of Neurology diagnostic criteria for ALS/MND require the presence of signs of upper and lower motor neurone degeneration, with evidence of progression in the absence of evidence of other disease processes.[8]

The selectivity of the pathological process for the motor system is now recognised to be relative rather than absolute. Detailed investigation has revealed involvement of extramotor parts of the CNS, including changes in other long tract systems — for example, sensory and spinocerebellar pathways — and cellular injury of neuronal groups including the substantia nigra neurones and dentate granule cells within the hippocampus.[9] Overt dementia is found in approximately 2–3% of MND patients,[10] and detailed neuropsychological testing shows more subtle neurophysiological changes, particularly affecting frontal function, in approximately 30% of patients.[11] Thus MND is now regarded as a multisystem disease in which motor neurones tend to be affected earliest and most prominently.[4]

WHAT DO WE KNOW ABOUT THE CAUSES OF MOTOR NEURONE DEGENERATION?

The primary pathogenic processes underlying MND are likely to be multifactorial, and the precise mechanisms underlying selective cell death in the disease are at present unknown. Current understanding of the neurodegenerative process in MND suggests that there may be a complex interplay between multiple mechanisms including genetic factors, oxidative stress, excitotoxicity, protein aggregation, and damage to critical cellular processes, including axonal transport and organelles such as mitochondria (Table 1). Recently there has been growing interest in the role play by non-neuronal neighbourhood cells in the pathogenesis of motor neurone injury and in dysfunction of particular molecular signalling pathways (Fig. 1). The relative importance of these different pathways may well vary in different subgroups of patients, and a very important task for clinicians and scientists in the future is to further delineate the subcategories of MND. Evidence has also accumulated that the final process of motor neurone

Table 1 Pathogenic mechanisms which may contribute to motor neurone injury and cell death in motor neurone disease.

A

- 1. Genetic factors
- 2. Oxidative stress
- 3. Protein aggregation
- 4. Glutamatergic toxicity
- 5. Mitochondrial dysfunction
- 6. Impairment of axonal transport
- 7. Inflammatory cascades/contribution of non-neuronal cells
- 8. Dysfunctional signalling pathways, e.g. through VEGF, Nrf2

The above factors alone or in combination may lead to a programmed cell death mechanism similar to apoptosis

death is likely to occur through a caspase dependent programmed cell death pathway resembling apoptosis.

GENETICS OF ALS/MND

MND is sporadic in 90–95% of cases and familial in approximately 5–10%. Inheritance in familial MND is usually autosomal dominant, though autosomal recessive and X-linked inheritance may be seen in some pedigrees. It is apparent that multiple abnormal gene products can set the scene for motor neurone degeneration. There are at least six dominantly inherited adult onset ALS genes (Table 2), of which only three have so far been identified.

Copper-Zinc Superoxide Dismutase (SOD1)

A major research breakthrough 11 years ago came from the finding that 20% of families with autosomal dominant MND showed mutations in the gene on chromosome 21q22.1 which encodes the free radical scavenging enzyme superoxide dismutase 1 (SOD1).[12] More than 100 different mutations have been identified throughout the SOD1 gene.[13,14] The majority of mutations in SOD1 are missense mutations, with a small number of deletion and insertion mutations resulting in truncated SOD1 polypeptides. SOD1 is a

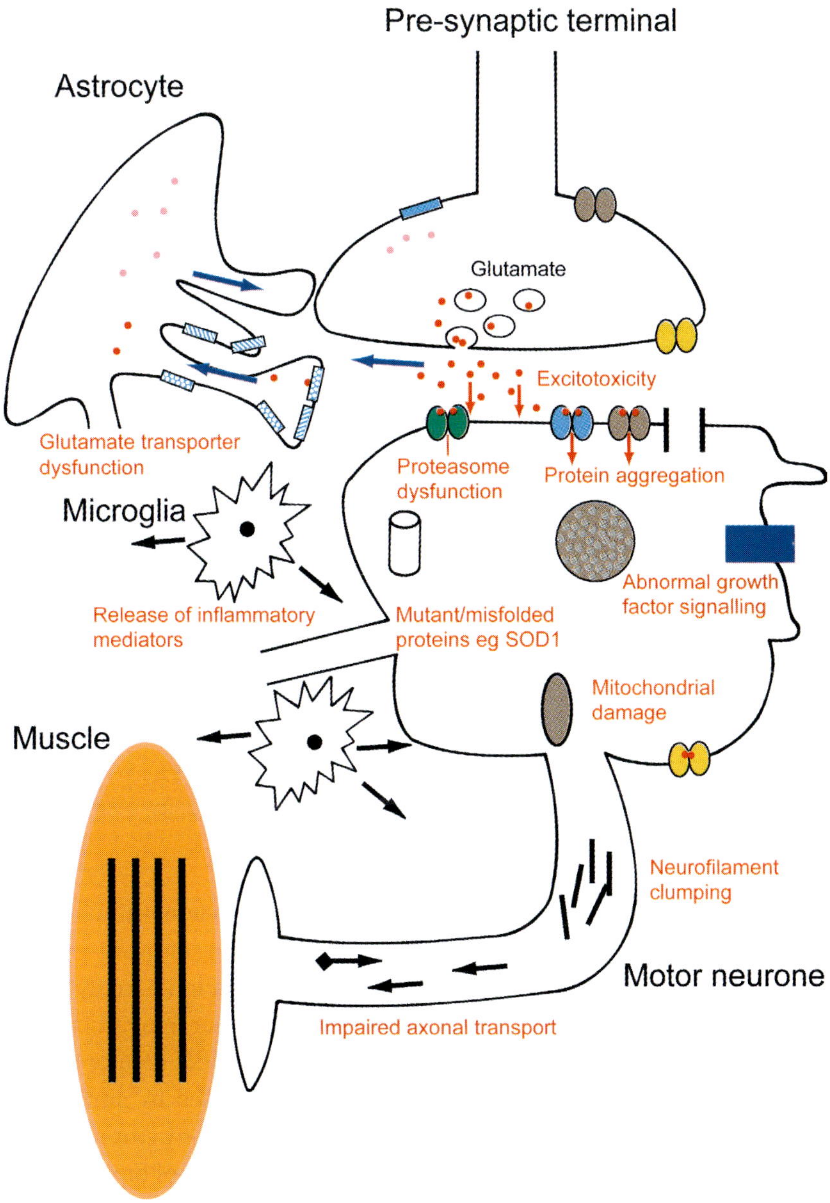

Fig. 1 Molecular mechanisms that may contribute to motor neurone injury in motor neurone disease.

Table 2 Genetic subtypes of ALS/MND.

E

Subtype of ALS/MND	Gene	Inheritance, chromosome
ALS1	*SOD1*	Dominant–adult onset, 21q22.1
ALS2	*ALSIN*	Recessive–juvenile onset, 2q33
ALS3	—	Dominant–adult onset, 18q21
ALS4	*Senataxin (SETX)*	Dominant–juvenile onset, 9q34
ALS5	—	Recessive–juvenile onset, 15q15.1–q21.1
ALS6	—	Dominant–adult onset, 16q12
ALS7	—	Dominant–adult onset, 20ptel–p13
ALS8	*Vesicle associated membrane protein (VAPB)*	Dominant–adult onset, 20q13.33
ALS-FTD	—	Dominant–adult onset, 9q21–22
ALS-X	—	Dominant–adult onset, Xp11–q12
ALS with parkinsonism and dementia	*Microtubule associated protein tau (MAPT)*	Dominant–adult onset, 17q21
Progressive LMN disease	*Dynactin p150 subunit (DCTN1)*	Dominant–adult onset, 2p13

ALS, amyotrophic lateral sclerosis; LMN, lower motor neurone; MND, motor neurone disease.

metalloenzyme of 153 amino acids which functions as a homodimer whose major function is to convert intracellular superoxide free radicals — a toxic waste product of mitochondrial oxidative phosphorylation — to hydrogen peroxide which is in turn removed by the action of other free radical scavenging enzymes. The SOD1 enzyme contains an essential copper atom at the active site which is alternately reduced and oxidised by superoxide. The presence of zinc is thought to stabilise the protein structure. SOD1 is an abundant protein in the CNS, accounting for about 1% of brain protein, but it is also ubiquitously expressed in all other tissues. SOD1 was initially thought to be confined to the cytosolic compartment of cells but it is now recognised that a small proportion of the protein is located in the intermembrane space of the mitochondria.[15] The reasons why motor neurones are especially vulnerable to injury in the presence of SOD1 mutations are not yet clear.

Despite 11 years of intensive research effort, the pathways leading to the cell death of motor neurones in the presence of SOD1 mutations have not yet been fully identified, though there is a convincing body of evidence that the mutant SOD1 protein exerts its detrimental effects through a toxic gain of function rather than a loss of function. Most of our current level of understanding of disease mechanisms in ALS/MND has come from the study of the effects of SOD1 mutations, but even in this defined genetic subgroup of disease the pathways to neurodegeneration appear to be complex and multifactorial.[16]

There is considerable variation in disease phenotype in terms of age of onset and rate of disease progression in human SOD1 related MND. It is apparent that the clinical phenotype must be modified by other genetic or environmental factors, or both. There has been much interest in the D90A SOD1 mutation, which has a dominant inheritance in some genetic backgrounds but is recessively inherited, with two mutated copies of the gene required to cause disease, in Scandinavian populations, implying a co-inherited protective factor.[17] Some intensively studied SOD1 mutations such as the A4V mutation do not show 100% penetrance[18] and the disease phenotype in mice can vary significantly according to the background strain of mouse employed.[19]

Mutant SOD1 transgenic mice that develop a disease which clinically and pathologically resembles human MND have been developed. The most extensively studied are SOD1 G93A, SOD1 G37R, and SOD1 G85R.[20–22] Transgenic rats, carrying G93A or H46R SOD1, also develop an MND phenotype.[23,24] In addition, cellular models of SOD1 related MND have been generated which have helped to elucidate cellular mechanisms of disease.[25–27]

The toxic gain of function of mutant SOD1 has not yet been fully defined, but several pathophysiological processes may be involved, including oxidative stress, mitochondrial dysfunction, excitotoxicity, protein aggregation, and inflammation. These mechanisms are not mutually exclusive and it is possible that all of them play a role in the development of motor neurone injury. These potential mechanisms will be discussed in subsequent sections.

The genetic alterations underlying the remaining 80% of cases of autosomal dominant MND at present remain unknown. However, three other genes have recently been identified as causative in rare cases of familial MND.

ALS 2: Alsin

In 2001, two groups identified alsin as the causative gene for an autosomal recessive form of juvenile ALS linked to chromosome 2q33.[28,29] Mutations in alsin can also cause a motor neurone degenerative disorder with a predominant upper motor neurone phenotype, infantile onset ascending hereditary spastic paralysis,[30,31] and one family has been described with autosomal recessive complicated hereditary spastic paraplegia.[32] Thus, ALS2 mutations account for several juvenile onset autosomal recessive neurodegenerative disorders of motor neurones.

ALS2 encodes a 184 kDa protein which contains three putative guanine nucleotide exchange factor (GEF) domains. GEFs are known to activate small GTPase proteins by stimulating the release of guanosine diphosphate (GDP) in exchange for guanosine triphosphate (GTP).[33] Given the conserved GEF domains of ALS2, it is predicted to function as an activator of particular small GTPases. The small GTPases control a range of important cellular processes including nuclear transport, cytoskeletal reorganisation, transcription, cell migration, and membrane trafficking. They function as binary switches — alternating between inactive GDP bound and active GTP bound states. The alsin protein is widely expressed, but enriched within the CNS, where it is localised to the cytoplasmic face of endosomal membranes.[34] The functions of alsin are still being worked out, but to date it has been shown to bind specifically to the small GTPase, Rab5, and to function as a guanine nucleotide exchange factor or activator of Rab5.[35,36] This implies that alsin is important in endosomal dynamics, and the working hypothesis is that it normally regulates trafficking of signalling molecules important for proper development or maintenance of the health of motor neurones. ALS2 knockout mice have been generated but no major motor system disease phenotype has yet been reported.[37,38]

Alsin/ALS2 is alternatively spliced to generate a short and a long transcript. It has recently been reported that the long isoform of alsin specifically binds, through its RhoGEF domain, to mutant SOD1 and protects cultured motor neurones from mutant SOD1 mediated toxicity.[39] Further examination of this interaction may help elucidate motor neurone specific pathways of neurodegeneration.

ALS4: Senataxin

The ALS4 locus linked to chromosome 9q34 was originally identified in a single large pedigree with juvenile onset, autosomal dominant ALS/MND. The disease course in this family was indolent and did not reduce life expectancy. Chen and colleagues identified three different missense mutations (L3095, R2136H, and T3I) in three families with this subtype of ALS/MND.[40] The SETX gene encodes senataxin, a large 302.8 kDa protein of unknown function. Much of the protein has no homology with other known proteins, but there is one DNA/RNA helicase domain. DNA/RNA helicase proteins are known to have roles in processes such as repair, replication, recombination or transcription of DNA and RNA processing, RNA transcript stability, and the initiation of translation. Recessive loss of function mutations in SETX are associated with ataxia-oculomotor apraxia type 2.[41] It is predicted that the different phenotype of dominantly inherited ALS4 is likely to be caused by a toxic gain of function of the mutated senataxin protein.

ALS8: VAPB (Vesicle Associated Membrane Protein/Synaptobrevin Associated Membrane Protein)

Very recently Nishimura and coworkers described a novel missense mutation (P565) in the VAPB gene at chromosome 20q 13.3 in a Brazilian family with ALS8, an autosomal dominant slowly progressive disorder characterised by fasciculation, cramps, and postural tremor.[42] They subsequently found the same mutation in six further families with different clinical phenotypes, including late onset spinal muscular atrophy and classical rapidly progressive ALS/MND. Thus, modifier genes or environmental factors are likely to play an important role in modulating the clinical course of disease in individuals carrying the same mutation. Vesicle associated proteins are intracellular membrane proteins that can associate with microtubules and have been shown to function in membrane transport. The VAPB protein has three identifiable structural domains. The first 150 residues form an MSP domain conserved between all members of this protein family; the central region contains an amphipathic helical structure predicted to form a coiled/coil protein–protein interaction motif and at the carboxy terminus is a hydrophobic region that acts as a membrane anchor. Preliminary cell biological studies have indicated that the wild-type VAPB protein localises predominantly to the endoplasmic reticulum. The P56S mutation dramatically disrupts the subcellular distribution and induces the formation of intracellular protein aggregates.[42]

Dynactin Mutation

Puls and colleagues identified a mutation (G595) substitution in the gene encoding the P150 subunit of dynactin (DCTN1) in a single family with a slowly progressive lower motor neurone degenerative disorder.[43] The described family had a highly unusual and characteristic phenotype, presenting in early adulthood with respiratory difficulties from vocal cord paralysis, progressive facial weakness, weakness and atrophy of the hands, and the later development of lower motor neurone signs distally in the lower limbs. The amino acid change caused by the mutation would be predicted to distort the folding of the microtubule binding domain of dynactin.

The dynactin–protein complex is required for dynein mediated retrograde axonal transport of vesicles and organelles along the microtubule system. It provides the link between the specific cargo, the microtubule, and cytoplasmic dynein during vesicle transport. Interestingly, it has also been shown that overexpression of the P50 subunit of dynactin also disrupts the function of this protein complex and causes late onset progressive motor neurone degeneration in genetically engineered mice.[44]

Other ALS/MND Loci

The genes for several other subtypes of ALS remain to be identified, as indicated in Table 2. Three separate families have shown linkage to chromosome 16, allowing significant refinement of the region of interest. ALS/MND with fronto-temporal dementia has been mapped to a 17-cM interval chromosome 9q21,[45] and one Swedish family with a similar phenotype without linkage to the chromosome 9 locus has recently been identified, suggesting genetic heterogeneity for this subtype of disease.[46] Motor neurone degeneration may sometimes occur in patients with fronto-temporal dementia and Parkinson's disease, associated with mutations in the microtubule associated protein tau.[47,48] The mutant tau protein forms filamentous inclusions and insoluble aggregates that are associated with neurodegeneration. Some patients with familial fronto-temporal dementia, parkinsonism, and ALS/MND do not have identified mutations in tau,[49,50] suggesting that further genes causing this triad of features remain to be identified.

Possible Genetic Risk Factors in Sporadic ALS/MND

There have been reports of genetic variants found in individuals with apparently sporadic MND (Table 3). Deletions or insertions have been

Table 3 Potential genetic risk factors in sporadic ALS/MND and genetic modifiers in familial ALS/MND.

D

Gene	**Chromosome**	**Genetic variant**	**ALS association**
Neurofilament heavy (NEFH)	22q12.1–q13.1	KSP deletion/insertion	Sporadic
Apolipoprotein E (Apo E Σ4)	19q13.2	Σ4 genotype	Sporadic
Cytochrome c oxidase subunit 1	Mt	Microdeletion	Sporadic
Excitatory amino acid transporter 2 ($EAAT_2$)	11p13–p12	Decreased expression	Familial/sporadic
AMPA receptor subunit (GluR2)	4q32–q33	Altered RNA editing	Sporadic
? Survival motor neurone 1 (SMN1)	5q12.2–q13.3	Copy number	Sporadic
? Survival motor neurone 2 (SMN2)	5q12.2–q13.3	Copy number	Sporadic
Ciliary neurotrophic factor (CNTF)	11q 12.2	Null allele	Familial
Vascular endothelial growth factor (VEGF)	6p12	Promotor polymorphisms	Sporadic

ALS, amyotrophic lateral sclerosis; MND, motor neurone disease.

described in the KSP repeat region of the gene encoding the neurofilament heavy protein, which is a major component of the neuronal cytoskeleton.[51,52] The apolipoprotein E Σ4 genotype, which correlates with an earlier age of onset in Alzheimer's disease, has been examined in ALS/MND and the results obtained are not clear. Some studies have indicated that the apolipoprotein Σ4 genotype may be a risk factor for the development of bulbar onset ALS.[53] Another study provided evidence that this genotype correlates with a shortened survival in patients with ALS, though it did not obviously influence age of onset.[54] However, several other groups have been unable to confirm these findings. A single patient with an ALS/MND phenotype has been described with a mutation in the mitochondrially encoded subunit 1 of cytochrome C oxidase, which is an important component of the mitochondrial respiratory chain.[55] Decreased expression of the glial glutamate transporter $EAAT_2$ is observed in the spinal cord of human ALS/MND patients and in mouse and rat SOD1 transgenic murine models.[24,56–58] Initially it was suggested that aberrant splicing of $EAAT_2$ RNA might underlie this alteration in transporter expression,[59] but subsequent studies have indicated that $EAAT_2$ splice variants are observed as frequently in the CNS of controls as in patients with ALS/MND.[60,61] A single individual with sporadic ALS/MND has been found to have a mutation (N2065) within a putative glycosylation site of EAAT2, which appeared functionally significant in causing aberrant targeting to the cell membrane and reduced glutamate transport.[62] A further component of the glutamate neurotransmitter system has been found to be altered in ALS/MND. Kawahara and colleagues reported that the editing of the GluR2 AMPA receptor subunit appeared defective in motor neurones isolated from ALS spinal cord.[63] The normal RNA editing process of this receptor subunit changes a glutamine to an arginine residue in almost 100% of transcripts, with the important functional consequence that the encoded AMPA receptor channel is impermeable to calcium.

Homozygous deletions of the survival motor neurone gene (SMN1) located on chromosome 5 cause autosomal recessive proximal spinal muscular atrophy, usually of childhood onset. A second adjacent gene, SMN2, has five nucleotide differences between intron 6 and exon 8 which distinguish it from SMN1. One of these polymorphic variants causes frequent skipping of exon 7, which in turn results in low expression of full length SMN2 protein. Two early studies failed to provide evidence for homozygous deletion of SMN1 in sporadic ALS.[64,65] Several studies have been conducted to determine whether alterations of SMN1 or SMN2 are associated as genetic risk factors for ALS. The results are conflicting. Veldink and coworkers found that SMN2 gene deletions were overrepresented in patients with sporadic ALS (16% ALS patients *v* 4% of controls).[66] Corcia *et al.* reported no difference in SMN2 copy number, but that patients with ALS were significantly more likely to have an abnormal copy number of SMN1 (one or three copies).[67] Further work is necessary to determine whether alterations in SMN copy number are true risk factors for the development of ALS.

Ciliary neurotrophic factor (CNTF) is a potent survival factor for motor neurones. Inactivation of the CNTF gene causes mild progressive motor neurone loss in adult mice but does not result in an ALS/MND phenotype. A splice site acceptor mutation in the CNTF gene causes a null mutant allele lacking biological activity, and 1–2% of the normal human population are homozygous for this null allele. A study of SOD1 related familial ALS/MND with a V148G mutation and marked intrafamilial variation in the clinical phenotype indicated that a homozygous null mutation in CNTF might correlate with early onset and rapid disease progression.[68] In addition, G93A SOD1 mice crossed with CNTF knockout mice developed disease at a significantly earlier age.

Vascular endothelial growth factor (VEGF) is an angiogenic factor essential for the formation of new blood vessels during embryogenesis and in many pathological conditions, and has also recently been found to have a significant role as a neurotrophic factor.[69] The expression of VEGF is normally upregulated as a response to tissue hypoxia. Lambrechts *et al.* have recently shown that VEGF is a modifier of ALS/MND in both humans and mice.[70] Deletion of the hypoxia response element (HRE) from the VEGF promoter in mice unexpectedly caused a late onset motor neurone disease which resembled ALS.[71] Cross breeding of these VEGF mice with G93A SOD1 transgenic mice resulted in a worsening of the phenotype with earlier onset of disease. No mutations of the HRE of VEGF have been identified in human ALS/MND.[70] In a meta-analysis of a pooled population of 1900 individuals from four different European populations, two "at risk" haplotypes of promoter (5′) polymorphisms were identified which conferred an overall 1.8-fold increased risk of ALS/MND. These haplotypes were associated with lower serum VEGF levels in both control and ALS/MND populations and with reduced *in vitro* transcription of a luciferase reporter gene.[68] There was substantial heterogeneity in terms of the risk of ALS/MND associated with VEGF promoter polymorphisms between the different geographical groups of patients included in this study, and some studies with smaller numbers of patients and controls have shown that these polymorphisms do not confer an increased risk of ALS/MND.[72,73] Thus further exploration of the role of VEGF in the pathogenesis of human ALS/MND is required.

OXIDATIVE STRESS

The effects of oxidative stress within non-dividing cells such as neurones may be cumulative, and cellular injury by free radical species is a major potential cause of the age related deterioration in neuronal function that occurs in neurodegenerative diseases. There has been particular interest in the role of oxidative stress in ALS/MND, given that mutations in SOD1 — which encodes a key cellular antioxidant defence protein — underlie around 20% of familial ALS/MND cases. The close clinical and pathological similarity between sporadic and SOD1 related familial subtypes of MND suggest that common pathophysiological mechanisms may be operating. Studies of CSF and human postmortem CNS tissue have shown the presence of biochemical changes which represent the effects of free radical damage or abnormal free radical metabolism, and these changes are more pronounced in ALS/MND cases than in controls.[74–77] Fibroblasts cultured from the skin of patients with both familial and sporadic MND show increased sensitivity to oxidative insults compared with those from control cases.[78]

In relation to the toxic gain of function of the mutant SOD1 protein, oxidative damage or metal mishandling, or both, have been strongly implicated. The main hypotheses have been that mutations alter the structure of the SOD1 protein, allowing greater access of abnormal substrates to the active copper site of the dimeric enzyme, resulting in the production of damaging free radical species including peroxynitrite and hydroxyl radicals. Nitration of tyrosine residues on cellular proteins by peroxynitrite can have damaging consequences.[79] Some mutations in SOD1 render the protein more likely to form a zinc deficient variant,[80,81] which in turn makes the copper site more accessible to abnormal substrates. *In vitro* studies have shown that zinc deficient SOD1 causes peroxynitrite dependent cell death.[81] However, several experiments have raised questions as to whether the toxicity of mutant SOD1 can be explained by copper dependent oxidative mechanisms. Thus SOD1 that has been engineered not to bind copper by mutating the four histidine residues for copper binding still causes ALS/MND in transgenic mice.[82] Also, knock out of the gene encoding the copper chaperone protein normally required for insertion of copper into SOD1 has no effect on the disease phenotype in SOD1 transgenic mice.[83] Finally, reduction in nitric oxide (NO) synthesis by pharmacological inhibition of neuronal nitric oxide synthase (nNOS) or genetic manipulation of nNOS would be expected to ameliorate the disease phenotype in mutant SOD1 transgenic mice if peroxynitrite is indeed a key contributor to motor neurone injury. However, these interventions have not been shown to have a significant effect on the murine disease,[84,85] nor did deletion of inducible NOS, which is normally expressed within astrocytes and microglia.[86]

EXCITOTOXICITY

Glutamate is the major excitatory transmitter in the human CNS, and tremendous complexity has been uncovered in the molecular structure of the repertoire of receptors for this neurotransmitter system. Excitotoxicity is the term coined for neuronal injury induced by excessive stimulation of glutamate receptors, by mechanisms which include derangement of intracellular calcium homeostasis and excessive free radical production. Motor neurones are particularly susceptible to toxicity through activation of cell surface AMPA receptors.[87] A body of evidence, which is still circumstantial, has implicated glutamatergic toxicity as a contributory factor to motor neurone injury (reviewed by Heath and Shaw[88]). The key findings are that the expression and function of the major glial glutamate reuptake transporter protein $EAAT_2$ may be impaired in the CNS of MND patients and that CSF (and therefore CNS extracellular fluid) levels of glutamate appear to be abnormally raised at least in a proportion of MND patients.[56,58,89,90] The balance of evidence does not favour RNA mis-splicing as the cause of reduced $EAAT_2$ expression as discussed above.

Excitotoxicity has provided one of the few examples of a mechanistic link between mutant SOD1 mediated MND and the sporadic form of the disease. The presence of mutant SOD1 increases the sensitivity of motor neurones to glutamate toxicity,[27,91] causes alteration in AMPA receptor subunit expression,[92] and causes reduced expression of the major glutamate reuptake transporter $EAAT_2$.[24,93]

Whether as a primary or a propagating process, it appears that glutamate toxicity plays a contributory role to the injury of motor neurones in ALS/MND. This is supported by the finding that anti-glutamate treatment with riluzole has some effect, albeit modest, in prolonging survival in human ALS patients and in mutant SOD1 mouse models.[94,95]

MITOCHONDRIAL DYSFUNCTION

Important properties of mitochondria include the generation of intracellular ATP, the buffering of intracellular calcium, the generation of intracellular free radicals, and involvement in the initiation of apoptotic cell death. Age related deterioration in mitochondrial function is considered a potentially important factor contributing to late onset neurodegenerative diseases.

There is a body of evidence emerging from investigation of human material and cellular and animal models indicating that mitochondrial dysfunction may contribute to motor neurone injury in ALS/MND, and this has been reviewed.[96,97] The key evidence for mitochondrial dysfunction in human ALS/MND includes the following:

- alteration in the morphology of mitochondria in hepatocytes, muscle, and motor neurones;
- increased mitochondrial volume and calcium levels within motor axon terminals in muscle biopsies from sporadic ALS/MND cases[98];
- reduced complex IV activity in spinal motor neurones in sporadic ALS[99];
- high frequency of mitochondrial DNA mutations in motor cortex tissue in sporadic ALS[100];
- multiple mutations and decreased mitochondrial DNA in muscle and spinal cord in sporadic ALS[101];
- ALS-like phenotype in one patient with a deletion in the cytochrome oxidase c subunit/gene.[55]

Further evidence for the role of mitochondrial dysfunction as a contributory factor to motor neurone injury has come from the examination of cellular models of SOD1 related ALS/MND. Expression of mutant (G93A) SOD1 in the NSC34 motor neurone cell line results in the development of morphologically swollen mitochondria, impaired activity of complexes II and IV of the mitochondrial respiratory chain, impaired cellular bioenergetic status, and alteration in the mitochondrial proteome.[102,103] Takeuchi and colleagues showed that molecular targeting of mutant SOD1 to the mitochondria but not to the nucleus or endoplasmic reticulum leads to activation of the apoptosis cascade and cell death.[104]

Mitochondrial dysfunction has also been studied in mutant SOD1 transgenic mice. At least in some strains (for example, G93A) mitochondrial vacuolation within motor neurones is an early feature of the pathology.[20] Whereas SOD1 was previously considered to be an exclusively cytosolic protein, it is now recognised also to reside in the intermembrane space of mitochondria.[105] SOD1 has been shown to accumulate in vacuolated mitochondria in mutant SOD1 mice.[106] It has been shown that the activities of several complexes of the mitochondrial respiratory chain are reduced before disease onset and that these changes increase with age.[107] Mattiazzi and colleagues reported the presence of oxidative damage to mitochondrial protein and lipids and decreased ATP synthesis at the onset of the murine disease.[108] Several groups have shown translocation of cytochrome C, an initiator of apoptosis, from the mitochondrial to the cytosol during disease progression in the mouse.[109,110] Partial deficiency of the mitochondrial form of SOD (MnSOD) exacerbates disease in transgenic SOD1 mice.[111] Recently Lui *et al.* reported that mutant SOD1 is selectively and aberrantly recruited to the cytoplasmic face of mitochondria in spinal cord tissue from mutant SOD1 transgenic mice. Covalently damaged adducts of mutant SOD accumulated on the cytoplasmic face of mitochondria in the spinal cord.[112] This tissue specific recruitment raises the possibility that mitochondrial abnormalities may be involved in the initiation of motor neurone injury. Pasinelli *et al.* also recently showed that the anti-apoptotic protein Bcl2 may be entrapped within large protein aggregates of SOD1 in spinal cord tissue, which may result in reduction of the availability of this protein to regulate apoptosis.[113]

Therapeutic effects of compounds which modulate mitochondrial function have begun to be investigated in SOD1 transgenic mouse models. Creatine buffers energy levels within the cell, maintains ATP levels, and stabilises mitochondrial creatine kinase, which inhibits opening of the mitochondrial permeability transition pore. Administration of creatine to G93A transgenic mice improved motor function and extended survival in a dose dependent manner, as well as causing a reduction in biochemical indices of oxidative damage in the spinal cord.[114] Minocycline, a tetracycline derivative which inhibits microglial activation and blocks release of cytochrome c from mitochondria, also slows disease in mutant SOD1 mice.[110]

CYTOSKELETAL ELEMENTS AND AXONAL TRANSPORT

Neurofilament proteins form a major component of the cytoskeleton of neurones, and important functions include maintenance of cell shape and axonal calibre, as well as axonal transport. Neurofilaments are the most abundant structural proteins in large cells with long axons such as motor neurones. Neurofilament subunits are assembled in the motor neurone cell body, and transported down the axon by slow axonal transport, with progressive phosphorylation during movement down the axon.

Neurofilament proteins are potential subcellular targets for injury in ALS/MND and other forms of motor neurone degeneration. Accumulation and abnormal assembly of neurofilaments are common pathological hallmarks of ALS/MND. Ubiquitinated inclusions with compact or Lewy body-like morphology within surviving motor neurones in ALS/MND may show immunoreactivity for neurofilament epitopes. In some cases of SOD1 related ALS, large argryophilic hyaline conglomerate inclusions expressing both phosphorylated and non-phosphorylated neurofilament epitopes have been observed in the cell bodies and axons of motor neurones.[4] The importance of neurofilaments in the normal functioning of motor neurones is demonstrated by the finding that approximately 1% of sporadic ALS/MND cases have deletions of insertions in the KSP repeat region of the neurofilament heavy (NFH) gene.[51,52] In addition, pathological changes within motor neurones develop in mice overexpressing NF-light or NF-heavy subunits, or in mice expressing mutations in the NF-light gene.[115–117] Transgenic mice which carry mutations in SOD1 also show alterations in neurofilament organisation, with the development of neurofilament spheroids, as well as reduced neurofilament protein and decreased transport rate in the ventral root axons.[118,119] Genetic manipulations to alter the expression of neurofilament proteins have been shown to alter the disease course in SOD1 transgenic mice. Increased expression of NF-heavy, which traps most neurofilaments within the cell body, robustly improves the disease course — by as much as six months in mutant SOD1 mice.[120]

The reasons for this somewhat counterintuitive effect are not understood, though it has been suggested that excess neurofilaments within the cell body may function as a buffer for some other deleterious process, for example offering phosphorylation sites for dysregulated intracellular kinases, or reducing the burden of axonal transport.

Another intermediate filament protein, peripherin, may play a role in motor neurone degeneration. Genetically engineered mice which over-express the major peripherin isoform (peripherin 58) develop late onset motor neurone degeneration accompanied by disruption of neurofilament assembly.[121] Another isoform, peripherin 61, is toxic when expressed in primary motor neurones and this toxic isoform is detectable in the spinal cord of sporadic ALS/MND cases.[122] However, manipulating the level of expression of peripherin in SOD1 transgenic mice does not appear to have any effect on the disease phenotype.[123]

Motor neurones, which in the human nervous system may have axons up to one metre in length, are highly reliant on an efficient intracellular transport system with anterograde and retrograde components. It is interesting that in SOD1 mutant mice, axonal transport is demonstrably impaired several months before clinical disease onset.[124] The kinesin complex of proteins are important molecular motors for anterograde axonal transport on the microtubule system. Mutations of genes encoding several kinesin proteins have been shown to cause various types of motor neurone degeneration including hereditary spastic paraplegia (SPG10) and type 2A Charcot–Marie–Tooth disease,[125,126] though they have not yet been associated with ALS/MND. The dynein–dynactin complex is the important motor for retrograde transport on the microtubule system, returning components (for example, multivesicular bodies and neurotrophic factors) back to the cell body. Mutations in dynein and the dynactin complex, which is an activator of cytoplasmic dynein, cause progressive motor neurone disease in mice.[44,127] As discussed in the genetics section, a dominant point mutation is the P150 subunit of dynactin, which causes a lower motor neurone disorder with vocal cord paresis in human subjects.[43]

PROTEIN AGGREGATION

A recurring theme highlighted in research into neurodegenerative diseases has been the misfolding of mutant proteins with the formation of intracellular aggregates. There is continuing debate as to whether such aggregated proteins play a key role in disease pathogenesis, whether they represent harmless bystanders, or whether they could be beneficial to the cell by sequestering potentially toxic abnormal proteins. In the SOD1 transgenic mouse model of familial ALS, the mutant SOD1 protein forms conspicuous cytoplasmic inclusions in motor neurones and sometimes in astrocytes, which develop before the onset of motor dysfunction. Several hypotheses have been put forward to explain how mutant SOD1 aggregates could produce cellular toxicity. First, there might be sequestration of other proteins required for normal motor neurone function. Several additional proteins have been found present in SOD1 aggregates including CCS (copper chaperone for SOD1), ubiquitin neurofilaments, glial fibrillary acidic protein, two neuronal glutamate transporters, BCl2, and proteins involved in chaperone and proteosome functions.[113,128] Second, by repeatedly misfolding, the SOD1 aggregates may reduce the availability of chaperone proteins required for the folding and function of other essential intracellular proteins.[129] Third, the SOD1 mutant protein aggregates may reduce proteasome activity needed for normal protein turnover.[130,131] Fourth, there could be inhibition of the function of specific organelles (for example, mitochondria) by aggregation on or within these organelles. Overexpression of chaperone proteins can reduce mutant SOD1 aggregation and enhances the survival and function of motor neurones in culture.[132] In addition, arimoclamol — a drug which enhances the expression of heat shock proteins — increases the life span of G93A SOD1 mice by 22%.[133] Clearly protein aggregates, which can be identified by ubiquitin immunostaining, are a feature of sporadic as well as familial SOD1 related ALS. SOD1 containing aggregates are not a characteristic feature of sporadic ALS, and determining the nature of the protein inclusions in sporadic ALS is a key research goal.

INFLAMMATORY CASCADES AND THE ROLE OF NON-NEURONAL CELLS

Recently there has been increasing interest in the possibility that non-neuronal cells, including activated microglia and astrocytes, may contribute to the pathogenesis or propagation of the disease process in ALS/MND. Several studies in genetically engineered mouse models have indicated that expression of mutant SOD1 in neurones alone is insufficient to cause motor neurone degeneration and that participation of non-neuronal cells may be required.[134,135] More recently Clement and colleagues produced several sets of chimeric mice which have both normal and mutant SOD1 expressing cells.[136] Motor neurones expressing mutant SOD1 could escape disease if surrounded by a sufficient number of normal non-neuronal cells. Conversely normal motor neurones surrounded by mutant SOD1 containing non-neuronal cells developed signs of abnormality, with the development of ubiquitinated intraneuronal deposits. Thus mutant SOD1 may cause neurotoxicity indirectly by disturbing the function of non-neuronal cells, for example microglia. Microglia play a critical role as resident immunocompetent and phagocytic cells within the CNS. Activation is associated with transformation to phagocytic cells capable of releasing potentially cytotoxic molecules including reactive oxygen species, nitric oxide, proteases, and proinflammatory cytokines such as interleukin-1B, tumour necrosis factor α (TNFα), and interleukin 6 (IL-6).[137] Given this, there is little doubt that activated microglia can inflict significant damage on neurones, but their role is complex and they are capable of stimulating neuroprotective as well as neurotoxic effects.

Proliferation of activated microglia is a prominent histological feature in the spinal ventral horn both in mutant SOD1 transgenic mice and in human ALS/MND.[138,139] In the mice, microglial activation is present before the onset of significant motor neurone loss or motor weakness. Various inflammatory cytokines or enzymes are upregulated in the spinal cord or CSF of ALS/MND patients (IL-6, IL-1β, cyclo-oxygenase 2 (COX2), and prostaglandin E2 (PGE2)) or, in the spinal cord of mutant SOD1 mice (IL-1β, TNFα, COX2, PGE2).[140–143] Microglia appear to mediate the toxicity to neurones in culture of CSF from patients with ALS/MND by releasing factors that enhance glutamate toxicity.[144] Minocycline, which inhibits microglial activation, ameliorates disease progression in mutant SOD1 mice.[110,145]

There is a tendency in ALS/MND for the disease to start focally and to spread "like a bush fire" to contiguous groups of motor neurones.[146] It would be very relevant to identify molecules that contribute to this propagation and those released from activated microglia would clearly be plausible candidates.

APOPTOSIS

Apoptosis describes the controlled removal of cells by an energy dependent cell death programme. Key molecular players contributing to the control of apoptosis include: the caspase family of proteolytic enzymes which orchestrate cell destruction by destroying several intracellular targets including structural and regulatory proteins; the Bcl2 family of oncoproteins, where the balance and subcellular distribution between pro- and anti-apoptotic members is important in regulating cell survival or destruction; and the apoptosis inhibitor family of proteins which suppress apoptosis by preventing proteolytic activation of specific caspases. Three main pathways triggering caspase activation have been identified including: release of proapoptotic factors (for example cytochrome c) from mitochondria; activation of cell surface ligand receptor systems of the tumour necrosis factor family including Fas-Fas ligand, with subsequent recruitment of cytosotic adaptor proteins; and stress to the endoplasmic reticulum with activation of caspase 12.

The evidence that motor neurones may die in ALS/MND by a programmed cell death pathway has been reviewed.[147,148] Key evidence from investigation of human necropsy material includes the following:

- evidence of structural morphology of degenerating motor neurones compatible with the apoptosis as well as internucleosomal DNA fragmentation detected by TUNEL staining[149];
- increased expression of specific apoptosis related molecules, for example Le^{y} antigen and prostate apoptosis response-4 protein in spinal cord;
- alteration in the balance of expression and subcellular compartmental localisation of pro- and anti-apoptotic members of the Bcl2 family in a direction favouring apoptosis[149,150];
- significant increases in the activities of caspases 1 and 3 in the spinal cord.[149]

Study of cellular models of SOD1 related ALS/MND has indicated that motor neuronal cells expressing mutant SOD1 are more likely to die by apoptosis when oxidatively stressed.[25] In addition, under unstressed basal culture conditions, these mutant SOD1 containing cells appear to be "primed" for apoptosis by expressing increased amounts compared with control cells of phosphatidyl serine on the cell surface and increased cleavage/activation of the initiator caspase 9.[151] In the mutant SOD1 mouse model, there is evidence of DNA laddering, increased expression and activation of caspase 1 and caspase 3 in the spinal cord of symptomatic mice, and alterations in the balance of key members of the Bcl2 protein family in a direction favouring apoptosis.[152,153] Cross breeding experiments between G93A SOD1 transgenic mice and mice genetically engineered to overexpress antiapopototic molecules results in amelioration of the murine disease. The administration of caspase inhibitors has a partial neuroprotective effect in cellular models,[151] and intraventricular administration of a broad spectrum caspase inhibitor to mutant SOD1 mice prolongs life span by approximately 20%.[152]

CELL SPECIFIC FEATURES OF MOTOR NEURONES WHICH MAY PREDISPOSE TO NEURODEGENERATION

One of the unsolved enigmas in neurodegenerative diseases in general, and in motor neurone degeneration in particular, is the selective vulnerability of certain neuronal groups to the neurodegenerative process. This vulnerability is relative rather than absolute. SOD1 is a ubiquitously distributed antioxidant defence protein, yet when the protein is mutated, it is motor neurones that are most susceptible to injury. The cell specific features of motor neurones that may predispose to age related degeneration have been reviewed[154,155] and are outlined in Table 4. Key features are likely to include the cell size of motor neurones, which has consequences for intracellular transport, energy metabolism, and neurofilament content. The neurones vulnerable to degeneration in ALS have a particular sensitivity to glutamatergic toxicity through AMPA receptor activation and differ from most other neuronal groups in expressing a high preponderance of calcium permeable AMPA receptors, which lack the GluR2 subunit.[156] Motor neurones also have a relative lack of expression of calcium buffering proteins[157] and appear to have a high threshold for mounting a protective heat shock response.[158] Recent studies suggest that the properties of mitochondria from the spinal cord may differ from those of mitochondria from other tissues.[112,159]

Table 4 Cell specific features of motor neurones predisposing to neurodegeneration.

E
1 • Cell size and axonal length
2 • High metabolic rate
3 • ? Specific features of motor neurone mitochondria
4 • Cytoskeleton, neurofilament content and reliance on efficient intracellular transport system
5 • Characteristic profile of cell surface glutamate receptors, with high relative expression of calcium permeable AMPA receptors, lacking the GluR2 subunit
6 • High expression of glutamate transporters in the vicinity of vulnerable motor neurone groups
7 • Low expression of specific calcium binding proteins
8 • High expression of SOD1 protein
9 • High threshold for mounting a heat shock response/ upregulation of chaperone proteins

CONCLUSIONS

The process of motor neurone degeneration in ALS/MND is complex and mutifactorial. Several genetic alterations can set the scene for motor neurone injury in familial ALS, but much remains to be learned about the genetic and environmental factors predisposing to the commoner sporadic form of the disease. Most has been learned about the mechanisms of motor neurone degeneration in the subtype of disease caused by SOD1 mutations, but even here there appears to be a complex interplay between multiple pathogenic processes including oxidative stress, protein aggregation, mitochondrial dysfunction excitotoxicity, and impaired axonal transport. New evidence is emerging that non-neuronal cells in the vicinity of motor neurones may contribute to neuronal injury. Evidence has accumulated that the final demise of motor neurones is likely to occur by a programmed cell death pathway resembling apoptosis.

To date only the antiglutamate agent riluzole has been shown reproducibly to prolong the survival of patients with ALS/MND, and this is a modest effect. In the recent past, robust cellular and animal models of motor neurone degeneration have emerged which are being used to evaluate new potential therapeutic strategies. New technologies including gene expression profiling using microarray platforms,[160,161] analysis of the repertoire of cellular proteins using proteomic approaches,[103,130] and the ability to subdissect motor neurones from complex tissues using laser capture microdissection[162] are likely to lead to clarification of our knowledge of the cellular mechanisms of disease in ALS/MND over the next few years. Important priorities for future research include the search for other genes associated with familial MND and for genetic and environmental factors predisposing to the sporadic form of the disease. In addition, further probing for insights into the cell specific biochemistry and physiology of motor neurones and the cellular pathways deranged during motor neurone degeneration are likely to lead to the development of more effective neuroprotective treatments for patients. The space constraints of this article have not permitted a detailed discussion of potential therapies targeted to the outlined molecular mechanisms of motor neurone injury. This topic has recently been reviewed in relation both to experimental models and to human MND.[16,163] Future treatment of ALS/MND is likely to involve a cocktail of neuroprotective compounds akin to chemotherapeutic combinations for malignant disease, which interfere with several molecular pathways that lead to neuronal injury.

ACKNOWLEDGEMENTS

PJS is supported by the Wellcome Trust, the Motor Neurone Disease Association, and the Medical Research Council.

REFERENCES

1. Maasilta P, Jokelainen M, Loytonen M, *et al.* Mortality from amyotrophic lateral sclerosis in Finland, 1986–1995. *Acta Neurol Scand* 2001; **104**:232–235.
2. Rowland LP. Clinical aspects of sporadic ALS/MND. In: Shaw PJ, Strong MJ, eds. *Blue Book of Practical Neurology: Motor Neuron Disorders*. Philadelphia: Heinemann, 2003:111–144.
3. Charcot JM, Joffroy A. Deux cas d'atrophie musculaire progressive avec lesions de la substance grise et des faisceaux antero-lateraux de la moelle epiniere. *Arch Physiol Neurol Pathol* 1869; **2**:744.
4. Ince PG, Tomkins J, Slade JY, *et al.* Amyotrophic lateral sclerosis associated with genetic abnormalities in the gene encoding Cu/Zn superoxide

dismutase: Molecular pathology of five new cases, and comparison with previous reports and 73 sporadic cases of ALS. *J Neuropathol Exp Neurol* 1998; **57**: 895–904.

5. Mannen T, Iwata M, Toyokura Y, *et al.* Preservation of a certain motoneurone group of the sacral cord in amyotrophic lateral sclerosis: Its clinical significance. *J Neurol Neurosurg Psychiatry* 1977; **40**:464–469.
6. del Aguila MA, Longstreth WT, McGuire V, *et al.* Prognosis in amyotrophic lateral sclerosis: A population-based study. *Neurology* 2003; **60**:813–819.
7. Turner MR, Parton MJ, Shaw CE, *et al.* Prolonged survival in motor neuron disease: A descriptive study of the King's database 1990–2002. *J Neurol Neurosurg Psychiatry* 2003; **74**:995–997.
8. Brooks BR, Miller RG, Swash M, *et al.* El Escorial revisited: Revised criteria for the diagnosis of amyotrophic lateral sclerosis. *Amyotroph Lateral Scler Other Motor Neuron Disord* 2000; **1**:293–299.
9. Wharton SP. Pathology of motor neuron disorders. In: Shaw PJ Strong MJ, eds. *Blue Books of Practical Neurology: Motor Neuron Disorders*. Philadelphia: Butterworth-Heinemann, 2003:17–50.
10. Hudson AJ. Amyotrophic lateral sclerosis and its association with dementia, parkinsonism and other neurological disorders: A review. *Brain* 1981; **104**: 217–247.
11. Massman PJ, Sims J, Cooke N, *et al.* Prevalence and correlates of neuropsychological deficits in amyotrophic lateral sclerosis. *J Neurol Neurosurg Psychiatry* 1996; **61**:450–455.
12. Rosen DR, Siddique T, Patterson D, *et al.* Mutations in Cu/Zn superoxide dismutase gene are associated with familial amyotrophic lateral sclerosis. *Nature* 1993; **364**:362.
13. Andersen PM, Sims KB, Xin WW, *et al.* Sixteen novel mutations in the Cu/Zn superoxide dismutase gene in amyotrophic lateral sclerosis: A decade of discoveries, defects and disputes. *Amyotroph Lateral Scler Other Motor Neuron Disord* 2003; **4**:62–73.
14. http://www.alsod.org.
15. Okado-Matsumoto A, Fridovich I. Subcellular distribution of superoxide dismutases (SOD) in rat liver: Cu,Zn-SOD in mitochondria. *J Biol Chem* 2001; **276**:38388–38393.
16. Bruijn LI, Miller TM, Cleveland DW. Unraveling the mechanisms involved in motor neuron degeneration in ALS. *Annu Rev Neurosci* 2004; **27**:723–749.
17. Andersen PM, Forsgren L, Binzer M, *et al.* Autosomal recessive adult-onset amyotrophic lateral sclerosis associated with homozygosity for Asp90Ala CuZn-superoxide dismutase mutation. A clinical and genealogical study of 36 patients. *Brain* 1996; **119**:1153–1172.

18. Cudkowicz ME, McKenna-Yasek D, Sapp PE, *et al.* Epidemiology of mutations in superoxide dismutase in amyotrophic lateral sclerosis. *Ann Neurol* 1997; **41**:210–221.
19. Kunst CB, Messer L, Gordon J, *et al.* Genetic mapping of a mouse modifier gene that can prevent ALS onset. *Genomics* 2000; **70**:181–189.
20. Wong PC, Pardo CA, Borchelt DR, *et al.* An adverse property of a familial ALS-linked SOD1 mutation causes motor neuron disease characterized by vacuolar degeneration of mitochondria. *Neuron* 1995; **14**:1105–1116.
21. Gurney ME, Pu H, Chiu AY, *et al.* Motor neuron degeneration in mice that express a human Cu/Zn superoxide dismutase mutation. *Science* 1994; **264**: 1772–1775.
22. Bruijn LI, Cleveland DW. Mechanisms of selective motor neuron death in ALS: Insights from transgenic mouse models of motor neuron disease. *Neuropathol Appl Neurobiol* 1996; **22**:373–387.
23. Nagai M, Aoki M, Miyoshi I, *et al.* Rats expressing human cytosolic copper-zinc superoxide dismutase transgenes with amyotrophic lateral sclerosis: Associated mutations develop motor neuron disease. *J Neurosci* 2001; **21**:9246–9254.
24. Howland DS, Liu J, She Y, *et al.* Focal loss of the glutamate transporter EAAT2 in a transgenic rat model of SOD1 mutant-mediated amyotrophic lateral sclerosis (ALS). *Proc Natl Acad Sci USA* 2002; **99**:1604–1609.
25. Cookson MR, Menzies FM, Manning P, *et al.* Cu/Zn superoxide dismutase (SOD1) mutations associated with familial amyotrophic lateral sclerosis (ALS) affect cellular free radical release in the presence of oxidative stress. *Amyotroph Lateral Scler Other Motor Neuron Disord* 2002; **3**:75–85.
26. Pasinelli P, Borchelt DR, Houseweart MK, *et al.* Caspase-1 is activated in neural cells and tissue with amyotrophic lateral sclerosis-associated mutations in copper-zinc superoxide dismutase. *Proc Natl Acad Sci USA* 1998; **95**:15763–15768.
27. Roy J, Minotti S, Dong L, *et al.* Glutamate potentiates the toxicity of mutant Cu/Zn-superoxide dismutase in motor neurons by postsynaptic calcium-dependent mechanisms. *J Neurosci* 1998; **18**:9673–9684.
28. Hadano S, Hand CK, Osuga H, *et al.* A gene encoding a putative GTPase regulator is mutated in familial amyotrophic lateral sclerosis 2. *Nat Genet* 2001; **29**:166–173.
29. Yang Y, Hentati A, Deng HX, *et al.* The gene encoding alsin, a protein with three guanine-nucleotide exchange factor domains, is mutated in a form of recessive amyotrophic lateral sclerosis. *Nat Genet* 2001; **29**:160–165.
30. Eymard-Pierre E, Lesca G, Dollet S, *et al.* Infantile-onset ascending hereditary spastic paralysis is associated with mutations in the alsin gene. *Am J Hum Genet* 2002; **71**:518–527.

31. Devon RS, Helm JR, Rouleau GA, *et al.* The first nonsense mutation in alsin results in a homogeneous phenotype of infantile-onset ascending spastic paralysis with bulbar involvement in two siblings. *Clin Genet* 2003; **64**: 210–215.
32. Gros-Louis F, Meijer IA, Hand CK, *et al.* An ALS2 gene mutation causes hereditary spastic paraplegia in a Pakistani kindred. *Ann Neurol* 2003; **53**:144–145.
33. Vetter IR, Wittinghofer A. The guanine nucleotide-binding switch in three dimensions. *Science* 2001; **294**:1299–1304.
34. Yamanaka K, Vande Velde C, Eymard-Pierre E, *et al.* Unstable mutants in the peripheral endosomal membrane component ALS2 cause early-onset motor neuron disease. *Proc Natl Acad Sci USA* 2003; **100**:16041–16046.
35. Otomo A, Hadano S, Okada T, *et al.* ALS2, a novel guanine nucleotide exchange factor for the small GTPase Rab5, is implicated in endosomal dynamics. *Hum Mol Genet* 2003; **12**:1671–1687.
36. Kunita R, Otomo A, Mizumura H, *et al.* Homo-oligomerization of ALS2 through its unique carboxyl-terminal regions is essential for the ALS2-associated Rab5 guanine nucleotide exchange activity and its regulatory function on endosome trafficking. *J Biol Chem* 2004; **279**:38626–38635.
37. Kris J, Millecamps S, Zhu Q, *et al.* Creation of a mouse model for juvenile amyotrophic lateral sclerosis. *Amyotroph Lateral Scler Other Motor Neuron Disord* 2003; **4**(suppl 1):11.
38. Cai H, Wen H, Chaing HC, *et al.* Physiological role of ALS2: Selective vulnerability and generation of ALS2 knockout mice. *Amyotroph Lateral Scler Other Motor Neuron Disord* 2003; **4**(suppl 1):11–12.
39. Kanekura K, Hashimoto Y, Niikura T, *et al.* Alsin, the product of ALS2 gene, suppresses SOD1 mutant neurotoxicity through RhoGEF domain by interacting with SOD1 mutants. *J Biol Chem* 2004; **279**:19247–19256.
40. Chen YZ, Bennett CL, Huynh HM, *et al.* DNA/RNA helicase gene mutations in a form of juvenile amyotrophic lateral sclerosis (ALS4). *Am J Hum Genet* 2004; **74**:1128–1135.
41. Moreira MC, Klur S, Watanabe M, *et al.* Senataxin, the ortholog of a yeast RNA helicase, is mutant in ataxia-ocular apraxia 2. *Nat Genet* 2004; **36**:225–227.
42. Nishimura AL, Mitne-Neto M, Silva HC, *et al.* A mutation in the vesicle-trafficking protein VAPB causes late-onset spinal muscular atrophy and amyotrophic lateral sclerosis. *Am J Hum Genet* 2004; **75**:822–831.
43. Puls I, Jonnakuty C, LaMonte BH, *et al.* Mutant dynactin in motor neuron disease. *Nat Genet* 2003; **33**:455–456.
44. LaMonte BH, Wallace KE, Holloway BA, *et al.* Disruption of dynein/dynactin inhibits axonal transport in motor neurons causing late-onset progressive degeneration. *Neuron* 2002; **34**:715–727.

45. Hosler BA, Siddique T, Sapp PC, *et al.* Linkage of familial amyotrophic lateral sclerosis with frontotemporal dementia to chromosome 9q21–q22. *JAMA* 2000; **284**:1664–1669.
46. Ostojic J, Axelman K, Lannfelt L, *et al.* No evidence of linkage to chromosome 9q21–22 in a Swedish family with frontotemporal dementia and amyotrophic lateral sclerosis. *Neurosci Lett* 2003; **340**:245–247.
47. Clark LN, Poorkaj P, Wszolek Z, *et al.* Pathogenic implications of mutations in the tau gene in pallido-ponto-nigral degeneration and related neurodegenerative disorders linked to chromosome 17. *Proc Natl Acad Sci USA* 1998; **95**:13103–13107.
48. Hutton M, Lendon CL, Rizzu P, *et al.* Association of missense and 5′-splice-site mutations in tau with the inherited dementia FTDP-17. *Nature* 1998; **393**: 702–705.
49. Kowalska A, Konagaya M, Sakai M, *et al.* Familial amyotrophic lateral sclerosis and parkinsonism-dementia complex–tauopathy without mutations in the tau gene? *Folia Neuropathol* 2003; **41**:59–64.
50. Wilhelmsen KC, Forman MS, Rosen HJ, *et al.* 17q-linked frontotemporal dementia-amyotrophic lateral sclerosis without tau mutations with tau and alpha-synuclein inclusions. *Arch Neurol* 2004; **61**:398–406.
51. Figlewicz DA, Krizus A, Martinoli MG, *et al.* Variants of the heavy neurofilament subunit are associated with the development of amyotrophic lateral sclerosis. *Hum Mol Genet* 1994; **3**:1757–1761.
52. Tomkins J, Usher P, Slade JY, *et al.* Novel insertion in the KSP region of the neurofilament heavy gene in amyotrophic lateral sclerosis (ALS). *Neuroreport* 1998; **9**:3967–3970.
53. Al-Chalabi A, Enayat ZE, Bakker MC, *et al.* Association of apolipoprotein E epsilon 4 allele with bulbar-onset motor neuron disease. *Lancet* 1996; **347**:159–160.
54. Drory VE, Birnbaum M, Korczyn AD, *et al.* Association of APOE epsilon4 allele with survival in amyotrophic lateral sclerosis. *J Neurol Sci* 2001; **190**:17–20.
55. Comi GP, Bordoni A, Salani S, *et al.* Cytochrome c oxidase subunit I microdeletion in a patient with motor neuron disease. *Ann Neurol* 1998; **43**:110–116.
56. Rothstein JD, Van Kammen M, Levey AI, *et al.* Selective loss of glial glutamate transporter GLT-1 in amyotrophic lateral sclerosis. *Ann Neurol* 1995; **38**:73–84.
57. Rao SD, Yin HZ, Weiss JH. Disruption of glial glutamate transport by reactive oxygen species produced in motor neurons. *J Neurosci* 2003; **23**: 2627–2633.
58. Fray AE, Ince PG, Banner SJ, *et al.* The expression of the glial glutamate transporter protein EAAT2 in motor neuron disease: An immunohistochemical study. *Eur J Neurosci* 1998; **10**:2481–2489.

59. Lin CL, Bristol LA, Jin L, *et al.* Aberrant RNA processing in a neurodegenerative disease: The cause for absent EAAT2, a glutamate transporter, in amyotrophic lateral sclerosis. *Neuron* 1998; **20**:589–602.
60. Meyer T, Fromm A, Munch C, *et al.* The RNA of the glutamate transporter EAAT2 is variably spliced in amyotrophic lateral sclerosis and normal individuals. *J Neurol Sci* 1999; **170**:45–50.
61. Flowers JM, Powell JF, Leigh PN, *et al.* Intron 7 retention and exon 9 skipping EAAT2 mRNA variants are not associated with amyotrophic lateral sclerosis. *Ann Neurol* 2001; **49**:643–649.
62. Trotti D, Aoki M, Pasinelli P, *et al.* Amyotrophic lateral sclerosis-linked glutamate transporter mutant has impaired glutamate clearance capacity. *J Biol Chem* 2001; **276**:576–582.
63. Kawahara Y, Ito K, Sun H, *et al.* Glutamate receptors: RNA editing and death of motor neurons. *Nature* 2004; **427**:801.
64. Jackson M, Morrison KE, Al-Chalabi A, *et al.* Analysis of chromosome 5 q13 genes in ALS: Homozygous NAIP deletion in a sporadic case. *Ann Neurol* 1996; **39**:796–800.
65. Orrell RW, Habgood JJ, de Belleroche JS, *et al.* The relationship of spinal muscular atrophy to motor neuron disease: Investigation of SMN and NAIP gene deletions in sporadic and familial amyotrophic lateral sclerosis. *J Neurol Sci* 1997; **145**:55–61.
66. Veldink JH, van den Berg LH, Cobben JM, *et al.* Homozygous deletion of the survival motor neuron 2 gene is a prognostic factor in sporadic ALS. *Neurology* 2001; **56**:749–752.
67. Corcia P, Mayeux-Portas V, Khoris J, *et al.* Abnormal SMN1 gene copy number is a susceptibility factor for amyotrophic lateral sclerosis. *Ann Neurol* 2002; **51**:243–246.
68. Giess R, Holtmann B, Braga M, *et al.* Early onset of severe familial amyotrophic lateral sclerosis with a SOD-1 mutation: Potential impact of CNTF as a candidate modifier gene. *Am J Hum Genet* 2002; **70**:1277–1286.
69. Brockington A, Lewis C, Wharton S, *et al.* Vascular endothelial growth factor and the nervous system. *Neuropathol Appl Neurobiol* 2004; **30**: 427–446.
70. Lambrechts D, Storkebaum E, Morimoto M, *et al.* VEGF is a modifier of amyotrophic lateral sclerosis in mice and humans and protects motoneurons against ischemic death. *Nat Genet* 2003; **34**:383–394.
71. Oosthuyse B, Moons L, Storkebaum E, *et al.* Deletion of the hypoxia-response element in the vascular endothelial growth factor promoter causes motor neuron degeneration. *Nat Genet* 2001; **28**:131–138.

72. Brockington A, Kirby J, Eggitt D, *et al.* Screening of the regulatory and coding regions of vascular endothelial growth factor in amyotrophic lateral sclerosis. *Neurogenetics* 2005, March 18 [Epub ahead of print].
73. Gros-Louis F, Laurent S, Lopes AA, *et al.* Absence of mutations in the hypoxia response element of VEGF in ALS. *Muscle Nerve* 2003; **28**:774–775.
74. Shaw PJ, Ince PG, Falkous G, *et al.* Oxidative damage to protein in sporadic motor neuron disease spinal cord. *Ann Neurol* 1995; **38**:691–695.
75. Ferrante RJ, Browne SE, Shinobu LA, *et al.* Evidence of increased oxidative damage in both sporadic and familial amyotrophic lateral sclerosis. *J Neurochem* 1997; **69**:2064–2074.
76. Smith RG, Henry YK, Mattson MP, *et al.* Presence of 4-hydroxynonenal in cerebrospinal fluid of patients with sporadic amyotrophic lateral sclerosis. *Ann Neurol* 1998; **44**:696–699.
77. Tohgi H, Abe T, Yamazaki K, *et al.* Remarkable increase in cerebrospinal fluid 3-nitrotyrosine in patients with sporadic amyotrophic lateral sclerosis. *Ann Neurol* 1999; **46**:129–131.
78. Aguirre T, Van Den Bosch L, Goetschalckx K, *et al.* Increased sensitivity of fibroblasts from amyotrophic lateral sclerosis patients to oxidative stress. *Ann Neurol* 1998; **43**:452–457.
79. Beckman JS, Carson M, Smith CD, *et al.* ALS, SOD and peroxynitrite. *Nature* 1993; **364**:584.
80. Crow JP, Sampson JB, Zhuang Y, *et al.* Decreased zinc affinity of amyotrophic lateral sclerosis-associated superoxide dismutase mutants leads to enhanced catalysis of tyrosine nitration by peroxynitrite. *J Neurochem* 1997; **69**:1936–1944.
81. Estevez AG, Crow JP, Sampson JB, *et al.* Induction of nitric oxide-dependent apoptosis in motor neurons by zinc-deficient superoxide dismutase. *Science* 1999; **286**:2498–2500.
82. Wang J, Slunt H, Gonzales V, *et al.* Copper-binding-site-null SOD1 causes ALS in transgenic mice: Aggregates of non-native SOD1 delineate a common feature. *Hum Mol Genet* 2003; **12**:2753–2764.
83. Subramaniam JR, Lyons WE, Liu J, *et al.* Mutant SOD1 causes motor neuron disease independent of copper chaperone-mediated copper loading. *Nat Neurosci* 2002; **5**:301–307.
84. Facchinetti F, Sasaki M, Cutting FB, *et al.* Lack of involvement of neuronal nitric oxide synthase in the pathogenesis of a transgenic mouse model of familial amyotrophic lateral sclerosis. *Neuroscience* 1999; **90**:1483–1492.
85. Upton-Rice MN, Cudkowicz ME, Mathew RK, *et al.* Administration of nitric oxide synthase inhibitors does not alter disease course of amyotrophic lateral sclerosis SOD1 mutant transgenic mice. *Ann Neurol* 1999; **45**:413–414.

86. Son M, Fathallah-Shaykh HM, Elliott JL. Survival in a transgenic model of FALS is independent of iNOS expression. *Ann Neurol* 2001; **50**:273.
87. Carriedo SG, Yin HZ, Weiss JH. Motor neurons are selectively vulnerable to AMPA/kainate receptor-mediated injury *in vitro*. *J Neurosci* 1996; **16**: 4069–4079.
88. Heath PR, Shaw PJ. Update on the glutamatergic neurotransmitter system and the role of excitotoxicity in amyotrophic lateral sclerosis. *Muscle Nerve* 2002; **26**:438–458.
89. Shaw PJ, Forrest V, Ince PG, *et al.* CSF and plasma amino acid levels in motor neuron disease: Elevation of CSF glutamate in a subset of patients. *Neurodegeneration* 1995; **4**:209–216.
90. Spreux-Varoquaux O, Bensimon G, Lacomblez L, *et al.* Glutamate levels in cerebrospinal fluid in amyotrophic lateral sclerosis: A reappraisal using a new HPLC method with coulometric detection in a large cohort of patients. *J Neurol Sci* 2002; **193**:73–78.
91. Kruman II, Pedersen WA, Springer JE, *et al.* ALS-linked Cu/Zn-SOD mutation increases vulnerability of motor neurons to excitotoxicity by a mechanism involving increased oxidative stress and perturbed calcium homeostasis. *Exp Neurol* 1999; **160**:28–39.
92. Spalloni A, Albo F, Ferrari F, *et al.* Cu/Zn-superoxide dismutase (GLY93→ALA) mutation alters AMPA receptor subunit expression and function and potentiates kainate-mediated toxicity in motor neurons in culture. *Neurobiol Dis* 2004; **15**:340–350.
93. Bendotti C, Tortarolo M, Suchak SK, *et al.* Transgenic SOD1 G93A mice develop reduced GLT-1 in spinal cord without alterations in cerebrospinal fluid glutamate levels. *J Neurochem* 2001; **79**:737–746.
94. Lacomblez L, Bensimon G, Leigh PN, *et al.* Dose-ranging study of riluzole in amyotrophic lateral sclerosis. Amyotrophic Lateral Sclerosis/Riluzole Study Group II. *Lancet* 1996; **347**:1425–1431.
95. Gurney ME, Cutting FB, Zhai P, *et al.* Benefit of vitamin E, riluzole, and gabapentin in a transgenic model of familial amyotrophic lateral sclerosis. *Ann Neurol* 1996; **39**:147–157.
96. Beal MF. Mitochondria and the pathogenesis of ALS. *Brain* 2000; **123**: 1291–1292.
97. Menzies FM, Ince PG, Shaw PJ. Mitochondrial involvement in amyotrophic lateral sclerosis. *Neurochem Int* 2002; **40**:543–551.
98. Siklos L, Engelhardt J, Harati Y, *et al.* Ultrastructural evidence for altered calcium in motor nerve terminals in amyotropic lateral sclerosis. *Ann Neurol* 1996; **39**:203–216.

99. Borthwick GM, Johnson MA, Ince PG, *et al.* Mitochondrial enzyme activity in amyotrophic lateral sclerosis: Implications for the role of mitochondria in neuronal cell death. *Ann Neurol* 1999; **46**:787–790.
100. Dhaliwal GK, Grewal RP. Mitochondrial DNA deletion mutation levels are elevated in ALS brains. *Neuroreport* 2000; **11**:2507–2509.
101. Wiedemann FR, Manfredi G, Mawrin C, *et al.* Mitochondrial DNA and respiratory chain function in spinal cords of ALS patients. *J Neurochem* 2002; **80**: 616–625.
102. Menzies FM, Cookson MR, Taylor RW, *et al.* Mitochondrial dysfunction in a cell culture model of familial amyotrophic lateral sclerosis. *Brain* 2002; **125**:1522–1533.
103. Fukada K, Zhang F, Vien A, *et al.* Mitochondrial proteomic analysis of a cell line model of familial amyotrophic lateral sclerosis. *Mol Cell Proteomics* 2004; **3**:1211–1223.
104. Takeuchi H, Kobayashi Y, Ishigaki S, *et al.* Mitochondrial localization of mutant superoxide dismutase 1 triggers caspase-dependent cell death in a cellular model of familial amyotrophic lateral sclerosis. *J Biol Chem* 2002; **277**: 50966–50972.
105. Higgins CM, Jung C, Ding H, *et al.* Mutant Cu, Zn superoxide dismutase that causes motoneuron degeneration is present in mitochondria in the CNS. *J Neurosci* 2002; **22**:RC215.
106. Jaarsma D, Rognoni F, van Duijn W, *et al.* CuZn superoxide dismutase (SOD1) accumulates in vacuolated mitochondria in transgenic mice expressing amyotrophic lateral sclerosis-linked SOD1 mutations. *Acta Neuropathol (Berl)* 2001; **102**:293–305.
107. Jung C, Higgins CM, Xu Z. A quantitative histochemical assay for activities of mitochondrial electron transport chain complexes in mouse spinal cord sections. *J Neurosci Methods* 2002; **114**:165–172.
108. Mattiazzi M, D'Aurelio M, Gajewski CD, *et al.* Mutated human SOD1 causes dysfunction of oxidative phosphorylation in mitochondria of transgenic mice. *J Biol Chem* 2002; **277**:29626–29633.
109. Guegan C, Vila M, Rosoklija G, *et al.* Recruitment of the mitochondrial-dependent apoptotic pathway in amyotrophic lateral sclerosis. *J Neurosci* 2001; **21**:6569–6576.
110. Zhu S, Stavrovskaya IG, Drozda M, *et al.* Minocycline inhibits cytochrome c release and delays progression of amyotrophic lateral sclerosis in mice. *Nature* 2002; **417**:74–78.
111. Andreassen OA, Ferrante RJ, Klivenyi P, *et al.* Partial deficiency of manganese superoxide dismutase exacerbates a transgenic mouse model of amyotrophic lateral sclerosis. *Ann Neurol* 2000; **47**:447–455.

112. Liu J, Lillo C, Jonsson PA, *et al.* Toxicity of familial ALS-linked SOD1 mutants from selective recruitment to spinal mitochondria. *Neuron* 2004; **43**:5–17.
113. Pasinelli P, Belford ME, Lennon N, *et al.* Amyotrophic lateral sclerosis-associated SOD1 mutant proteins bind and aggregate with Bcl-2 in spinal cord mitochondria. *Neuron* 2004; **43**:19–30.
114. Klivenyi P, Ferrante RJ, Matthews RT, *et al.* Neuroprotective effects of creatine in a transgenic animal model of amyotrophic lateral sclerosis. *Nat Med* 1999; **5**:347–350.
115. Xu Z, Cork LC, Griffin JW, *et al.* Increased expression of neurofilament subunit NF-L produces morphological alterations that resemble the pathology of human motor neuron disease. *Cell* 1993; **73**:23–33.
116. Cote F, Collard JF, Julien JP. Progressive neuronopathy in transgenic mice expressing the human neurofilament heavy gene: A mouse model of amyotrophic lateral sclerosis. *Cell* 1993; **73**:35–46.
117. Lee MK, Marszalek JR, Cleveland DW. A mutant neurofilament subunit causes massive, selective motor neuron death: Implications for the pathogenesis of human motor neuron disease. *Neuron* 1994; **13**:975–988.
118. Tu PH, Raju P, Robinson KA, *et al.* Transgenic mice carrying a human mutant superoxide dismutase transgene develop neuronal cytoskeletal pathology resembling human amyotrophic lateral sclerosis lesions. *Proc Natl Acad Sci USA* 1996; **93**:3155–3160.
119. Zhang B, Tu P, Abtahian F, *et al.* Neurofilaments and orthograde transport are reduced in ventral root axons of transgenic mice that express human SOD1 with a G93A mutation. *J Cell Biol* 1997; **139**:1307–1315.
120. Couillard-Despres S, Zhu Q, Wong PC, *et al.* Protective effect of neurofilament heavy gene overexpression in motor neuron disease induced by mutant superoxide dismutase. *Proc Natl Acad Sci USA* 1998; **95**:9626–9630.
121. Beaulieu JM, Nguyen MD, Julien JP. Late onset of motor neurons in mice overexpressing wild-type peripherin. *J Cell Biol* 1999; **147**:531–544.
122. Robertson J, Doroudchi MM, Nguyen MD, *et al.* A neurotoxic peripherin splice variant in a mouse model of ALS. *J Cell Biol* 2003; **160**:939–949.
123. Lariviere RC, Beaulieu JM, Nguyen MD, *et al.* Peripherin is not a contributing factor to motor neuron disease in a mouse model of amyotrophic lateral sclerosis caused by mutant superoxide dismutase. *Neurobiol Dis* 2003; **13**:158–166.
124. Williamson TL, Cleveland DW. Slowing of axonal transport is a very early event in the toxicity of ALS-linked SOD1 mutants to motor neurons. *Nat Neurosci* 1999; **2**:50–56.
125. Reid E, Kloos M, Ashley-Koch A, *et al.* A kinesin heavy chain (KIF5A) mutation in hereditary spastic paraplegia (SPG10). *Am J Hum Genet* 2002; **71**:1189–1194.

126. Zhao C, Takita J, Tanaka Y, *et al.* Charcot-Marie-Tooth disease type 2A caused by mutation in a microtubule motor KIF1Bbeta. *Cell* 2001; **105**:587–597.
127. Hafezparast M, Klocke R, Ruhrberg C, *et al.* Mutations in dynein link motor neuron degeneration to defects in retrograde transport. *Science* 2003; **300**: 808–812.
128. Watanabe M, Dykes-Hoberg M, Culotta VC, *et al.* Histological evidence of protein aggregation in mutant SOD1 transgenic mice and in amyotrophic lateral sclerosis neural tissues. *Neurobiol Dis* 2001; **8**:933–941.
129. Bruening W, Roy J, Giasson B, *et al.* Up-regulation of protein chaperones preserves viability of cells expressing toxic Cu/Zn-superoxide dismutase mutants associated with amyotrophic lateral sclerosis. *J Neurochem* 1999; **72**:693–699.
130. Allen S, Heath PR, Kirby J, *et al.* Analysis of the cytosolic proteome in a cell culture model of familial amyotrophic lateral sclerosis reveals alterations to the proteasome, antioxidant defenses, and nitric oxide synthetic pathways. *J Biol Chem* 2003; **278**:6371–6383.
131. Johnston JA, Dalton MJ, Gurney ME, *et al.* Formation of high molecular weight complexes of mutant Cu, Zn-superoxide dismutase in a mouse model for familial amyotrophic lateral sclerosis. *Proc Natl Acad Sci USA* 2000; **97**: 12571–12576.
132. Takeuchi H, Kobayashi Y, Yoshihara T, *et al.* Hsp70 and Hsp40 improve neurite outgrowth and suppress intracytoplasmic aggregate formation in cultured neuronal cells expressing mutant SOD1. *Brain Res* 2002; **949**:11–22.
133. Kieran D, Kalmar B, Dick JR, *et al.* Treatment with arimoclomol, a coinducer of heat shock proteins, delays disease progression in ALS mice. *Nat Med* 2004; **10**:402–405.
134. Pramatarova A, Laganiere J, Roussel J, *et al.* Neuron-specific expression of mutant superoxide dismutase 1 in transgenic mice does not lead to motor impairment. *J Neurosci* 2001; **21**:3369–3374.
135. Lino MM, Schneider C, Caroni P. Accumulation of SOD1 mutants in postnatal motoneurons does not cause motoneuron pathology or motoneuron disease. *J Neurosci* 2002; **22**:4825–4832.
136. Clement AM, Nguyen MD, Roberts EA, *et al.* Wild-type nonneuronal cells extend survival of SOD1 mutant motor neurons in ALS mice. *Science* 2003; **302**:113–117.
137. Gonzalez-Scarano F, Baltuch G. Microglia as mediators of inflammatory and degenerative diseases. *Annu Rev Neurosci* 1999; **22**:219–240.
138. Alexianu ME, Kozovska M, Appel SH. Immune reactivity in a mouse model of familial ALS correlates with disease progression. *Neurology* 2001; **57**: 1282–1289.

139. Kawamata T, Akiyama H, Yamada T, *et al.* Immunologic reactions in amyotrophic lateral sclerosis brain and spinal cord tissue. *Am J Pathol* 1992; **140**: 691–707.
140. Sekizawa T, Openshaw H, Ohbo K, *et al.* Cerebrospinal fluid interleukin 6 in amyotrophic lateral sclerosis: Immunological parameter and comparison with inflammatory and non-inflammatory central nervous system diseases. *J Neurol Sci* 1998; **154**:194–199.
141. Hensley K, Floyd RA, Gordon B, *et al.* Temporal patterns of cytokine and apoptosis-related gene expression in spinal cords of the G93A-SOD1 mouse model of amyotrophic lateral sclerosis. *J Neurochem* 2002; **82**:365–374.
142. Almer G, Teismann P, Stevic Z, *et al.* Increased levels of the pro-inflammatory prostaglandin PGE2 in CSF from ALS patients. *Neurology* 2002; **58**:1277–1279.
143. Elliott JL. Cytokine upregulation in a murine model of familial amyotrophic lateral sclerosis. *Brain Res Mol Brain Res* 2001; **95**:172–178.
144. Tikka TM, Vartiainen NE, Goldsteins G, *et al.* Minocycline prevents neurotoxicity induced by cerebrospinal fluid from patients with motor neurone disease. *Brain* 2002; **125**:722–731.
145. Zhang W, Narayanan M, Friedlander RM. Additive neuroprotective effects of minocycline with creatine in a mouse model of ALS. *Ann Neurol* 2003; **53**: 267–270.
146. Brooks BR, Shodis KA, Lewis DH, *et al.* Natural history of amyotrophic lateral sclerosis. Quantification of symptoms, signs, strength, and function. *Adv Neurol* 1995; **68**:163–184.
147. Sathasivam S, Ince PG, Shaw PJ. Apoptosis in amyotrophic lateral sclerosis: A review of the evidence. *Neuropathol Appl Neurobiol* 2001; **27**:257–274.
148. Guegan C, Przedborski S. Programmed cell death in amyotrophic lateral sclerosis. *J Clin Invest* 2003; **111**:153–161.
149. Martin LJ. Neuronal death in amyotrophic lateral sclerosis is apoptosis: Possible contribution of a programmed cell death mechanism. *J Neuropathol Exp Neurol* 1999; **58**:459–471.
150. Mu X, He J, Anderson DW, *et al.* Altered expression of bcl-2 and bax mRNA in amyotrophic lateral sclerosis spinal cord motor neurons. *Ann Neurol* 1996; **40**:379–386.
151. Sathasivam S, Grierson AJ, Shaw PJ. Characterisation of the caspase cascade in a cell culture model of SOD1-related FALS: Expression, activation and therapeutic effects of inhibition. *Neuropath Appl Neurobiol* (in press).
152. Li M, Ona VO, Guegan C, *et al.* Functional role of caspase-1 and caspase-3 in an ALS transgenic mouse model. *Science* 2000; **288**:335–339.

153. Vukosavic S, Dubois-Dauphin M, Romero N, *et al.* Bax and Bcl-2 interaction in a transgenic mouse model of familial amyotrophic lateral sclerosis. *J Neurochem* 1999; **73**:2460–2468.
154. Shaw PJ, Eggett CJ. Molecular factors underlying selective vulnerability of motor neurons to neurodegeneration in amyotrophic lateral sclerosis. *J Neurol* 2000; **247**(suppl 1):I17–I27.
155. Durham HD. Factors underlying the selective vulnerability of motor neurons to neurodegeneration. In: Shaw PJ, Strong MJ, eds. *Blue Books of Practical Neurology: Motor Neuron Disorders*. Philadelphia: Butterworth Heinemann, 2003:379–400.
156. Williams TL, Day NC, Ince PG, *et al.* Calcium-permeable alpha-amino-3-hydroxy-5-methyl-4-isoxazole propionic acid receptors: A molecular determinant of selective vulnerability in amyotrophic lateral sclerosis. *Ann Neurol* 1997; **42**:200–207.
157. Ince P, Stout N, Shaw P, *et al.* Parvalbumin and calbindin D-28k in the human motor system and in motor neuron disease. *Neuropathol Appl Neurobiol* 1993; **19**:291–299.
158. Batulan ZB, Minotti S, Figlewicz DA, *et al.* High threshold for activation of the stress response in motor neurons: A source of vulnerability in ALS? *Society of Neuroscience Abstracts* 2000; **26**:85.15.
159. Sullivan PG, Rabchevsky AG, Keller JN, *et al.* Intrinsic differences in brain and spinal cord mitochondria: Implication for therapeutic interventions. *J Comp Neurol* 2004; **474**:524–534.
160. Kirby J, Halligan E, Baptista MJ, *et al.* Mutant SOD1 alters the motor neuronal transcriptome: Implications for familial ALS. *Brain* (in press).
161. Olsen MK, Roberds SL, Ellerbrock BR, *et al.* Disease mechanisms revealed by transcription profiling in SOD1-G93A transgenic mouse spinal cord. *Ann Neurol* 2001; **50**:730–740.
162. Heath PR, Tomkins J, Ince PG, *et al.* Quantitative assessment of AMPA receptor mRNA in human spinal motor neurons isolated by laser capture microdissection. *Neuroreport* 2002; **13**:1753–1757.
163. Turner MR, Leigh PN. Disease-modifying therapies in motor neuron disorders: The present position and potential future developments. In: Shaw PJ, Strong MJ, eds. *Blue Books of Practical Neurology: Motor Neuron Disorders*. Philadelphia: Butterworth-Heinemann, 2003:497–544.

7

Neurodegenerative Disorders: Parkinson's Disease and Huntington's Disease

*S M Hague, S Klaffke and O Bandmann**

Parkinson's disease and Huntington's disease are both model diseases. Parkinson's disease is obviously only one of several akinetic-rigid syndromes (albeit by far the most common one) and Huntington's disease is only one of an ever growing number of trinucleotide repeat disorders. Molecular genetic studies and subsequent molecular biological studies have provided fascinating new insights into the pathogenesis of both disorders and there is now real hope for disease modifying treatment in the not too distant future for patients with Parkinson's disease or Huntington's disease.

PARKINSON'S DISEASE

Parkinson's disease is the second most common neurodegenerative disorder after Alzheimer's disease. The age adjusted prevalence for Parkinson's

*Academic Neurology Unit, Division of Genomic Medicine, University of Sheffield, UK

Dr Oliver Bandmann, Academic Neurology Unit, E Floor Medical School, Beech Hill Road, Sheffield S10 2RX, UK; o.bandmann@sheffield.ac.uk disease and other types of parkinsonism in the United Kingdom is 254/100 000, but the prevalence rises with age from 0.143% in the 50 to 59 year old population to 1.75% in the population aged 80 years or older.[1] The pathological hallmarks are dopaminergic cell loss in the substantia nigra and the presence of Lewy bodies and Lewy neurites. Lewy bodies and dystrophic Lewy neurites are cytoplasmic accumulations of aggregated proteins.

Research into the pathogenesis of this disorder in the 1980s and early 1990s predominantly focused on oxidative stress and impaired function of the mitochondrial respiratory chain. Since the mid-1990s, scientific progress has been mainly the result of molecular genetic research and further studies investigating the physiological role of the mutated genes/proteins, the functional consequences of the disease causing mutations, and a subsequent investigation of the affected pathways in sporadic Parkinson's disease (Table 1).

Table 1 Familial forms of Parkinson's disease: genes, chromosomal loci, and mode of inheritance.

E

Locus/gene	Chromosomal location	Inheritance pattern	Clinical features
PARK-1/α-synuclein	4q21–q23	AD	Early onset, rapid progression, cognitive impairment
PARK-2/parkin	6q25.2–q27	AR	Juvenile onset, slow progression, focal dystonia
PARK-3	2p13	AD	Late onset
Formerly PARK-4/α-synuclein triplication	4q21–q23	AD	Early onset, rapid progression, postural tremor, late dementia
PARK-5//UCHL-1	4p14	AD	Late onset
PARK-6/PINK-1	1p35–p36	AR	Early onset, slow progression
PARK-7/DJ-1	1p36	AR	Early onset, slow progression
PARK-8/LRRK-2	12p11.2–q13.1	AD	Late onset
PARK-10/unknown	1p32		Late onset
PARK-11/unknown	2q36–q37		Late onset
NA/synphilin-1	5q23.1–q23.3	AD	Late onset
NA/NR4A2	2q22–q23	AD	Late onset

AD, autosomal dominant; AR, autosomal recessive.

α-Synuclein (A-S) was the first identified Parkinson's disease gene (PARK1). To date, only three autosomal dominantly inherited point mutations (Ala53Thr, Ala30Pro, and Glu46Lys) have been described.[2–4] The Ala53Thr mutation has been detected in several families of Mediterranean origin with autosomal dominantly inherited Parkinson's disease, but the Ala30Pro mutation has been discovered only in one small German family.[2,3] Within the families with either the Ala53Thr or the Ala30Pro mutation, the ratio of normal (wild type) to mutant A-S correlates with the disease severity: the more severely patients are affected, the less mutant A-S is expressed (because PARK1 is autosomal dominantly inherited, both a normal and a mutant copy of the gene (allele) are present). Thus the ratio between normal and mutant A-S (haploinsufficiency) may be important for disease progression and severity.[5] The average age of onset in the famous Contursi kindred (in which the Ala53Thr mutation was first identified) was 45.6 years. Affected family members presented with typical parkinsonian features such as resting tremor, bradykinesia, and gait disturbance, but progressed more rapidly than typical Parkinson patients.[6] Subsequently, additional features such as central hypoventilation, orthostatic hypotension, prominent myoclonus, and urinary incontinence have been described in a different Ala53Thr family, indicating a wide range of phenotypes for Ala53Thr mutation carriers.[7] Cognitive impairment is a frequent and early symptom in Ala30Pro mutation carriers, but the phenotype is otherwise similar to sporadic Parkinson's disease, with an age of onset ranging from 54 to 76 years.[8] The most recently discovered Glu46Lys mutation not only causes parkinsonism but also results in clinical and pathological features characteristic of Lewy body dementia.[4] It has been suggested that single A-S polymorphisms or haplotypes formed by a combination of several A-S sequence variants may be a risk factor for isolated Parkinson's disease, but this has not been consistently confirmed by others.[9–11] A-S, but not the closely related β and γ synucleins, is consistently found in Lewy bodies, not only in those rare families with a genomic A-S mutation but in all cases of Lewy-body Parkinson's disease.[12] A-S is also present in the glial cytoplasmic inclusion bodies typical for multiple system atrophy.[13]

The physiological role of A-S is still largely unknown, but its localisation at presynaptic terminals and some functional studies indicate a possible role in synaptic plasticity and vesicular transport.[12] A-S knockout mice have a defect in dopamine release and reuptake, supporting a role for A-S in the regulation of dopamine transmission.[14] How does A-S contribute to the cell death observed in Parkinson's disease and how does it fit in with previously recognised pathogenic mechanisms such as oxidative stress? Wild type A-S expression confers an increased resistance to various apoptotic insults, whereas mutant A-S results in increased apoptotic response and enhanced susceptibility to oxidative stress.[12] More importantly, different mechanisms lead to the aggregation of A-S with subsequent formation of protofibrils, fibrils, and eventually conversion to Lewy bodies.[15] Wild type A-S is natively unfolded but forms fibrils at increasing concentrations, and its overexpression in marmosets results in dopaminergic cell loss as well as A-S positive cytoplasmic inclusions.[12,16] Fibril formation is accelerated by the A53T mutation of A-S, but also by oxidative stress, heavy metals, pesticides, and 1-methyl-4-phenyl-1,2,3,6 tetrahydropyridine (MPTP).[12,15] Fascinatingly, A-S knockout mice show resistance to MPTP but increased susceptibility to a different neurotoxin, rotenone.[17] Aggregation is also facilitated by nitration, indicating the relevance of post-translational modification processes.[18] Why does A-S cause selective neurodegeneration in Parkinson's disease? This may at least partly reflect the fact that A-S renders endogenous levels of dopamine toxic in cultured human dopaminergic neurones, this toxicity being mediated by reactive oxygen species. In contrast, A-S is not toxic in non-dopaminergic human cortical neurones, but rather shows neuroprotective activity.[19] The recent discovery of A-S triplication as a further cause for Parkinson's disease indicates that not only mutated but also non-physiological amounts of A-S can lead to nigral cell death.[20]

Parkin is the second Parkinson's disease gene (PARK2). Parkin mutations are typically detected in autosomal recessively inherited young onset parkinsonism with dystonia, sleep benefit, and the tendency to develop L-dopa induced dyskinesias,[21] but some patients with Parkin mutations have a later age of onset (up to 65 years) and are clinically indistinguishable from patients with idiopathic Parkinson's disease.[21,22] Disease progression is considerably slower in patients with parkin mutations than in idiopathic, Lewy body Parkinson's disease.[23] Parkin localises to the Lewy bodies of Parkinson's disease in those patients who do not carry Parkin mutations. In contrast, Lewy bodies are absent in patients with homozygous Parkin mutations. Unusual presentations such as levodopa unresponsive parkinsonism combined with cerebellar and pyramidal tract dysfunction have also been reported.[24] Interestingly, heterozygote mutations can also be found in some patients with young onset parkinsonism, and some asymptomatic heterozygote mutation carriers also show significant striatal dopaminergic dysfunction on ^{18}F-dopa positron emission tomography (PET).[23] A polymorphism (−258 T/G) in the promoter region of the parkin gene may be associated with idiopathic Parkinson's disease, but this finding awaits confirmation by others.[25]

A mutation in the gene encoding ubiquitin carboxy-terminal hydrolase 1 (UCHL1) has been described in a single small German kindred with autosomal dominantly inherited Parkinson's disease (PARK5) and the S18Y polymorphism in the UCHL1 gene may be associated with decreased risk for Parkinson's disease.[26,27]

Both parkin and UCHL1 are involved in the ubiquitin dependent degradation of intracellular misfolded, unassembled, or damaged proteins by the proteasome, a multicatalytic complex. Parkin is an E3 ligase responsible for the attachment of ubiquitin to substrates, whereas UCHL-1 removes polyubiquitin chains once the substrate has been attached to the proteasome.[28] Glycosylated A-S and the A-S interacting protein synphilin-1 are among the recognised substrates of parkin (Fig. 1). Mutations in either the parkin gene or UCHL1 lead to impaired protein degradation.[29] Expression of A53T mutant A-S also induces disruption of the ubiquitin dependent proteolytic system in cell culture models.[30] Subsequent work showed reduced enzymatic activity and impaired structural integrity of the proteasome in the substantia nigra of patients with sporadic Parkinson's disease.[31] These findings suggest that failure of the ubiquitin–proteasome system to achieve adequate clearance of unwanted proteins may underlie vulnerability and degeneration of the substantia nigra, even in sporadic Parkinson's disease. The expression of proteasome subunits in the mesencephalon of rats declines with age. One could therefore speculate that the age dependent prevalence of Parkinson's disease reflects a lifetime accumulation of abnormal intracellular proteins (for example, from oxidative damage) and an increasingly incompetent ubiquitin–proteasomal complex. However, proteasome subunits are also selectively vulnerable to oxidative stress, and the structural and functional integrity of the proteasome depends on sufficient ATP production. Thus the observed impairment of the proteasome in Parkinson's disease may merely be secondary, because of oxidative stress and impaired complex I activity of the mitochondrial respiratory chain (see McNaught *et al.*[31] for more detailed discussion).

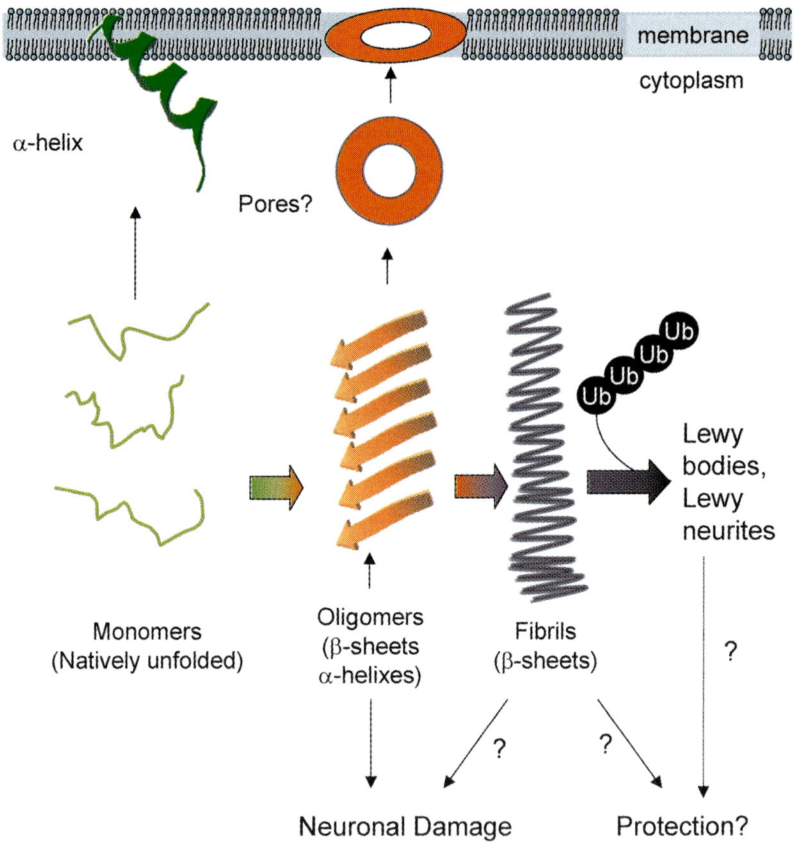

Fig. 1 The ubiquitin–proteasome system. Parkin, shown in green, consists of several of distinct domains, a ubiquitin-like domain (Ubl) and two RING fingers at the C-terminus, separated by an in between ring (IBR) domain. Parkin has been shown to bind to several substrates (shown in red), which are oriented with the PARKIN domains to which they bind. The RING-IBR-RING domain also binds various non-substrate interactors (yellow), namely the heat shock protein Hsp 70, the C-terminal HSP interacting protein CHIP, and hsel10. Additional non-substrate interactors bind to the C-terminus of Parkin, including, CASK/lin10, and a protein in the cap of the proteasome, Rpn10. The N terminus of Parkin is responsible for binding to the proteasome. RING2 recruits E2 enzymes (blue), which themselves carry ubiquitin. Several rounds of ubiquitin addition build a polyubiquitin chain on the substrate, which is subsequently degraded by the proteasome. Mutant and wild type α-synuclein (to a lesser degree) inhibit the proteasome. (Reproduced with kind permission of Mark Cookson and Sage Publications, Beverly Hills, California, USA).

Autosomal recessive mutations in the DJ-1 gene (PARK7) were originally identified in two consanguineous families from genetically isolated communities in the Netherlands and Italy, but subsequently also in other populations.[32] Behavioural and psychiatric disturbances at onset or early in the disease course, and focal dystonia including blepharospasm, were noticed in both families.[33] DJ-1 mutations lead to reduced DJ-1 protein stability, and the mutant protein is rapidly degraded through the ubiquitin–proteasome system.[34] Knockdown of DJ-1 in cell culture systems leads to increased susceptibility to oxidative stress and proteasome inhibition.[35] DJ-1 co-localises with tau inclusions, but Lewy bodies are DJ-1 negative and necropsy reports on patients with a DJ-1 mutation are not yet available.[36,37]

Autosomal recessively inherited mutations in the PTEN induced kinase 1 on chromosome 1p36 (PARK6) were also initially discovered in Parkinson patients from consanguineous families, but subsequently reported in sporadic patients with early onset Parkinson's disease as well.[38,39] PINK1 is localised to the mitochondria. Wild type, but not mutant, PINK1 protein protects against stress induced mitochondrial dysfunction and apoptosis.[38] Neither DJ-1 nor PINK1 polymorphisms (naturally occurring DNA sequence variants) confer increased susceptibility to sporadic Parkinson's disease.[40,41]

Most recently, leucine-rich repeat kinase 2 (LRRK2) or dardarin was identified as the causative gene in families linked to the autosomal dominantly inherited PARK8 locus on chromosome 12p11.2–q13.1.[42,43] The predicted product of the LRRK2 gene is a large protein with 2527 amino acids; sequence comparison suggests that it may function as a protein kinase. Necropsy diagnoses of six mutation carriers included abnormalities consistent with Lewy body Parkinson's disease, diffuse Lewy body Parkinson's disease, nigral degeneration without distinctive histopathology, progressive supranuclear palsy-like pathology, and clinical diagnoses of parkinsonism with dementia or amyotrophy, or both, with their associated pathology, were also noted.[43] Thus LRRK2 may not only play a role in the pathogenesis of Parkinson's disease as such, but also of other neurodegenerative disorders. Subsequent studies have described the Gly2019Ser mutations in both familial and sporadic forms of Parkinson's disease in several distinct populations, with a frequency ranging from 1% to 6%.[44–46]

More information about the penetrance and other clinically relevant aspects of this mutation are needed, but genetic testing for the Gly2019Ser mutation may be the first diagnostic genetic test for Parkinson's disease to enter clinical practice.

The NR4A2 (also known as NURR1) gene encodes a member of the nuclear receptor superfamily and is essential for the differentiation of nigral dopaminergic neurones. Heterozygote mutations in NR4A2 have been detected in 10 of 107 patients with autosomal dominantly inherited Parkinson's disease, but not in sporadic disease or controls. Age of onset and clinical features were not different from typical Parkinson's disease. The mutations result in a marked decrease of NR4A2 mRNA levels and downregulate the transcription of the tyrosine hydroxylase gene.[47] Numerous subsequent studies failed to detect any sequence variants in other Parkinson patient cohorts, and there is now at least the suspicion that the observed sequence changes of NR4A2 in the original study may simply reflect a haplotype co-segregating with the disease in that particular population rather than a disease causing mutation as such.

In addition to genetic factors, exposure to toxins is the only other universally recognised risk factor for Parkinson's disease. Systemic application of the herbicide paraquat has been shown to kill dopaminergic neurones in the substantia nigra in rodents.[48] The pesticide rotenone, a specific inhibitor of mitochondrial complex I, is highly lipophilic, easily crosses biological membranes, and does not selectively accumulate in nigral neurones. Nevertheless, chronic systemic administration of rotenone has been reported to result in selective nigral degeneration.[49] Others have, however, described a multisystem degeneration in rats treated with rotenone, indicating non-selective neurotoxicity.[50] Interestingly, non-toxic concentrations of rotenone and the inflammogen lipopolysaccharide synergistically induced cell death in a dopaminergic mesencephalic cell culture model.[51] Microglial generation of reactive oxygen species appeared to be a key contributor to this synergistic neurotoxicity. Other microglia originated factors such as nitric oxide, tumour necrosis factor α, and interleukin 1β may further contribute to the neurodegenerative process observed in Parkinson's disease.[52] Inflammatory processes associated with an increased expression of cyclo-oxygenase 2 (COX2), the rate limiting enzyme in prostaglandin E_2 synthesis, and raised levels of prostaglandin E_2 have been implicated in the pathogenesis of several neurodegenerative disorders, and upregulation of COX2 has now also been observed in dopaminergic neurones of both Parkinson's disease and MPTP mice. Interestingly,

COX-2 inhibition does not protect against MPTP induced dopaminergic neurodegeneration by mitigating inflammation. Rather, COX2 inhibition prevents the formation of the oxidant species dopamine-quinone. COX-2 inhibitors can penetrate the blood–brain barrier and it was thus initially hoped that these drugs might be of future therapeutic use in Parkinson's disease.[53] The recently reported side effect profile of COX-2 inhibitors, however, has obviously cast some doubt on this.

The discovery of the different gene defects described above highlighted the relevance of the ubiquitin-proteasome pathway for neuronal cell death in Parkinson's disease. The function of this pathway can, however, also be influenced by naturally or synthetic proteasome inhibitors. Treatment of rats with proteasome inhibitors closely mimics Parkinson's disease in rodents.[54] The treated rats develop progressive parkinsonism with dykinesia, rigidity, and tremor. Necropsy analysis showed striatal dopamine depletion, dopaminergic cell death with inflammation, and apoptosis in the substantia nigra pars compacta. Lewy bodies were additionally described in a subset of the remaining neurones. This model potentially offers a substantial improvement over the previously described model systems as it appears to resemble the cardinal features of Parkinson's disease more closely.

Currently, only symptomatic treatment is available for Parkinson's disease. The study of the underlying genetic and cellular defects in both familial and sporadic disease provides an opportunity to identify novel targets and tools to ameliorate disease progression and possibly even provide a cure. Overexpression of molecular chaperones such as HSP40 or HSP70 markedly reduces the formation of inclusion bodies.[55] Administration of galdenamycin induces the expression of heat shock proteins and protects against A-S toxicity in *Drosophila* Parkinson's disease models.[56] Thus the induction of such molecular chaperones may become an exciting new type of disease modifying treatment for Parkinson's disease.

Too much A-S may not only lead to Parkinson's disease in those comparatively rare families with duplication or triplication of the A-S gene, but also in those sporadic patients who produce too much A-S because of the presence of a particular sequence variant in the promoter region of A-S.[57] A reduction of the A-S protein levels in the affected individuals may be a further promising therapeutic avenue. Indeed, a reduction in A-S protein levels has been achieved in a rodent model with overexpression of virus delivered parkin.[58] The common mutations of Gly2019 and Ile2020 in PARK8/LRRK2 may alter the kinase activity of this protein. The modification of kinase activity with specific inhibitors provides a very attractive and achievable treatment strategy, which could become useful in the treatment of both familial and sporadic disease.

HUNTINGTON'S DISEASE

Polyglutamine diseases such as Huntington's disease, Kennedy's disease, dentatorubro-pallidoluysian atrophy (DRPLA), and some of the autosomal dominantly inherited spinocerebellar ataxias result from an increased number of CAG nucleotide repeats that encode polyglutamine tracts within the corresponding gene products. The various proteins show no sequence homology outside the polyglutamine tract, span different lengths, have different cellular localisations and, where known, different functions.[59] A relatively modest quantitative change of approx 10–20% in repeat length differentiates between normal (in Huntington's disease up to 35 repeats) and progressive neurodegeneration (in Huntington's disease 40 repeats or more). Subjects with 36–39 repeats have reduced penetrance for Huntington's disease. Longer expansions correlate with earlier onset and more severe disease. Rare cases of Huntington patients homozygous for an expansion (pathological expansion on *both* alleles) develop a more severe phenotype, but the presence of two expanded alleles rather than one does not seem to influence the age of onset.[60]

Proteins with elongated polyglutamine tract misfolds aggregate as antiparallel β strands termed "polar zippers" and form intranuclear inclusions. These inclusions are typically but not exclusively found in those brain regions that are predominantly affected. They are not limited to those neurones that are most likely to degenerate and can also be found in non-neuronal tissue.[61] Furthermore, nuclear and cytosolic aggregates of huntingtin, the protein product of the Huntington's disease gene IT15, can also be found in non-neuronal tissue. This indicates that these aggregates are neither specific nor sufficient for cell death. Very recent evidence actually suggests that inclusion body formation reduces the levels of mutant huntingtin and the risk of neuronal cell death. Thus inclusion body formation could be interpreted as a "coping response" of the cell to toxic mutant huntingtin.[62]

Oligomerisation of expanded polyglutamine is not only a critical step in the formation of these inclusions, but it also stimulates different important cell death mechanisms previously identified in Huntington's disease, such as apoptosis and disturbed energy metabolism. Proapoptotic enzymes such as caspase 1 or 8 are activated in Huntington's disease and required for polyglutamine toxicity in cell culture models. Caspases can be inhibited pharmacologically by minocycline, and an influence of minocycline treatment on disease progression in Huntington mice was initially reported but subsequently not confirmed by others.[63,64] Lactate levels are raised and mitochondrial respiratory chain function is impaired in Huntington brain tissue, whereas creatine administration — which increases phosphocreatine levels and normalises mitochondrial function — leads to increased survival and delays motor symptoms in Huntington mice.[65] Inhibition of the oligomerisation of expanded huntingtin by Congo red prevents ATP depletion and caspase activation, preserves normal cellular protein synthesis and degradation functions *in vitro*, and promotes the clearance of expanded huntingtin *in vivo*.[66]

Mutant huntingtin is more resistant to proteolysis, and aggregation of abnormal huntingtin is thus further promoted by insufficient breakdown of this protein by the proteasome pathway.[67] Impaired function of the ubiquitin proteasome system cannot only be observed in brain tissue, but also in skin fibroblasts of Huntington's disease patients.[68] Altered proteasomal function is also associated with disrupted mitochondrial membrane potential, released cytochrome c from mitochondria into the cytosol, and caspase activation.[69] Normal function of the proteasome is closely linked to the machinery of molecular chaperones which mediate proper folding of other proteins and facilitate the transfer of misfolded proteins to the proteasome for degradation. Overexpression of chaperones such as Hsp 70 and Hsp 40 suppresses the aggregation and toxicity of polyglutamine containing proteins. For example, overexpression of Hsp 70 in a mouse model of spinocerebellar ataxia 1 (SCA1) not only reduced pathological changes but also ameliorated the phenotype.[67]

A further "toxic gain of function" of mutant huntingtin is its interaction with transcriptional factors such as Sp1 and its coactivator TAFII130 which in turn bind to a whole variety of genes such as neurotransmitter receptors and intracellular signalling systems. Coexpression of Sp1 and TAFII130 in cultured striatal cells from wild type and Huntington transgenic mice reverses the transcriptional inhibition of the dopamine D2 receptor gene caused by mutant huntingtin and protects neurones from huntingtin induced cellular toxicity.[62] Huntingtin, as well as other polyglutamine containing proteins, can also interact directly with other transcription factors such as the CREB binding protein (CBP), rendering them inactive.[61] CBP is one of several histone acetylases sequestered by polyglutamine inclusions. Histone acetylases are important gene expression regulators, and acetylation leads to increased mRNA transcription. CBP regulates the nuclear responses to a variety of cell signalling cascades including the neuronal response to neurotrophins, and overexpression of CBP rescues cells from polyglutamine toxicity. The incorporation of CBP into nuclear inclusions and resulting inactivation might therefore lead to a reduced capacity of the cells to respond to trophic factors essential for their

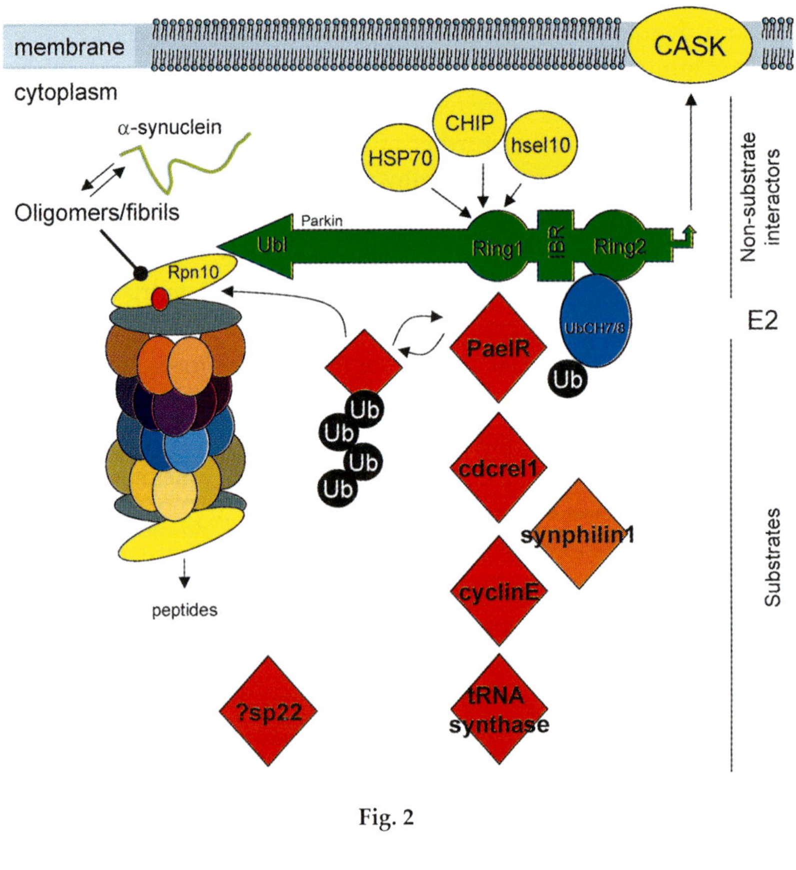

Fig. 2

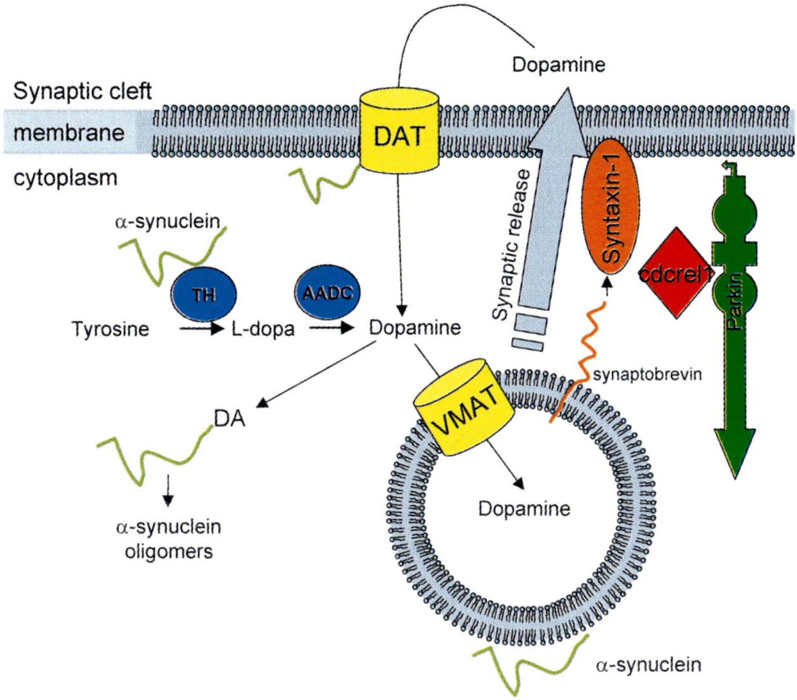

Fig. 3

survival. The incorporation and subsequent inactivation of other histone acetylases in the Huntington inclusions will further disturb the complicated gene expression network vital for the normal function and survival on neuronal cells.[71] Histone acetylation itself can not currently be promoted pharmacologically but inhibitors of the physiological antagonist histone deacetylase markedly reduce polyglutamine induced toxicity.[72]

Loss of physiological function of huntingtin might also contribute to the pathogenesis of Huntington's disease, and increasing expression of wild type huntingtin in transgenic mice protects against the toxic effects of mutant huntingtin. Wild type huntingtin also shows antiapoptotic properties, which may be because wild type but not mutant huntingtin interacts with the proapoptotic protein HIP1.[72] Wild type huntingtin also increases vesicular transport of brain derived neurotrophic factor (BDNF) along microtubules, but BDNF transport is impaired in the presence of mutant huntingtin or if the levels of wild type huntingtin are reduced.

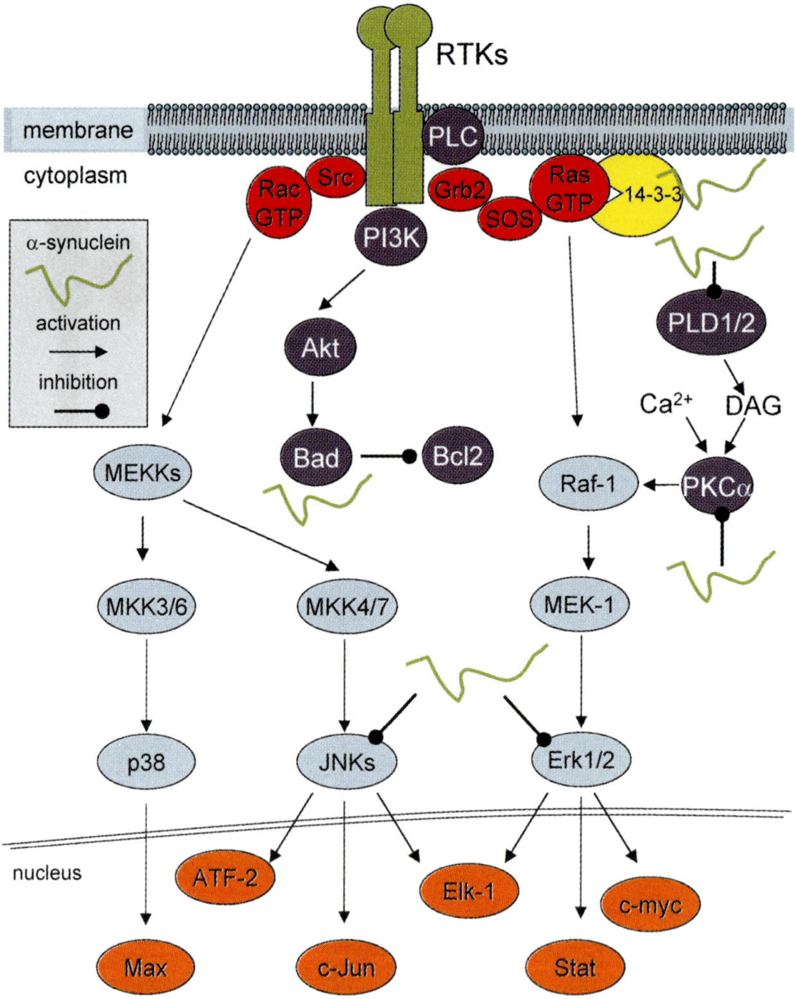

Fig. 4

CONCLUSIONS

Molecular genetic studies of familial Parkinson's disease have identified misfolding of proteins and failure of the proteasome to degrade such proteins as being key events in the pathogenesis of Parkinson's disease. Significant challenges remain — namely, to extrapolate these findings to encompass the possible role of these identified genes and their gene products in the more common sporadic forms of the disease, but increasing knowledge and understanding of the identified genes and pathways are already being used to develop novel strategies for the treatment of this disease. Astounding progress has also been made in our understanding of the underlying mechanisms leading to cell death in Huntington's disease. A European-wide network, EURO-Huntington's disease, has now been established which will facilitate drug trials aiming to identify disease modifying compounds.

REFERENCES

1. Schrag A, Ben-Shlomo Y, Quinn NP. Cross sectional prevalence survey of idiopathic Parkinson's disease and Parkinsonism in London. *BMJ* 2000; **321**:21–22.
2. Polymeropoulos MH, Lavedan C, Leroy E, *et al.* Mutation in the alpha-synuclein gene identified in families with Parkinson's disease. *Science* 1997; **276**:2045–2047.
3. Kruger R, Kuhn W, Muller T, *et al.* Ala30Pro mutation in the gene encoding alpha-synuclein in Parkinson's disease. *Nat Genet* 1998; **18**:106–108.
4. Zarranz JJ, Alegre J, Gomez-Esteban JC, *et al.* The new mutation, E46K, of alpha-synuclein causes Parkinson and Lewy body dementia. *Ann Neurol* 2004; **55**: 164–173.
5. Kobayashi H, Kruger R, Markopoulou K, *et al.* Haploinsufficiency at the alpha-synuclein gene underlies phenotypic severity in familial Parkinson's disease. *Brain* 2003; **126**:32–42.
6. Golbe LI, Di Iorio G, Sanges G, *et al.* Clinical genetic analysis of Parkinson's disease in the Contursi kindred. *Ann Neurol* 1996; **40**:767–775.
7. Spira PJ, Sharpe DM, Halliday G, *et al.* Clinical and pathological features of a Parkinsonian syndrome in a family with an Ala53Thr alpha-synuclein mutation. *Ann Neurol* 2001; **49**:313–319.
8. Kruger R, Kuhn W, Leenders KL, *et al.* Familial parkinsonism with synuclein pathology: clinical and PET studies of A30P mutation carriers. *Neurology* 2000; **56**:1355–1362.
9. Kruger R, Vieira-Saecker AM, Kuhn W, *et al.* Increased susceptibility to sporadic Parkinson's disease by a certain combined alpha-synuclein/apolipoprotein E genotype. *Ann Neurol* 1999; **45**:611–617.
10. Khan N, Graham E, Dixon P, *et al.* Parkinson's disease is not associated with the combined alpha-synuclein/apolipoprotein E susceptibility genotype. *Ann Neurol* 2001; **49**:665–668.

11. Farrer M, Maraganore DM, Lockhart P, *et al.* Alpha-synuclein gene haplotypes are associated with Parkinson's disease. *Hum Mol Genet* 2001; **10**:1847–1851.
12. Alves da Costa C. Recent advances on alpha-synuclein cell biology: functions and dysfunctions. *Curr Mol Med* 2003; **3**:17–24.
13. Papp MI, Kahn JE, Lantos PL. Glial cytoplasmic inclusions in the CNS of patients with multiple system atrophy (striatonigral degeneration, olivopontocerebellar atrophy and Shy-Drager syndrome). *J Neurol Sci* 1989; **94**:79–100.
14. Abeliovich A, Schmitz Y, Farinas I, *et al.* Mice lacking alpha-synuclein display functional deficits in the nigrostriatal dopamine system. *Neuron* 2000; **25**: 239–252.
15. Lansbury PT, Brice A. Genetics of Parkinson's disease and biochemical studies of implicated gene products. *Curr Opin Cell Biol* 2002; **14**:653–660.
16. Kirik D, Annett LE, Burger C, *et al.* Nigrostriatal alpha-synucleinopathy induced by viral vector-mediated overexpression of human alpha-synuclein: a new primate model of Parkinson's disease. *Proc Natl Acad Sci USA* 2003; **100**:2884–2889.
17. Dauer W, Kholodilov N, Vila M, *et al.* Resistance of alpha-synuclein null mice to the parkinsonian neurotoxin MPTP. *Proc Natl Acad Sci USA* 2002; **99**: 14524–14529.
18. Souza JM, Giasson BI, Chen Q, *et al.* Dityrosine cross-linking promotes formation of stable alpha-synuclein polymers. Implication of nitrative and oxidative stress in the pathogenesis of neurodegenerative synucleinopathies. *J Biol Chem* 2000; **275**:18344–18349.
19. Xu J, Kao SY, Lee FJ, *et al.* Dopamine-dependent neurotoxicity of alpha-synuclein: a mechanism for selective neurodegeneration in Parkinson disease. *Nat Med* 2002; **8**:600–606.
20. Singleton AB, Farrer M, Johnson J, *et al.* Alpha-synuclein locus triplication causes Parkinson's disease. *Science* 2003; **302**:841.
21. Lucking CB, Durr A, Bonifati V, *et al.* Association between early-onset Parkinson's disease and mutations in the parkin gene. French Parkinson's Disease Genetics Study Group. *N Engl J Med* 2000; **342**:1560–1567.
22. Klein C, Pramstaller PP, Kis B, *et al.* Parkin deletions in a family with adult-onset, tremor-dominant parkinsonism: expanding the phenotype. *Ann Neurol* 2000; **48**:65–71.
23. Khan NL, Brooks DJ, Pavese N, *et al.* Progression of nigrostriatal dysfunction in a parkin kindred: an [18F]dopa PET and clinical study. *Brain* 2002; **125**: 2248–2256.
24. Kuroda Y, Mitsui T, Akaike M, *et al.* Homozygous deletion mutation of the parkin gene in patients with atypical parkinsonism. *J Neurol Neurosurg Psychiatry* 2001; **71**:231–234.

25. West AB, Maraganore D, Crook J, *et al.* Functional association of the parkin gene promoter with idiopathic Parkinson's disease. *Hum Mol Genet* 2002; **11**: 2787–2792.
26. Leroy E, Boyer R, Auburger G, *et al.* The ubiquitin pathway in Parkinson's disease. *Nature* 1998; **395**:451–452.
27. Maraganore DM, Farrer MJ, Hardy JA, *et al.* Case–control study of the ubiquitin carboxy-terminal hydrolase L1 gene in Parkinson's disease. *Neurology* 1999; **53**:1858–1860.
28. Ciechanover A, Orian A, Schwartz AL. The ubiquitin-mediated proteolytic pathway: mode of action and clinical implications. *J Cell Biochem Suppl* 2000; **34**:40–51.
29. Dawson TM, Dawson VL. Rare genetic mutations shed light on the pathogenesis of Parkinson disease. *J Clin Invest* 2003; **111**:145–151.
30. Stefanis L, Larsen KE, Rideout HJ, *et al.* Expression of A53T mutant but not wild-type alpha-synuclein in PC12 cells induces alterations of the ubiquitin-dependent degradation system, loss of dopamine release, and autophagic cell death. *J Neurosci* 2001; **21**:9549–9560.
31. McNaught KS, Belizaire R, Isacson O, *et al.* Altered proteasomal function in sporadic Parkinson's disease. *Exp Neurol* 2003; **179**:38–46.
32. Hague S, Rogaeva E, Hernandez D, *et al.* Early-onset Parkinson's disease caused by a compound heterozygous DJ-1 mutation. *Ann Neurol* 2003; **54**: 271–274.
33. Bonifati V, Breedveld GJ, Squitieri F, *et al.* Localization of autosomal recessive early-onset parkinsonism to chromosome 1p36 (PARK7) in an independent dataset. *Ann Neurol* 2002; **51**:253–256.
34. Miller DW, Ahmad R, Hague S, *et al.* L166P mutant DJ-1, causative for recessive Parkinson's disease, is degraded through the ubiquitin-proteasome system. *J Biol Chem* 2003; **278**:36588–36595.
35. Yokota T, Sugawara K, Ito K, *et al.* Down regulation of DJ-1 enhances cell death by oxidative stress, ER stress, and proteasome inhibition. *Biochem Biophys Res Commun* 2003; **312**:1342–1348.
36. Rizzu P, Hinkle DA, Zhukareva V, *et al.* DJ-1 colocalizes with tau inclusions: a link between parkinsonism and dementia. *Ann Neurol* 2004; **55**: 113–118.
37. Neumann M, Muller V, Gorner K, *et al.* Pathological properties of the Parkinson's disease-associated protein DJ-1 in alpha-synucleinopathies and tauopathies: relevance for multiple system atrophy and Pick's disease. *Acta Neuropathol (Berl)* 2004; **107**:489–496.
38. Valente EM, Abou-Sleiman PM, Caputo V, *et al.* Hereditary early-onset Parkinson's disease caused by mutations in PINK1. *Science* 2004; **304**:1158–1160.

39. Rohe CF, Montagna P, Breedveld G, *et al.* Homozygous PINK1 C-terminus mutation causing early-onset parkinsonism. *Ann Neurol* 2004; **56**:427–431.
40. Eerola J, Hernandez D, Launes J, *et al.* Assessment of a DJ-1 (PARK7) polymorphism in Finnish PD. *Neurology* 2003; **61**:1000–1002.
41. Healy DG, Abou-Sleiman PM, Ahmadi KR, *et al.* The gene responsible for PARK6 Parkinson's disease, PINK1, does not influence common forms of parkinsonism. *Ann Neurol* 2004; **56**:329–335.
42. Paisan-Ruiz C, Jain S, Evans EW, *et al.* Cloning of the Gene Containing Mutations that Cause PARK8-Linked Parkinson's Disease. *Neuron* 2004; **44**:595–600.
43. Zimprich A, Biskup S, Leitner P, *et al.* Mutations in LRRK2 Cause Autosomal-Dominant Parkinsonism with Pleomorphic Pathology. *Neuron* 2004; **44**:601–607.
44. Nichols WC, Pankratz N, Hernandez D, *et al.* Genetic screening for a single common LRRK2 mutation in familial Parkinson's disease. *Lancet* 2005; **365**: 410–412.
45. Di Fonzo A, Rohe CF, Ferreira J, *et al.* A frequent LRRK2 gene mutation associated with autosomal dominant Parkinson's disease. *Lancet* 2005; **365**:412–415.
46. Gilks WP, Abou-Sleiman PM, Gandhi S, *et al.* A common LRRK2 mutation in idiopathic Parkinson's disease. *Lancet* 2005; **365**:415–416.
47. Le WD, Xu P, Jankovic J, *et al.* Mutations in NR4A2 associated with familial Parkinson disease. *Nat Genet* 2003; **33**:85–89.
48. McCormack AL, Thiruchelvam M, Manning-Bog AB, *et al.* Environmental risk factors and Parkinson's disease: selective degeneration of nigral dopaminergic neurons caused by the herbicide paraquat. *Neurobiol Dis* 2002; **10**:119–127.
49. Betarbet R, Sherer TB, MacKenzie G, *et al.* Chronic systemic pesticide exposure reproduces features of Parkinson's disease. *Nat Neurosci* 2000; **3**:1301–1306.
50. Hoglinger GU, Feger J, Prigent A, *et al.* Chronic systemic complex I inhibition induces a hypokinetic multisystem degeneration in rats. *J Neurochem* 2003; **84**:491–502.
51. Gao HM, Hong JS, Zhang W, *et al.* Synergistic dopaminergic neurotoxicity of the pesticide rotenone and inflammogen lipopolysaccharide: relevance to the etiology of Parkinson's disease. *J Neurosci* 2003; **23**:1228–1236.
52. Orr CF, Rowe DB, Halliday GM. An inflammatory review of Parkinson's disease. *Prog Neurobiol* 2002; **68**:325–340.
53. Teismann P, Tieu K, Choi DK, *et al.* Cyclooxygenase-2 is instrumental in Parkinson's disease neurodegeneration. *Proc Natl Acad Sci USA* 2003; **100**:5473–5478.
54. McNaught KS, Pearl DP, Brownell AL, *et al.* Systemic exposure to proteasome inhibitors causes a progressive model of Parkinson's disease. *Ann Neurol* 2004; **56**:149–162.

55. McLean PJ, Kawamata H, Hyman BT. α-synuclein-enhanced green fluorescent protein fusion proteins form proteasome sensitive inclusions in primary neurons. *Neuroscience* 2001; **104**:901–912.
56. Auluck PK, Meulener MC, Bonini NM. Mechanisms of Suppression of alpha synuclein Neurotoxicity by Geldanamycin in Drosophila. *J Biol Chem* 2005; **280**:2873–2878.
57. Farrer M, Maraganore DM, Lockhart P, *et al.* alpha-Synuclein gene haplotypes are associated with Parkinson's disease. *Hum Mol Genet* 2001; **10**:1847–1851.
58. Yamada M, Mizuno Y, Mochizuki H. Parkin gene therapy for alpha-synucleinopathy: a rat model of Parkinson's disease. *Hum Gene Ther* 2005; **16**:262–270.
59. Temussi PA, Masino L, Pastore A. From Alzheimer to Huntington: why is a structural understanding so difficult? *Embo J* 2003; **22**:355–361.
60. Squitieri F, Gellera C, Cannella M, *et al.* Homozygosity for CAG mutation in Huntington disease is associated with a more severe clinical course. *Brain* 2003; **126**:946–955.
61. Ross CA, Poirier MA, Wanker EE, *et al.* Polyglutamine fibrillogenesis: the pathway unfolds. *Proc Natl Acad Sci USA* 2003; **100**:1–3.
62. Arrasate M, Mitra S, Schweitzer ES, *et al.* Inclusion body formation reduces levels of mutant huntingtin and the risk of neuronal death. *Nature* 2004; **431**: 805–810.
63. Hughes RE, Olson JM. Therapeutic opportunities in polyglutamine disease. *Nat Med* 2001; **7**:419–423.
64. Smith DL, Woodman B, Mahal A, *et al.* Minocycline and doxycycline are not beneficial in a model of Huntington's disease. *Ann Neurol* 2003; **54**:186–196.
65. Andreassen OA, Dedeoglu A, Ferrante RJ, *et al.* Creatine increase survival and delays motor symptoms in a transgenic animal model of Huntington's disease. *Neurobiol Dis* 2001; **8**:479–491.
66. Sanchez I, Mahlke C, Yuan J. Pivotal role of oligomerization in expanded polyglutamine neurodegenerative disorders. *Nature* 2003; **421**:373–379.
67. Muchowski PJ. Protein misfolding, amyloid formation, and neurodegeneration: a critical role for molecular chaperones? *Neuron* 2002; **35**:9–12.
68. Seo H, Sonntag KC, Isacson O. Generalized brain and skin proteasome inhibition in Huntington's disease. *Ann Neurol* 2004; **56**:319–328.
69. Jana NR, Zemskov EA, Wang G, *et al.* Altered proteasomal function due to the expression of polyglutamine-expanded truncated N-terminal huntingtin induces apoptosis by caspase activation through mitochondrial cytochrome c release. *Hum Mol Genet* 2001; **10**:1049–1059.
70. Dunah AW, Jeong H, Griffin A, *et al.* Sp1 and TAFII130 transcriptional activity disrupted in early Huntington's disease. *Science* 2002; **296**:2238–2243.

71. Hughes RE. Polyglutamine disease: acetyltransferases awry. *Curr Biol* 2002; **12**:R141–R143.
72. Cattaneo E, Rigamonti D, Goffredo D, *et al*. Loss of normal huntingtin function: new developments in Huntington's disease research. *Trends Neurosci* 2001; **24**:182–188.

8

Molecular Pathogenesis of Neuroinflammation

M Bradl and R Hohlfeld*

The last few years have seen significant progress towards understanding the mechanisms of immune surveillance and inflammation in the nervous system. In this review, the milestones of scientific discovery in this field are discussed, and the strengths and limitations of the different ways of examining the molecular pathogenesis of neuroinflammation examined. The review is limited to the inflammatory reactions of the central nervous system that occur in multiple sclerosis and experimental autoimmune encephalomyelitis.

A BRIEF HISTORY OF IMMUNE PRIVILEGE

Today we are well aware that there is substantial communication between the immune system and the central nervous system (CNS), and that interactions between these two systems occur both in the healthy organism and in pathological situations.[1] However, this was not always the case.

*Institute for Brain Research, Department of Neuroimmunology, Spitalgasse 4, A-1090 Vienna, Austria; monika.bradl@meduniwien.ac.at
Institute for Clinical Neuroimmunology, Klinikum Grosshadern, D-81366 München, Germany

Some 20 to 30 years ago, the CNS and the immune system were considered strictly separate structures, and the CNS was viewed as the classical example of an "immunoprivileged site." This assumption was based on the prolonged survival of tissue grafts within the CNS,[2] the presence of the blood–brain barrier (an endothelial barrier between blood and brain which forms tight junctions and prevents an uncontrolled influx of molecules and cells into the CNS[3]), the lack of classical lymphatic drainage pathways,[3] the lack of professional antigen presenting cells such as dendritic cells,[4] and the infrequent expression of MHC (major histocompatibility complex) molecules needed to present CNS antigens to infiltrating T cells.[4]

All these observations seemed to indicate that the immune system and the CNS are separate domains. However, over the years novel experimental evidence challenged this traditional viewpoint: the intact CNS parenchyma regularly contains small numbers of T cells,[5,6] T cells may also enter the CNS in the course of CNS inflammation and degeneration,[7,8] and interstitial/cerebrospinal fluid and proteins drain from the brain to the blood and may affect immune responses in the draining lymph nodes.[9]

It is now clear that the CNS is under constant immune surveillance, and that it interacts with cells of the immune system in both health and disease.

EXPERIMENTAL AUTOIMMUNE ENCEPHALOMYELITIS — AN ANIMAL MODEL TO STUDY INTERACTIONS OF THE IMMUNE SYSTEM WITH THE CNS

Most of our current knowledge about neuroinflammation has been obtained from experimental autoimmune encephalitis (EAE), which for decades has served as the best animal model to study the aspects of CNS inflammation and demyelination seen in human patients with multiple sclerosis. It is quite bizarre that this invaluable animal model had its origin in a series of neuroparalytic accidents in humans receiving anti-rabies treatments with killed carbolised virus isolated from infected animal brains. The discovery that brain material contaminating the vaccine was responsible for paralysis[10] paved the way for pioneering experiments in which CNS inflammation and demyelination were induced in experimental animals by immunisation with brain material.[11] The resulting disease model was termed EAE, and has been markedly refined over the years. At the start, crude spinal cord or brain extracts were used for immunisation[12]; later, purified CNS proteins were used, such as myelin basic protein (MBP),[13] proteolipid protein (PLP),[14] myelin oligodendrocyte glycoprotein (MOG),[15] S100β,[16] and now often only small peptides (of about 20 amino acids in length) derived from CNS proteins.[7]

A breakthrough came when it was observed that EAE can be transferred between animals by lymph node cells,[17] that pure CNS antigen specific T cells can be cultured and activated *in vitro*,[18] and that these cells were also able to initiate EAE upon transfer in naive recipient animals.[18] In fact, T cell transferred EAE has considerable advantages over immunisation induced EAE in that it starts earlier and shows less variation between animals.[7]

Other refinements of the EAE model applied to the choice of animals. As seen in human patients with multiple sclerosis, different strains of animal display different types of disease course, different symptoms, and different pathological changes within the CNS.[7] The most recent developments in the field are:

- the use of genetically modified animals with defined changes in cells of the immune or central nervous systems;
- the production and transfer of fluorescent CNS antigen specific T cells to trace their migration and cellular changes in the course of the disease;
- microdissection and molecular analyses of single cells involved in the disease process;
- spectratyping of disease relevant T cell receptors;
- the search for antigens recognised by CNS infiltrating T cells in biopsy/necropsy material from patients suffering from inflammatory CNS disease;
- the use of DNA microarrays to study simultaneously the interactions between many different genes in CNS inflammation;
- the development of new treatments based on the unique property of activated CNS antigen specific T cells to cross the blood–brain barrier.

HOW MODERN TECHNIQUES CONTRIBUTED TO OUR UNDERSTANDING OF NEUROINFLAMMATION IN EAE AND MULTIPLE SCLEROSIS

Genetically Manipulated Animals

Probably the most fruitful — and currently most popular — technology is the genetic manipulation of experimental animals so that factors or even entire cell populations are missing ("knock-out animals"), or so that proteins under investigation are produced in a temporally or spatially controllable way (conditionally "transgenic" animals). Such strategies have furthered our understanding of the different phases of autoimmune CNS inflammation.

The initiation of inflammatory lesions in the CNS

It is well established that CNS antigen specific T cells are normal components of the immune system of every healthy organism.[19,20] However, spontaneous CNS inflammation provoked by these cells is exceedingly rare. Why? To answer this question, transgenic mice were constructed which carry receptors for myelin basic protein — a major component of the myelin sheath in the CNS — on essentially all their T cells. Furthermore, the (rearranged) genes for these particular T cell receptors had been isolated from T cells that were found to be highly pathogenic in EAE experiments.[21] The first surprising observation was that the animals did not get sick as long as they were kept under specific pathogen-free (SPF) conditions. However, they did develop spontaneous autoimmune CNS inflammation when they were exposed to the ubiquitous pathogens of the environment. These experiments clearly showed that it is the activation and not the mere presence of CNS antigen specific T cells in the immune system that provokes the induction of neuroinflammation. As these animals had not been immunised with the CNS antigen recognised by the transgenic T cell receptor, T cell activation must have resulted from the contact with bacterial or viral proteins. Four recent observations suggest that this observation is not just intrinsic to the animal model of EAE, but may also be highly significant for human patients with multiple sclerosis. First, and perhaps trivially, the onset of disease or exacerbations of its course in human patients with multiple sclerosis or other disorders with an autoimmune component is sometimes associated with infections.[22] Second, some human MBP specific T cell clones can be activated by peptides derived from herpes simplex virus, Epstein-Barr virus, or influenza virus, which are presented by the same MHC molecule.[22] Third, even different MHC molecules loaded with different peptides may show such a high degree of structural equivalence that they are recognised by the same T cell receptor of a patient with multiple sclerosis (for example DRB1 * 1501 loaded with an MBP peptide and DRB5 * 0101 loaded with an Epstein-Barr virus peptide).[23] Finally, a "humanised mouse model" of multiple sclerosis has been developed using multiply transgenic mice.[24] These mice express three human components involved in T cell recognition of the multiple sclerosis related autoantigen MBP: HLA-DR2, an MHC class II antigen associated with multiple sclerosis in north Europeans; a T cell receptor from a T cell clone specific for the MBP peptide 84–102 derived from a multiple sclerosis patient; and the human coreceptor CD4. When backcrossed to Rag2 deficient mice, these transgenic humanised mice spontaneously develop inflammatory disease of the CNS.[24]

The EAE experiments described above clearly show that CNS antigen specific T cells which had been activated somewhere in the body — whether by direct antigen contact in the course of immunisation or by the antigen independent action of certain bacterial or viral proteins — are able to cross the intact blood–brain barrier and enter the CNS parenchyma. It is assumed that these cells find "their" specific antigen upon entry into the CNS, are further activated, and then interact with local glial cells, leading to a locally confined release of chemokines and cytokines. These proteins activate the blood–brain barrier and attract further cells from the immune system. A second wave of immune cell infiltration into the CNS follows, which seems to be independent of the antigen recognition or the activation status of the recruited cells. Inflammatory lesions develop.[1]

Progression of CNS inflammation and the resulting tissue damage

Like multiple sclerosis, EAE can take different clinical courses. In some animal strain or antigen combinations the course is monophasic; in others it is fulminant/lethal, relapsing/remitting, or chronically progressive. In animals with relapsing/remitting or chronically progressive types of EAE, and in human patients with isolated monosymptomatic demyelinating syndrome (IMDS), the proliferative response of T cells to the "priming" myelin protein often declines with time, while T cells with specificities for other parts of the same myelin protein or even for other myelin proteins increase in numbers. This effect is called "epitope spreading" and has been held responsible for the progression of chronic CNS inflammation in animals with EAE and for the progression of IMDS to clinically definite multiple sclerosis.[25] Moreover, the phenomenon of epitope spreading could present an obstacle to experimental "T cell vaccination treatments" aiming at the selective elimination of certain CNS antigen specific T cell clones from the immune repertoire of patients with multiple sclerosis.[26] However, a more recent study suggests that the possible relation between epitope spreading and disease progression may be more complicated.[27] To test whether progressive courses of EAE can also occur in the absence of epitope spreading, transgenic mice were created which carried T cells with only one defined CNS antigen specificity — which effectively excluded T cell reactivity to other CNS proteins. Despite a monospecific T cell response, these mice developed progressive and relapsing/remitting disease courses. This indicates that under certain conditions disease initiating T cells may also drive the subsequent progressive phase.[27]

CNS inflammation is commonly associated with some degree of tissue damage — that is, loss of myelin sheaths or loss of axons. This is observed in animal models of EAE and is a central theme in human patients with multiple sclerosis. Again, EAE models were instrumental in elucidating the different cellular and molecular pathways resulting in demyelination. One pathway was discovered in knock-out mice which had been rendered deficient in B cells by genetic disruption of the membrane bound antibody μ chain. When these animals were immunised with a myelin protein to induce EAE, they developed both inflammatory lesions and primary demyelination.[28] These data indicated that in mice, antibodies and B cells are not necessary for myelin loss, or that T cells and their products are sufficient for demyelination. What are these T cells products?

It is well established that not only macrophages but also T cells can produce tumour necrosis factor α (TNFα), and that activated T cells may also secrete γ interferon (IFNγ). Both these cytokines are found in inflammatory lesions of EAE animals. To learn more about the role of these cytokines in myelin loss, transgenic mice were created which overexpressed either TNFα[29] or IFNγ[30] in their CNS. Both types of transgenic animal developed demyelination in the absence of immunisation, which clearly showed that the presence of these cytokines in the CNS of mice is sufficient to trigger myelin loss.[29,30]

Another pathway leading to demyelination in the course of CNS inflammation was identified in rats. In these animals, T cells themselves are not capable of causing noticeable damage to myelin sheaths. Instead, loss of myelin is only observed when activated CNS antigen specific T cells open the blood–brain barrier and antibodies specifically recognising epitopes of proteins on the surface of myelin sheaths gain access to the inflammatory lesions.[31]

Yet another pathway to demyelination could be initiated by primary changes in the myelin sheaths themselves, as seen in transgenic mice and rats overexpressing PLP in their myelin forming oligodendrocytes.[32–34] When the PLP overdose exceeded a certain threshold, myelin sheaths were not produced at all owing to early oligodendrocyte death. Below the threshold, myelin sheaths were formed but were unstable. The resulting subclinical myelin degeneration paved the way for spontaneous demyelination in transgenic mice,[32,33] and for an increase in demyelination in EAE induced by T cells and antibodies directed against myelin proteins in transgenic rats (Bradl M, unpublished data). In summary, demyelination in the course of CNS inflammation may be caused by T cells and T cell products, by antibodies, and by an instability of myelin sheaths.

As mentioned above, multiple sclerosis and EAE are not only characterised by loss of myelin sheaths, but also by damage to axons. Transections of axons, axonal swellings, and spheroid formations in the CNS had already been described at the turn of last century, but were neglected for many decades and only recently attracted renewed attention.[35] These pathological changes are held responsible for the permanent clinical deficits in multiple sclerosis.[36] They can be caused by cytotoxic T cells[37] or by the action of inflammatory mediators such as nitric oxide.[38]

The resolution of inflammatory lesions within the CNS

Genetically modified animals are also essential for studying the termination of CNS inflammation — that is, the resolution of inflammatory lesions. Studies in conventional EAE models revealed unusually large numbers of apoptotic T cells in the late phases of CNS inflammation and it was concluded from these observations that inflammatory CNS lesions may resolve through the induction of apoptosis in T cells.[39,40] However, it remained unclear whether the induction of programmed cell death only affects autoantigen specific and hence disease initiating T cells, or whether it also affects recruited T cells non-specifically, and whether it occurs at some sites rather more often than at others. To address these questions, EAE was induced by injections of CNS antigen specific transgenic T cells into wild-type (non-transgenic) recipients. The transgenic T cells carried a nuclear marker and could therefore be easily identified in histological specimen of inflamed CNS tissue. It was found that apoptotic T cells are preferentially located within the CNS parenchyma, and almost never in the meninges or perivascular spaces. Moreover, all T cells are cleared from the CNS parenchyma, no matter whether they were disease initiating or non-specifically recruited, or whether they had previously been activated or were resting.[41]

Fluorescent T Cells

Based on the experiments described above it became firmly established that CNS antigen specific T cells have to be maximally activated before their transfer into naive recipients in order to induce EAE. Nevertheless, there was a need to explain the curious time window of about 72 hours between T cell transfer and clinical disease, and consequently to address the following questions. What happens to the disease inducing T cells during that time? Where do they reside? Is there a way of tracing and characterising such cells, and of discriminating them from lymphocytes with other antigen specificities which will never participate in the disease process?

In brief, there is indeed an elegant way to follow the fate of autoreactive T cells *in vivo*. This relies on the introduction of the gene for a green fluorescent protein (GFP) into CNS antigen specific T cells which can be achieved by retroviral gene transfer (Fig. 1) and which leads to highly fluorescent T cells with a stable, self replenishing, and selectable marker.[42] When such cells were injected into recipient animals to cause EAE, they could be reisolated for further studies at any time between disease induction (that is, T cell transfer) and CNS inflammation. These analyses provided very interesting information: it turns out that in the preclinical phase of EAE, freshly injected CNS antigen specific T cells migrate first to the perithymic lymph nodes

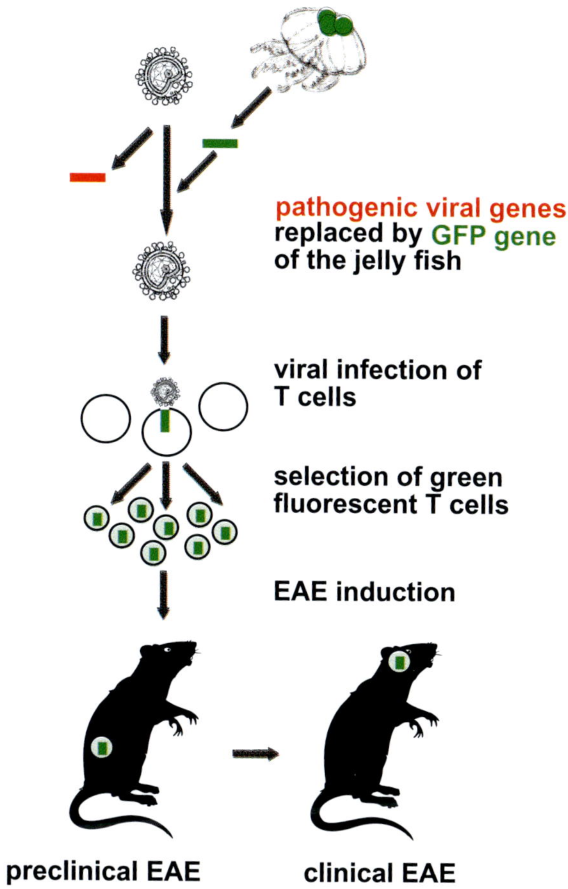

Fig. 1 Fluorescent T cells.[41–43] Viruses are genetically modified so that disease inducing viral genes are replaced by the gene encoding the green fluorescent protein derived from the jellyfish.[43] These viruses are then used to infect T cells and to introduce the GFP gene. In some cells the gene is expressed, the fluorescent protein produced, and the T cells turn green. Green fluorescent CNS antigen specific T cells can be selected, activated, and transferred to naive recipient animals. At different time points after T cell transfer, they can be isolated from the CNS, spleen, or lymph nodes for further analyses.

and then to the spleen, where they acquire a new "migratory" phenotype, characterised by downregulation of activation markers, upregulation of chemokine receptors, and an increase in MHC class II molecules on the cell surface. These T cells then change phenotype a second time when they enter the CNS. Here they are reactivated on encountering their specific antigen, which is presented by activated antigen presenting cells of the CNS.[44]

MICRODISSECTION AND MOLECULAR ANALYSES OF SINGLE CELLS INVOLVED IN THE DISEASE PROCESS

As mentioned above, T cells injected to induce EAE can easily be reisolated from the CNS or from lymphoid organs of the recipient animals. Their phenotype can be determined and their encephalitogenic potential can even be tested by further transfer into naive recipients. In humans the situation is quite different, for obvious reasons. Here, information must be gathered from the few T cells found in inflammatory lesions in necropsy or biopsy material, which demands more refined techniques such as microdissection and molecular analysis of individual T cells. The feasibility of this approach was recently demonstrated by Babbe and coworkers, when they analysed multiple sclerosis plaques.[45] They first stained frozen tissue sections for the presence of CD4 or CD8 positive T lymphocytes, and then mobilised and isolated single T cells from the surrounding tissue with the help of a micromanipulator. Afterwards, they analysed the T cell receptors of these cells by polymerase chain reaction (PCR). This approach revealed interesting differences between $CD8^+$ and $CD4^+$ T cells. $CD8^+$ T cells were the dominant T cell population in multiple sclerosis lesions but represented the descendents of only a few different T cell clones. In contrast, $CD4^+$ T cells were much less numerous but much more heterogeneous, representing progeny of many different T cell clones.[45]

The dominance of T cell clones in inflammatory lesions could have at least two different causes. It could reflect a preferential infiltration of certain clones into the affected area, or a random recruitment of clones which are overrepresented in the peripheral immune repertoire. One way to address this problem involves the isolation, cellular cloning, and characterisation of T cells *in vitro*, and the molecular cloning and characterisation of their receptors by sequencing analyses, which would be a rather laborious task. Instead, it is significantly more efficient to use a different technique that is especially suited to screening T cell receptor repertoires. This is called "CDR3 spectratyping."

CDR3 Spectratyping

This technique is based on the fact that individual T cells and their clonal descendants carry unique T cell receptors with which they recognise "their" specific antigen. The high degree of diversity of T cell receptors is caused by a random combination of gene segments and by inaccuracy in the joining process of these segments. This leads to the arbitrary addition or deletion of nucleotides and to the formation within each T cell receptor gene of areas which are hypervariable both in their length and in their molecular sequence. One of these regions of the T cell receptor is called the complementary determining (CDR)-3 region. For each T cell receptor gene segment, amplification of this region by PCR yields a spectrum of PCR products of differing lengths (the "spectratype"), which reveals the expansion of T cell clones in the peripheral blood,[46] or the persistence of CNS antigen specific T cells in blood[47] and cerebrospinal fluid of patients as strong signals ("peaks").

NOVEL TECHNIQUES AND THEIR CONTRIBUTION TO CLINICAL NEUROLOGY — A FUTURE PERSPECTIVE

The Search for the Autoantigen of CNS Inflammation

As mentioned above, in the animal model of EAE, activated T cells with specificity for neural antigens were shown to initiate CNS inflammation and autoimmune disease. However, the antigen specificity of CNS infiltrating T cells in multiple sclerosis lesions remains largely unknown. What do these T cells recognise?

If the (auto)antigens are unknown, the unique (clonotypic) T cell receptor of the presumably autoaggressive T cells might provide a clue. A potentially promising approach combines the cloning of T cell receptors found on individual cells within inflammatory lesions, and the expression of the cloned receptors in cells which contain the complete machinery for signalling through T cell receptors, but do not themselves express full receptors. If such cells are grafted with a cloned receptor derived from an autoaggressive T cell, they will secrete cytokines, and possibly even proliferate upon recognition of an antigen for which the T cell receptor is specific. The feasibility of this approach was recently demonstrated with autoreactive T cells derived from polymyositis lesions.[48] Once the antigen is known, antigen specific T cells can be tracked with soluble MHC–peptide tetramers. These synthetic structures behave in a similar way to the natural MHC–peptide complexes presented on the surface of antigen presenting cells: They are recognised by T cells carrying the corresponding specific T cell receptors, and bind only to these specific cells. Principally, tetramers can be constructed using MHC class I[49] or class II[50] molecules and peptides, thus allowing the tracking of $CD8^+$ and $CD4^+$ T cells. This approach works in FACS (fluorescent antibody cell sorting) analyses of peripheral blood cells, where it was used to visualise antigen specific T cells, for example in Epstein-Barr virus infected[49] or HIV infected patients,[51] or in cancer patients.[52,53] Unfortunately, it does not yet work reliably on histological sections. However, as there are ongoing efforts in many different laboratories to optimise the MHC–peptide tetramer techniques (see for example[54,55]), it is probably just a matter of time until this problem is solved.

The Search for Therapeutic Target Structures with DNA Microarrays

As stated earlier, inflammation of the CNS is a very complex process, requiring the spatial and temporal regulation of T cell activation, coordinated interaction between T lymphocytes and other components of the immune system, and intercellular interactions between T cells and local glial cells within the CNS parenchyma. Hence it is not surprising that there is also a complex interaction between many different genes and gene products of individual cell populations. To study these interactions and to identify novel molecules involved in the different phases of neuroinflammation, DNA microarrays (also called DNA chips) are increasingly used. They consist of hundreds or thousands of DNA sequences representing defined genes, which are attached to a glass surface and can be hybridised to the complete RNA (or cDNA) from cells of interest (Fig. 2). The RNA (or cDNA) probes bind to matching gene spots on the DNA microarray and identify all expressed genes of a given sample. To date, microarrays have been used to investigate multiple sclerosis or EAE tissue samples[57–59] to characterise molecules involved in specific

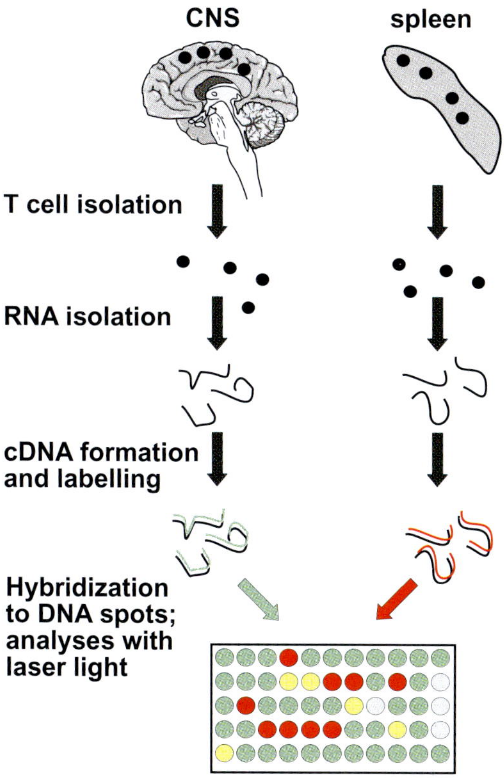

Fig. 2 DNA microarrays.[56] Messenger RNA is isolated from T cells found within the CNS or the spleen and is converted to fluorescence labelled, and hence coloured, cDNA. The cDNA of CNS derived T cells is labelled green, the cDNA of spleen derived T cells red. The two different sets of cDNA are mixed and hybridised to the microarray. A microarray (also called DNA chip) contains spots of thousands of DNA sequences which all represent different genes and which can be simultaneously tested with the labelled cDNA. If the cDNA finds corresponding sequences on the microarray, it will bind to it. Unbound cDNA is washed off. The microarray is analysed with laser light (red and green). The resulting images are analysed. Green spots represent genes that are expressed in CNS derived T cells only; red spots represent genes that are only expressed in spleen resident T cells; yellow spots represent genes that are expressed in both sets of T cells, and the absence of coloured spots indicates that a particular gene is not expressed at all.

biological processes (such as cellular activation, intercellular signalling, cell adhesion, transmigration, or cell–cell interactions), with the aim of ultimately identifying new targets for therapeutic intervention of CNS inflammation.

T Cells as Therapeutic Agents

All the scenarios described above clearly show that T cells and their products may provoke damage or even loss of myelin sheaths and axons. Other observations, however, indicate that the role of inflammatory cells, especially T cells, might be considerably more complex. For example, T cells were found to produce neuroprotective molecules such as neurotrophin-3 (NT-3)[60] or brain derived neurotrophic factor (BDNF).[61] Neurotrophins and neuroprotective factors promote neuronal survival and may also mediate anti-inflammatory effects.[62] In addition, depending on the subset studied T cells may also produce cytokines such as interleukin-4 or interleukin-10, which can also suppress autoimmune responses.[63]

These examples clearly show that T cells may also have beneficial functions in CNS inflammation.[1] Under natural conditions this "beneficial" side of inflammation might be too weak to outweigh the harmful and toxic effects. In certain experimental systems, however, the neuroprotective and anti-inflammatory effects of inflammation can be strengthened so that they become clinically apparent. An especially appealing approach is based on the following reasoning. If activated T cells can cross the endothelial blood–brain barrier whereas other factors are efficiently excluded from the CNS,[3] and if these cells accumulate at sites of neuronal degeneration[8] and can easily be manipulated to express foreign genes while maintaining all other cellular functions,[42,44,64,65] then why not use them as vehicles to transport therapeutic agents across the blood–brain barrier?[66] The feasibility of such an approach was demonstrated in animal models of EAE and experimental autoimmune neuritis (EAN). The gene for nerve growth factor (NGF) was retrovirally transduced into neural antigen specific T cells. When these T cells were transferred to experimental animals, an attenuation of the disease process was observed in the animal models of both EAE[65] and EAN.[64] These promising results encourage further efforts to refine and optimise this technique and to create better tools for the treatment of inflammatory CNS disease. Furthermore, it appears that some of the existing treatments[67] might — at least in part — even rely on the beneficial role of inflammation. For example, it is known that Copaxone (glatiramer acetate (GA)), which is approved for the treatment of multiple sclerosis,[67] induces a Th1 to Th2 cytokine shift in GA reactive T cells,[68] possibly because of its action on dendritic cells.[69] These findings indicate that activated GA-reactive T cells migrate into the CNS and produce immunomodulatory cytokines[68] or even BDNF[70] locally.

CONCLUSION

As we have discussed in this brief review, our understanding of the molecular pathogenesis of neuroinflammation is growing steadily. Several recent studies have revealed unexpected insights — for example, by providing hints of a potentially protective role of inflammation. New techniques are providing increasingly powerful research tools. Progress in different areas of basic research will help further to improve the treatment of patients with neuroimmunological diseases.

ACKNOWLEDGEMENTS

We thank DFG (SFB 571) and FWF (project P16047-B02) for financial support.

REFERENCES

1. Bradl M, Flügel A. The role of T cells in brain pathology. *Curr Top Microbiol Immunol* 2002; **265**:141–162.
2. Head JR, Griffin WST. Functional capacity of solid tissue implants in the brain: Evidence for immunological privilege. *Proc R Soc* 1985; **224**:375–387.
3. Barker CF, Billingham RE. Immunologically privileged sites. *Adv Immunol* 1977; **25**:1–54.
4. Wekerle H. Antigen presentation by CNS glia. In: Kettenmann H, Ransom B, eds. *Neuroglia*. USA: OUP USA, 1994.
5. Wekerle H, Linington C, Lassmann H, *et al.* Cellular immune reactivity within the CNS. *Trends Neurosci* 1986; **9**:271–277.
6. Hickey WF, Hsu BL, Kimura H. T lymphocyte entry into the central nervous system. *J Neurosci Res* 1991; **28**:254–260.
7. Wekerle H, Kojima K, Lannes-Vieira J, *et al.* Animal models. *Ann Neurol* 1994; **36**:S47–S53.
8. Raivich G, Jones LL, Kloss CU, *et al.* Immune surveillance in the injured nervous system: T lymphocytes invade the axotomized mouse facial motor nucleus and aggregate around sites of neuronal degeneration. *J Neurosci* 1998; **18**:5804–5816.
9. Cserr HF, Harling-Berg CJ, Knopf PM. Drainage of brain extracellular fluid into blood and deep cervical lymph and its immunological significance. *Brain Pathol* 1992; **2**:269–276.

10. Stuart G, Krikorian KS. The neuro-paralytic accidents of anti-rabies treatment. *Ann Trop Med Parasitol* 1928; **22**:327–377.
11. Rivers TM, Schwentker FF. Encephalomyelitis accompanied by myelin destruction experimentally produced in monkeys. *J Exp Med* 1935; **61**:689–702.
12. Kabat EA, Wolf A, Bezer AE. The rapid production of acute disseminated encephalomyelitis in rhesus monkeys by injection of heterologous and homologous brain tissue with adjuvants. *J Exp Med* 1947; **85**:117–130.
13. Einstein ER, Robertsons DM, DiCaprio JM, *et al.* The isolation from bovine spinal cord of a homogeneous protein with encephalitogenic activity. *J Neurochem* 1962; **9**:353–361.
14. Yamamura T, Namikawa T, Endoh M, *et al.* Experimental allergic encephalomyelitis induced by proteolipid apoprotein in Lewis rats. *J Neuroimmunol* 1986; **12**:143–153.
15. Adelmann M, Wood J, Benzel I, *et al.* The N-terminal domain of the myelin oligodendrocyte glycoprotein (MOG) induces acute demyelinating experimental autoimmune encephalomyelitis in the Lewis rat. *J Neuroimmunol* 1995; **63**:17–27.
16. Kojima K, Berger T, Lassmann H, *et al.* Experimental autoimmune panencephalitis and uveoretinitis transferred to the Lewis rat by T lymphocytes specific for the S100β molecule, a calcium binding protein of astroglia. *J Exp Med* 1994; **180**:817–829.
17. Paterson PY. Transfer of allergic encephalomyelitis in rats by means of lymph node cells. *J Exp Med* 1960; **111**:119–136.
18. Ben Nun A, Wekerle H, Cohen IR. The rapid isolation of clonable antigen-specific T lymphocyte lines capable of mediating autoimmune encephalomyelitis. *Eur J Immunol* 1981; **11**:195–199.
19. Schluesener HJ, Wekerle H. Autoaggressive T lymphocyte lines recognizing the encephalitogenic region of myelin basic protein: in vitro selection from unprimed rat T lymphocyte populations. *J Immunol* 1985; **135**:3128–3133.
20. Pette M, Fujita K, Wilkinson D, *et al.* Myelin autoreactivity in multiple sclerosis: Recognition of myelin basic protein in the context of HLA-DR2 products by T lymphocytes of multiple sclerosis patients and healthy donors. *Proc Natl Acad Sci USA* 1990; **87**:7968–7972.
21. Goverman J, Woods A, Larson L, *et al.* Transgenic mice that express a myelin basic protein-specific T cell receptor develop spontaneous autoimmunity. *Cell* 1993; **72**:551–560.
22. Wucherpfennig KW, Strominger JL. Molecular mimicry in T cell-mediated autoimmunity: viral peptides activate human T cell clones specific for myelin basic protein. *Cell* 1999; **80**:695–705.

23. Lang HL, Jacobsen H, Ikemizu S, *et al.* A functional and structural basis for TCR cross-reactivity in multiple sclerosis. *Nat Immunol* 2002; **3**:940–943.
24. Madsen LS, Andersson EC, Jansson L, *et al.* A humanized model for multiple sclerosis using HLA-DR2 and a human T-cell receptor. *Nat Genet* 1999; **23**: 343–347.
25. Tuohy VK, Yu M, Yin L, *et al.* The epitope spreading cascade during progression of experimental autoimmune encephalomyelitis and multiple sclerosis. *Immunol Rev* 1998; **164**:93–100.
26. Zhang J, Vandevyver C, Stinissen P, *et al.* In vivo clonotypic regulation in human myelin basic protein-reactive T cells by T cell vaccination. *J Immunol* 1995; **155**:5868–5877.
27. Jones RE, Bourdette D, Moes N, *et al.* Epitope spreading is not required for relapses in experimental autoimmune encephalomyelitis. *J Immunol* 2003; **170**:1690–1698.
28. Hjelmström P, Juedes AE, Fjell J, *et al.* Cutting edge: B cell-deficient mice develop experimental allergic encephalomyelitis with demyelination after myelin oligodendrocyte glycoprotein sensitization. *J Immunol* 1998; **161**:4480–4483.
29. Akassoglou K, Bauer J, Kassiotis G, *et al.* Oligodendrocyte apoptosis and primary demyelination induced by local TNF/p55TNF receptor signaling in the central nervous system of transgenic mice: Models for multiple sclerosis with primary oligodendrogliopathy. *Am J Pathol* 1998; **153**:801–813.
30. Horwitz MS, Evans CF, McGavern DB, *et al.* Primary demyelination in transgenic mice expressing interferon-gamma. *Nat Med* 1997; **3**:1037–1041.
31. Linington C, Bradl M, Lassmann H, *et al.* Augmentation of demyelination in rat acute allergic encephalomyelitis by circulating mouse monoclonal antibodies directed against a myelin/oligodendrocyte glycoprotein. *Am J Pathol* 1988; **130**:443–454.
32. Readhead C, Schneider A, Griffiths IR, *et al.* Premature arrest of myelin formation in transgenic mice with increased proteolipid protein gene dosage. *Neuron* 1994; **12**:583–595.
33. Kagawa T, Ikenaka K, Inoue Y, *et al.* Glial cell degeneration and hypomyelination caused by overexpression of myelin proteolipid protein gene. *Neuron* 1994; **13**:427–442.
34. Bradl M, Bauer J, Inomata T, *et al.* Transgenic Lewis rats overexpressing the proteolipid protein gene: myelin degeneration and its effect on T cell-mediated experimental autoimmune encephalomyelitis. *Acta Neuropathol (Berl)* 1999; **97**:595–606.
35. Kornek B, Lassmann H. Axonal pathology in multiple sclerosis. A historical note. *Brain Pathol* 1999; **9**:651–656.

36. Trapp BD, Peterson J, Ransohoff RM, *et al.* Axonal transection in the lesion of multiple sclerosis. *N Engl J Med* 1998; **338**:278–285.
37. Medana I, Martinic MA, Wekerle H, *et al.* Transection of major histocompatibility complex class I-induced neurites by cytotoxic T lymphocytes. *Am J Pathol* 2001; **159**:809–815.
38. Waxman SG. Nitric oxide and the axonal death cascade. *Ann Neurol* 2003; **53**:150–153.
39. Pender MP, Nguyen KB, McCombe PA, *et al.* Apoptosis in the nervous system in experimental allergic encephalomyelitis. *J Neurol Sci* 1991; **104**:81–87.
40. Schmied M, Breitschopf H, Gold R, *et al.* Apoptosis of T lymphocytes in experimental autoimmune encephalomyelitis: evidence for programmed cell death as a mechanism to control inflammation in the brain. *Am J Pathol* 1993; **143**: 446–452.
41. Bauer J, Bradl M, Klein M, *et al.* Endoplasmic reticulum stress in PLP-overexpressing transgenic rats: gray matter oligodendrocytes are more vulnerable than white matter oligodendrocytes. *J Neuropathol Exp Neurol* 2002; **61**:12–22.
42. Flügel A, Willem M, Berkowicz T, *et al.* Gene transfer into CD4+ T lymphocytes: green fluorescent protein-engineered, encephalitogenic T cells illuminate brain autoimmune responses. *Nat Med* 1999; **5**:843–847.
43. Miller AD, Rosman G. Improved retroviral vectors for gene transfer and expression. *BioTechniques* 1989; **7**:980–990.
44. Flügel A, Berkowicz T, Ritter T, *et al.* Migratory activity and functional changes of green fluorescent effector cells before and during experimental autoimmune encephalomyelitis. *Immunity* 2001; **14**:547–560.
45. Babbe H, Roers A, Waisman A, *et al.* Clonal expansions of CD8(+) T cells dominate the T cell infiltrate in active multiple sclerosis lesions as shown by micromanipulation and single cell polymerase chain reaction. *J Exp Med* 2000; **192**:393–404.
46. Hofbauer M, Wiesener S, Babbe H, *et al.* Clonal tracking of autoaggressive T cells in polymyositis by combining laser microdissection, single-cell PCR, and CDR3-spectratype analysis. *Proc Natl Acad Sci USA* 2003; **100**: 4090–4095.
47. Goebels N, Hofstetter H, Schmidt S, *et al.* Repertoire dynamics of autoreactive T cells in multiple sclerosis patients and healthy subjects: epitope spreading versus clonal persistence. *Brain* 2000; **123**:508–518.
48. Wiendl H, Malotka J, Holzwarth B, *et al.* An autoreactive gamma delta TCR derived from a polymyositis lesion. *J Immunol* 2002; **169**:515–521.
49. Callan MF, Tan L, Annels N, *et al.* Direct visualization of antigen-specific CD8+ T cells during the primary immune response to Epstein-Barr virus in vivo. *J Exp Med* 1998; **187**:1395–1402.

50. Reddy J, Bettelli E, Nicholson L, *et al.* Detection of autoreactive myelin proteolipid protein 139-151-specific T cells by using MHC II (IAs) tetramers. *J Immunol* 2003; **170**:870–877.
51. McCloskey TW, Haridas V, Pahwa R, *et al.* T cell receptor V beta repertoire of the antigen specific CD8 T lymphocyte subset of HIV infected children. *AIDS* 2002; **16**:1459–1465.
52. Pittet MJ, Speiser DE, Valmori D, *et al.* Ex vivo analysis of tumor antigen specific CD8+ T cell responses using MHC/peptide tetramers in cancer patients. *Int Immunopharmacol* 2001; **1**:1235–1247.
53. Cerundolo V. Use of major histocompatibility complex class I tetramers to monitor tumor-specific cytotoxic T lymphocyte response in melanoma patients. *Cancer Chemother Pharmacol* 2000; **46**:S83–S85.
54. Sato Y, Sahara H, Tsukahara T, *et al.* Improved generation of HLA class I/peptide tetramers. *J Immunol Methods* 2002; **271**:177–184.
55. Cunliffe SL, Wyer JR, Sutton JK, *et al.* Optimization of peptide linker length in production of MHC classII/peptide tetrameric complexes increases yield and stability, and allows identification of antigen-specific CD4+ T cells in peripheral blood mononuclear cells. *Eur J Immunol* 2002; **32**: 3366–3375.
56. Department of Biology, Davidson College, Davidson NC28036. DNA Microarray Methodology–Flash animation. http://www.bio.davidson.edu/courses/genomics/chip/chip.html
57. Whitney LW, Ludwin SK, McFarland HF, *et al.* Microarray analysis of gene expression in multiple sclerosis and EAE identifies 5-lipoxygenase as a component of inflammatory lesions. *J Neuroimmunol* 2001; **212**: 40–48.
58. Ibrahim SM, Mix E, Bottcher T, *et al.* Gene expression profiling of the nervous system in murine experimental autoimmune encephalomyelitis. *Brain* 2001; **124**:1927–1938.
59. Mycko MP, Papoian R, Boschert U, *et al.* cDNA microarray analysis in multiple sclerosis lesions: detection of genes associated with disease activity. *Brain* 2003; **126**:1048–1057.
60. Besser M, Wank R. Cutting edge: clonally restricted production of the neurotrophins brain-derived neurotrophic factor and neurotrophin-3 mRNA by human immune cells and Th1/Th2-polarized expression of their receptors. *J Immunol* 1999; **162**:6303–6306.
61. Kerschensteiner M, Gallmeier E, Behrens L, *et al.* Activated human T cells, B cells, and monocytes produce brain-derived neurotrophic factor in vitro and in inflammatory brain lesions: a neuroprotective role of inflammation? *J Exp Med* 1999; **189**:865–870.

62. Kerschensteiner M, Stadelmann C, Dechant G, *et al.* Neurotrophic cross-talk between the nervous and immune systems: implications for neurological diseases. *Ann Neurol* 2003; **53**:292–304.
63. Young DA, Lowe LD, Booth SS, *et al.* IL-4, IL-10, IL-13, and TGF-beta from an altered peptide ligand-specific Th2 cell clone down-regulate adoptive transfer of experimental autoimmune encephalomyelitis. *J Immunol* 2000; **164**: 3563–3572.
64. Kramer R, Zhang Y, Gehrmann J, *et al.* Gene transfer through the blood–nerve barrier: NGF-engineered neuritogenic T lymphocytes attenuate experimental autoimmune neuritis. *Nat Med* 1995; **1**:1162–1166.
65. Flügel A, Matsumoro K, Neumann H, *et al.* Anti-inflammatory activity of nerve growth factor in experimental autoimmune encephalomyelitis: inhibition of monocyte transendothelial migration. *Eur J Immunol* 2001; **31**:11–22.
66. Flügel A, Wekerle H. Genetic engineering of brain-specific T cells for treatment of diseases in the central nervous system. *T cell Autoimmunity and Multiple Sclerosis*. Georgetown: RG Landes company, 1999:37–57.
67. Hohlfeld R. Biotechnological agents for the immunotherapy of multiple sclerosis. *Brain* 1997; **120**:865–916.
68. Neuhaus O, Farina C, Yassouridis A, *et al.* Multiple sclerosis: comparison of copolymer-1-reactive T cell lines from treated and untreated subjects reveals cytokine shift from T helper 1 to T helper 2 cells. *Proc Natl Acad Sci USA* 2000; **97**:7452–7457.
69. Vieira PL, Heysteck HC, Wormmeester J, *et al.* Glatiramer acetate (copolymer-1, copaxone) promotes th2 cell development and increased IL-10 production through modulation of dendritic cells. *J Immunol* 2003; **170**: 4483–4488.
70. Ziemssen T, Kumpfel T, Klinkert WE, *et al.* Glatiramer acetate-specific T-helper1- and 2-type cell lines produce BDNF: implications for multiple sclerosis therapy. Brain-derived neurotrophic factor. *Brain* 2002; **125**: 2381–2391.

9

Neurovirology

*Peter Ge Kennedy**

INTRODUCTION

Neurovirology has emerged over the last 30 years as a major discipline which embraces a range of subject areas including virology, neurosciences and clinical neurology, molecular biology and immunology.[1] As Richard T Johnson, one of the pioneers in this field, has explained, the term Neurovirology is a broad one, and is used to cover the studies of the pathogenesis of viral infections of the nervous system, both in humans and experimental animal models, *in vitro* studies of the effects of viruses on defined neural cell types, the use of viruses as tools in neuroanatomical and developmental studies, and the use of genetically engineered viruses as vectors to deliver therapeutic genes into the human central nervous system (CNS).[1] It also includes the diagnosis and epidemiology of viral infections of the CNS, as well as the developmental of specific anti-viral agents to treat disease. The use of acyclovir for the effective treatment of herpes simplex encephalitis is a very good example of the latter.[2] Thus it is a discipline which has much

*Division of Clinical Neurosciences, Institute of Neurological Sciences, Southern General Hospital, Department of Neurology, Glasgow G51 4TF; Phone: 0141 201 2474; Fax: 0141 201 2993; E-mail: P.G.Kennedy@clinmed.gla.ac.uk

relevance to both human disease and many aspects of neuroscience. Clinical and basic neurovirological studies sometimes reveal complementary information.

Historically much of Neurovirology research has focused on Herpes Simplex Virus 1 and 2 (especially Herpes Simplex encephalitis [HSE] and herpesvirus latency), Varicella-Zoster Virus, Human Immunodeficiency Virus (both basic and clinical aspects), HTLV-1, JC Virus the cause of Progressive Multifocal Leukoencephalopathy [PML] in humans), Measles Virus, Poliovirus, Rabies Virus, Borna Disease virus, and animal models of demyelinating disease such as Theiler's virus, Semliki Forest Virus, Coronaviruses, and visna-maedi virus.[3] More recently there has been much interest in "emerging" viral infections of the nervous system such as Nipah virus and West Nile virus.[4,5] Research has also focussed on the host immune response to CNS virus infection, and work on virus-associated apoptosis, cytokines and chemokines is also flourishing.[3,6] The disease and financial burden of Neurovirological disease worldwide is considerable. For example, nervous system involvement in HIV infection is very common with 10% of AIDS patients presenting with neurological features and over 80% being neurologically affected at autopsy.[7] With the introduction of highly active retroviral therapy (HAART) clinicians have also had to deal with the challenge of a changing neurological disease profile of HIV infection. There are also a number of common viruses which affect the CNS such as influenza whose neuropathogenesis is not well understood. In addition, increasing immunosuppression with more effective anti-bacterial therapy has increased interest in better treatment of neurovirological disease, while more widespread international air travel have also raised the profile of viral infections of the nervous system arising in African and Asia such as Japanese Encephalitis.[8] Thus both host and viral factors may be contributing to the changing pattern of neurovirological disease.

Fortunately these changes in disease patterns have been accompanied by increasingly sophisticated investigative tools including the molecular analysis of both the virus and the host response to viral infection. In this brief overview of the field the approach will be to define briefly most of the major neurovirological techniques, both "classical" and more recent (Table 1), and then to indicate how these have been used to gain knowledge pertaining to pathogenesis, clinical investigation and in some cases treatment of viral infections of the CNS. The coverage is not exhaustive, but

Table 1 Summary of techniques described in this article.

Classical	Modern
Tissue culture of the nervous system	Polymerase Chain Reaction (PCR)
Virus isolation	Molecular hybridisation methods including *in situ* hybridisation
Serological studies	Gene microarrays
Viral antigen detection	Molecular analysis of viral isolates Use of viral mutants in neuropathogenesis studies Viral vectors for gene therapy in the CNS.

emphasis will be given to the newer approaches such as advanced molecular hybridisation techniques, gene microarrays and viral vectors to deliver therapeutic genes to the CNS.

SPECIFIC TECHNIQUES AND THEIR APPLICATIONS

Classical Techniques

For many years what I refer to as "classical" techniques have been used in Neurovirology, which will be mentioned here briefly.

(i) Tissue culture of the nervous system. Cultures of dissociated or explanted CNS or PNS tissues containing identifiable neural cells provide a useful *in vitro* tool for investigating the virus-neural cell interaction where the environmental conditions can be precisely controlled. While such a system can never mimic the complexities of the *in vivo* situation, such studies have provided a wealth of useful information. For example, the differential susceptibility of human glial cells and neurons to HSV-1 infection has been defined in this way,[9,10] infection of dissociated neural cells by HIV has been investigated,[11] *in vitro* models of HSV latency have been devised,[12] and the effects of either toxic viral products or drugs on neural cells can be assessed.

(ii) Virus isolation. Isolation of viruses from human tissues and body fluids is a classical method of establishing the viral aetiology of a neurological condition. Viruses can be isolated through their production of a characteristic cytopathic effect on susceptible indicator cells, or the virus can be identified using electron microscopy. The latter was used, for example, to demonstrate the papovavirus virus cause of PML,[13] and such methods have also been used to demonstrate HSV in brain biopsies of patients with HSE.[3,14] More recently molecular analysis of viral isolates has been important but this will be discussed below. Potential problems with interpretation of virus isolation results include the possibility that they may represent an irrelevant and/or co-existing virus infection, the presence of asymptomatic virus secretion by the patient, accidental viral contamination, or the presence of a viral reactivation rather than a primary infection.[15] Numerous different viruses have been isolated over the course of many years from the tissues of patients with Multiple Sclerosis (MS).[16] However, while there may well be a viral aetiology of MS, these reports have generally been viewed with considerable scepticism and there is, in the author's opinion, no hard evidence to date for there being a direct viral cause of MS.[17] It is quite possible, however, that this question may be clarified in the future by the use of recently developed molecular technology, and the current interest in the possible role of HHV-6 in MS[18] is certainly intriguing and worthy of further detailed study.

(iii) Serological analyses. A significantly rising viral antibody titre in paired serum or CSF samples provides convincing evidence of a recent virus infection as being the cause of a neurological illness, and has also been a useful tool in epidemiological studies. While the Polymerase Chain Reaction (PCR) (see below) has to a large extent superseded serology for CNS viral diagnosis, nevertheless the latter is still widely employed. Potential problems with interpretation of antibody measurements include non-specific polyclonal activation of virus resulting from a generalised immune response to infection, persisting viral antibody levels from a previous infection, inadequate test sensitivity and specificity, and the practical problems of obtaining the specimens and making important decisions in an acutely ill patient many days before the antibody results are available.[15]

(iv) Viral antigen detection. Identification of viral antigens in tissues and body fluids of patients provides strong evidence of the involvement of a virus in an illness, although some of the caveats mentioned in

(ii) and (iii) above also pertain here, especially the potential problems of viral contamination and a co-existing and aetiologically irrelevant viral infection. The techniques used include enzyme-linked immunosorbent assays (ELISA), Western blotting for viral proteins, immunofluorescence and immunocytochemistry.[15] Some of these techniques have been used, for example, for the diagnosis of HIV and HSV infections, and also in viral pathogenetic studies.[15,19,20] Particularly useful is double-label immunofluorescence or immunocytochemistry whereby two viral antigens can be co-localised within the same tissue region or cell or where a viral antigen can be visualised within an antigenic marker-identified neural cell type eg an HSV antigen within a human astrocyte during natural or experimental viral infection.[9] Evidence of viral protein expression is critical in assessing the nature of the infection such as a latent (with no or restricted protein expression) versus a productive infection.[21] Also, a knowledge of which viral proteins are expressed during infection may allow the development of non-live or subunit vaccines directed against those proteins in order to generate anti-viral immunity.

Polymerase Chain Reaction (PCR)

PCR has revolutionised both CNS viral diagnosis and pathogenetic studies (see article by Morrison in ref-22 for detailed description). PCR uses oligonucleotide primers, an ingenious thermal cycling method and a specific thermostable DNA polymerase to provide rapid and massive amplification of target nucleic acid (DNA or cDNA) PCR can identify a single viral genome in a tissue specimen containing many thousands of cells, and the specificity and sensitivity can be even further increased by various modifications.[22] A significant problem, however, with PCR is potential contamination during the procedure, and there are also a number of technical problems related to choice of the most appropriate primer sequences.[22] A further advantage is that reverse-transcription (RT)-PCR can be used to detect viral RNA which can also be quantitated as in the case of measuring viral load in patients with HIV infection.[23] A recently developed and extremely sensitive technique known as real-time PCR uses a special TaqMan PCR system to quantitate viral DNA or RNA in terms of viral copies.[24] High-throughput real-time RT-PCR is also proving of value in diagnostic studies where a range of viruses can be looked for in patient samples. PCR is also used in other molecular techniques relevant to Neurovirology such as DNA sequencing and cloning of genomic DNA.[22]

PCR has proved to be of great value in the rapid and specific diagnosis of CNS viral infections such as HSV-1, Cytomegalovirus (CMV), VZV and enteroviruses, and is now the viral diagnostic method of choice in such diseases. The utility of CSF PCR in the diagnosis of HSE has been particularly well studied with a specificity in experienced laboratories of over 95%.[19,25] False negative results are most likely to occur if the CSF sample is obtained within the first 24–48 hours or after 10–14 days from the onset of the illness, but the sensitivity of the test is otherwise about 95%.[25,26] PCR has also been a frequently used tool for examining patients' tissue samples for the possible presence of viruses.[3,20] While such an approach can reveal potentially valuable information about a possible viral component to pathogenesis, care must be taken to avoid contamination, and caution is also required in the interpretation of either negative or positive results.

Molecular Hybridisation Studies

Several techniques which are used in conventional molecular biology have also been of great value in Neurovirological studies, and these will be briefly summarised here.

(i) Southern and Northern blots. These techniques detect DNA and RNA respectively. In both cases target nucleic acids are prepared from the relevant tissues under test, then immobilised on a hybridisation membrane e.g. a nylon filter, and the nucleic acid fragments hybridised with a radiolabelled molecular probe, with the result visualised autoradiographically.[20,27] Quantitation of nucleic acids is possible with the use of careful controls; however, but Northern blots require relatively large quantities of RNA and both techniques are more research than clinical tools.

(ii) *In situ* hybridisation (ISH). This is one of the most useful techniques currently used in viral pathogenesis studies. Nucleic acid probes, either radiolabelled e.g. with ^{35}S, or chemically labelled e.g. with digoxigenin, are pre-treated and hybridised *in situ* with sections of relevant tissues and/or cell cultures fixed onto glass slides.[27,28] After rigorous washing, the hybridised tissues or cells are processed autoradiographically so that the presence, cellular distribution and location of viral DNA or RNA (depending on the particular ISH methodology) within the specimen can be determined. ISH is more sensitive than Southern and Northern blotting, being able to detect a few copies (typically 10 or less) of the viral genome per cell.[27] However, the technique is labour-intensive, difficult, and very prone to sampling error and artefacts so that both care and experience are necessary when interpreting the results. ISH has been used extensively in attempts to identify particular viral nucleic acid sequences e.g. for VZV and measles viruses,[28,29] within nervous tissues from both normal individuals and patients with neurological conditions. It has also been particularly useful in studies of VZV latency in both humans[30,31] and animals.[32]

(iii) PCR in situ amplification. This is a recently developed technique which combines the exquisite sensitivity of PCR with the cell localising ability of ISH. The technique uses a specially modified PCR machine which carries out thermocycling on glass slides. Due to increased sensitivity, tissues which are negative for the DNA of a particular virus using conventional ISH may test positive when analysed by PCR *in situ* as has recently been demonstrated in the case of latent VZV in human trigeminal ganglia.[28] Although the technique is far more sensitive than ISH alone, a major drawback with PCR *in situ* is that it is very prone to artefacts so that great caution must be taken to include rigorous control procedures. RT-PCR *in situ* amplification to detect RNA in specific tissue areas has also been described. In our experience both of these techniques are capricious and difficult, but when rigorously controlled may reveal very useful information.

(iv) Combined ISH and immunocytochemistry. This is another useful but technically difficult technique in which tissues or cells under test are first labelled immunocytochemically with specific antibodies against eg viral proteins or cellular constituents, and are then processed for ISH as described above.[33] Using this technique it is possible to colocalise viral nucleic acids and proteins in the same cell, and also to identify nucleic acids in marker-identified neural cell types. Both types of information can contribute to our understanding of the possible ways in which the virus produces neurological damage.

Gene Microarrays

The advent of gene microarrays for analysing thousands of different genes simultaneously represents one of the most exciting and promising advances in basic and applied molecular biology in recent years. The detailed technology and potential applications of microarrays will be covered in another article in this series, and has been the subject of extensive recent reviews (see for example Refs. 34 and 35), so only a brief outline of this will be given here, concentrating on its application to Neurovirology. Whereas the molecular hybridisation techniques described above are designed to use a single probe to detect one target gene at a time, the microarray system contains thousands of different test genes, consisting of virus or host-specific nucleic acid sequences, contained on a single and very small silica or glass slide.[34,35] The two main types of microarrays are (i) oligonucleotide arrays consisting of thousands of viral or other oligonucleotides corresponding to specific open reading frames (ORF's), representing different genes, which are printed onto a solid e.g. silica, slide, and (ii) DNA microarrays in which DNA in the form of cDNA, PCR products, or oligonucleotides, are spotted onto either glass slides or nylon membranes.[34,35] During this procedure, RNA is extracted from the tissue or other test sample, labelled with a fluorescent dye, and then exposed to the microarrays to allow hybridisation with the various target spots. The method includes a sophisticated computer-assisted analysis of the intensity of each array spot after hybridisation so that a quantitative read-out is obtained. Significant fold increases in array spot intensity are then computed allowing an analysis of the expression of thousands of different genes simultaneously. The potential applications of this methodology are numerous and array technology is rapidly becoming more sophisticated in its efficiency and scope.[35] In Neurovirology, there are several important applications of microarrays. Since the entire genome of a virus can be represented by hundreds or thousands of probe spots on a microarray, it is possible to study the function of multiple genes during various types of viral infection. For example, in the case of HSV-1, the expression of all the viral genes can be studied during different stages of an acute lytic infection of neural or other susceptible cells allowing a detailed transcriptional analysis leading to advances in our understanding of viral gene function. During a latent viral infection due to e.g. HSV-1 or VZV it should be possible to determine the full extent of viral gene expression using microarrays. Until the present time this had only been possible using conventional molecular hybridisation techniques to detect single viral genes at a time.[28,30] Knowledge of which viral genes are expressed during a latent infection should facilitate the development of anti-viral strategies against specific viral target genes and their products.[36] Microarrays can also be used to study multiple host cell functions during lytic and latent viral infections. The expression of many thousands of different host genes subserving a myriad of functions e.g. genes encoding cell replication and degradation, various neural functions, signal transduction, stress responses, and immune responses, can be studied simultaneously on the same specimen using microarrays. For example it has recently been possible to use arrays to analyse in great detail host cell gene transcription during acute VZV infection of human T cells and fibroblasts using a 40 000-spot human cDNA microarray system.[37] Similar experiments should be possible in lytic infection of neural cells and latent viral infections. Microarrays also have great promise in the diagnosis of infectious diseases including those caused in the CNS by viruses. A number of pathogen arrays have already been constructed including those for bacteria and viruses.[35] Using virus pathogen "chips" it should be possible to screen patients' sera and CSF for many different viruses simultaneously and quickly. At present array technology is extremely expensive, but over time it is to be hoped that the cost-effectiveness of this technology will increase and the basic costs decrease. Many advances in our understanding of the host and viral responses during viral infection of the CNS can be expected over the next decade (see also later). In addition, rapid advances in "proteomics", including the development of protein microarrays, are likely to greatly clarify many host-virus interactions and will also provide valuable information on a variety of protein–protein interactions[35] which will help us to understand in more detail how viruses affect neural cells. In a wider sense this technology should allow the detection of genetic and environmental factors which influence human susceptibility to viral infections of the nervous system. Further, in the future, microarray-based PCR analysis of samples obtained from, for example, encephalitis of unknown origin, may prove a previously unsuspected, or confirm a suspected, virus link to that condition-a form of "virus discovery" that can be applied to many different neurological conditions.

Molecular Analysis of Viral Isolates

The molecular analysis of viral isolates from patient's tissues has for many years been a useful tool in both viral diagnosis and pathogenesis. A seminal example was provided by the Whitley and colleagues[38] who used restriction enzyme analysis to characterise paired HSV-1 isolates from oral-labial and brain sites in patients with HSE. By comparing the molecular profiles of isolates from the two sites they were able to conclude that HSE can result from a primary HSV-1 infection, a reinfection with a second HSV-1 or a reactivation of latent HSV-1. Another remarkable study in this area was reported by Evans *et al.* who sequenced the genomes of daily faecal poliovirus isolates from infants who had received oral Sabin type 3 vaccine virus.[39] As the vaccine strain passed through the infant's gut it was found to be rapidly mutated to a neurovirulent strain which was attributed to a single nucleotide change of uridine to cytidine in the 5′ non-coding region of the genome which presumably altered the secondary structure of the virus.[39] Thus a very small change in the viral genome had a profound effect on the neurovirulence of the virus.[20] This type of approach has also been used to study the "molecular epidemiology" of virus infections. A very recent and elegant example of this was provided by the study by Quinlivan *et al.*[40] in which restriction enzyme analysis (using the BglI site of gene 54) of VZV isolates was carried out in patients from the UK, North and South America, Africa, Asia and the Far East. Significant differences were found in the positivity for this particular site in the various patient populations, with, for example, <22% of VZV strains from UK and American patients being positive compared with a positivity rate of 98.6% for patients in the other areas. Thus this analysis proved that there is strong evidence for geographical segregation of different VZV strains. Clearly this might account for differences in the clinical presentations of varicella and herpes zoster in different parts of the world. Restriction enzyme analysis has also been used to characterise viral isolates in the trigeminal and autonomic ganglia in individuals latently infected with HSV-1,[41] allowing conclusions to be made regarding viral neuropathogenesis.

Use of Viral Mutants in Pathogenesis Studies

The use of specially constructed or naturally occurring viral mutants has been a highly productive method of studying viral determinants of neuropathogenesis. Such mutants have included temperature-sensitive mutants which replicate at 31°C but not 38.5°C, deletion mutants with larger changes in their genomic structure, and viral antigenic variants produced by exposure to monoclonal antibodies of varying specificities.[20,27] A few examples will be given briefly here to indicate their utility. In all cases a general principle is that a specific molecularly characterised viral mutation is correlated with a loss or alteration of a neurological disease phenotype such as neurovirulence or latency, thereby allowing the attribution of the particular impaired or absent function to the deleted gene sequence.[20] It should then be possible to reinsert the missing viral gene to restore the specific biological function to prove a cause- and effect relationship. Such mutants have been very useful in studying HSV-1 latency. For example, HSV-1 mutants which have had the LAT (latency-associated transcripts) gene deleted show an altered phenotype in which the virus takes longer than usual to reactivate from latency in experimental mouse ganglia, this indicating the importance of the LAT gene in viral reactivation from latency.[42] In human latency, Brown *et al.*[43] used a number of temperature-sensitive HSV mutants to "superinfect" explanted human ganglionic tissues in order to detect latent HSV genetic information in the ganglia. This was due to interactions between the input superinfecting virus and the genome of the latent virus within the ganglia. Deletion mutants of certain strains of polioviruses have been used to show that the 5′ non-coding region of the viral genome is important in producing poliovirus neurovirulence in experimental monkeys.[27,44] Monoclonal antibody-derived viral variants of rabies virus have been used to show that the presence of a positively charged amino acid (arginine or lysine) in position 333 of the rabies virus glycoprotein is necessary for viral neurovirulence in mice.[27,45] It is remarkable how such small alterations in the viral genome can result in such a disrupted neurological phenotype. Fields *et al.* carried out extensive studies in the nervous system using reassortant mutants containing segmented gene regions from reovirus types 1 and 3.[46] The generation of these mutants with reassorted genomes allowed the correlation of defined genomic regions with particular neuropathogenic effects. For example, the reovirus S1 gene, which encodes the viral cell attachment protein known as sigma-1, is the key determinant of whether the virus spreads via neural or haematogenous pathways, and whether the virus infects neurons or ependymal cells in mouse brain.[46,47] A general principle which has emerged from this type of study, and others, is that the basis of viral neurotropism is the interaction of viral attachment proteins to specific cell surface receptors on the target neural cell.[27]

Viral Vectors for Gene Therapy in the CNS

This is one of the most important and rapidly developing areas in Neurovirolgy and some general principles and examples will be given (see Ref. 48 for a detailed discussion). The enormous advances in molecular genetics and biochemistry, virology and our understanding of the molecular basis of neurological diseases that have occurred over the last two decades underly the current interest and feasibility of gene therapy. Viral vectors represent one of the most prevalent methods of delivering exogenous genes into the nervous system because of the ability of certain viruses to spread along neural pathways to infect particular regions of the CNS. The general principle is to disable the virus using targeted gene deletions so that the virus retains its "useful" functions such as the ability to spread along neurons and attach to and enter target neural tissues, but is sufficiently "disabled" by the gene deletion to lose its ability to replicate in those cells, elicit an immunological response, and to produce disease. The viral vector, as it is called, is able to cross the cell membrane of the target cell to which the delivered foreign gene is inserted, something which enzymes by themselves cannot typically do. The foreign gene must then be transiently or stably expressed, with both transcription and translation of the genome occurring producing RNA and protein respectively. In this way the hope is that a therapeutic effect can be produced. There are various potential strategies for gene therapy which are summarised in Table 2, and some are probably more feasible than others. It is important to mention that some forms of what is termed "gene therapy" consist of direct cell killing of a tumour by a disabled virus which only grows in rapidly dividing tumour cells (see later). In some cases a viral vector can be used to insert a drug susceptibility gene into a tumour which is then killed when the drug is administered systemically. A good example of this is the insertion of the HSV-1 thymidine kinase gene into a tumour which is then destroyed after treatment with the anti-HSV drug acyclovir.[48] While numerous studies of viral vector-mediated gene therapy have been performed in experimental animals, there is now an increasing number of clinical trials of such treatment in patients.

There are many technical aspects of gene therapy which are becoming increasingly sophisticated and will not be discussed here. However, it should be emphasised that these problems are formidable, and it is still not known just how successful this new form of treatment is ultimately going to be, or how many neurological diseases will prove to be amenable to gene therapy. If one were to design the ideal viral vector for gene therapy then it would possess all the properties listed in Table 3. Two of the most problematic issues appear to be maintaining longterm expression of the delivered gene and controlling the level of expression of the gene in the target tissues. The main viruses which have been used to date for such therapy include HSV-1, adenovirus, adeno-associated virus and retroviruses, the relative merits of which have been described in detail elsewhere.[48] HSV-1 has the particular advantages of having a very large genome which can be

Table 2 Strategies for gene therapy in the CNS using viral vectors.

(1) Replace missing or defective gene
(2) Replace or enhance local growth factors or enzyme production
(3) Virus-directed enzyme pro-drug therapy
(4) Direct cell killing of CNS tumours
(5) Delivery of anti-sense sequences to particular genes
(6) Use of inserted DNA to increase cellular antigen expression, to boost immune response.

Reprinted by permission of Oxford University Press from Ref. 48.

Table 3 Ideal properties of viral vectors for gene therapy.

(1) Enough capacity of virus to package foreign gene
(2) Vector must be effectively delivered to target cells
    (i) Appropriate route of delivery
    (ii) No vector-induced damage along route or at target
    (iii) Vectors must remain intact during delivery
    (iv) Vector must target appropriate neural cells
(3) No viral replication and/or reactivation at target tissue site
(4) Foreign gene must be stably expressed in target tissue
(5) Must have ability to control level of expression of the gene products
(6) Vector should not elicit immune responses in the host
(7) Appropriate *in vitro* and *in vivo* models should be available prior to human therapy.

Reprinted by permission of Oxford University Press from Ref. 48.

easily "engineered" to remove relevant "harmful" regions, and the ability to spread along neural pathways and then produce a non-pathogenic latent infection in neurons.[21,48,49] Viral vectors need to contain molecular elements which drive transcription of the exogenous gene, and in practice the vectors contain both viral and cellular promoters, the former controlling transcription of the inserted gene and the latter, in principle, controlling its cell-type specificity,[48,49] although in practice these promoters do not necessarily function in this way as well as might be predicted.

There are several human neurological diseases which could potentially benefit from the gene therapy approach using viral vectors, including Parkinson's disease, malignant brain tumours, motor neuron disease, inherited conditions due to single gene defects, demyelinating diseases, cerebrovascular disease and acute CNS trauma.[48] However, in my view, the two conditions for which gene therapy is most likely to be successful are Parkinson's disease and malignant gliomas. In the former case, for example, the gene coding for tyrosine hydroxylase can be delivered to the affected basal ganglion region with a resultant increase in brain dopamine levels to produce clinical improvement.[48,50,51] Regarding brain tumours, clinical trials are now well under way to investigate the safety and efficacy of direct tumour killing by mutant HSV-1. In Glasgow, for instance, the HSV 1716 mutant, which is cell state- and cell type-specific, replicating in rapidly dividing tumour cells, but not normal cells,[52] has been used in clinical trials in patients with grade IV glioma since 1997.[53] The treatment was found to be safe, and 3 patients continue to survive for 3–4 years which is very promising, although further ongoing studies will be required to determine the future role of this treatment in such patients. It has already been established that HSV 1716 replicates in high-grade glioma without causing toxicity in both HSV-seropositive and seronegative patients,[53,54] and this approach may very well lead to important adjunct therapy.

Potential Applications of these Technologies over the Next 10 years

In Table 4, I have listed my best guess of the most likely applications of these various techniques in Neurovirology in both clinical and neuropathogenetic aspects. As can be seen these are somewhat interconnected and certainly not mutually exclusive.

Table 4 Possible major applications of techniques in Neurovirology over the next 10 years.

1. Greatly improved viral diagnostic advances, especially from high throughput gene microarrays.
2. Effective gene therapy for some neurological conditions using viral vectors.
3. Effective use of gene microarrays and proteomics to identify and utilise expressed viral genes as targets for anti-viral therapy.
4. Greatly increased understanding of neuropathogenesis of viral infections of the nervous system.
5. Greatly increased understanding of the alterations in host gene function in some neurovirological diseases.
6. Increased understanding of the molecular evolution of viruses affecting the CNS.
7. Major advances in virus discovery in nervous system diseases currently of unknown aetiology.
8. Discovery of host genetic susceptibility to virus infection by array analysis.

Glossary

Neuropathogenesis	The mechanism(s) by which a virus is able to cause disease in the nervous system.
PCR	Polymerase Chain Reaction. An extremely sensitive molecular technique for amplifying small amounts of specific DNA .
RT-PCR	Reverse-transcription Polymerase Chain Reaction. A type of PCR which amplifies RNA rather then DNA.
ISH	*In situ* hybridisation. A sensitive molecular technique which can detect small amounts of DNA or RNA in a localised region of tissue.
PCR-*In situ* amplification	A technique in which DNA is amplified by PCR within a localised region of tissue, giving greater sensitivity than ISH alone.
Gene microarray	A method of detecting thousands of viral or host genes simultaneously on a chip or slide, of great use in analysing gene expression.

Viral mutant	A virus which has had a part of its genetic structure altered leading to a detectable phenotype.
Neurotropism	The ability of a virus to bind to a particular neural cell type.
Neurovirulence	The ability of the virus to produce actual neurological disease
Gene Therapy	The delivery of an exogenous gene to an organ resulting in expression of the introduced gene leading to a therapeutic effect.
Viral Vector	A virus which has been genetically modified so that it can deliver an exogenous therapeutic gene to a target organ without producing deleterious effects.

Useful website

www.isnv.org; Official website of the International Society for Neurovirology, with details of current activities and the Journal of NeuroVirology.

ACKNOWLEDGEMENTS

I wish to thank Dr Randall Cohrs for critical reading of the manuscript. Personal research described here has been supported by the Wellcome Trust and the Chief Scientist's Office (Scotland).

REFERENCES

1. Johnson RT. Neurovirology: evolution of a new discipline. *J NeuroVirol* 1995; **1**:2–4.
2. Whitley RJ, Alford CA, Hirsch MS, *et al.* Vidarabine versus acyclovir therapy in herpes simplex encephalitis. *N Engl J Med* 1986; **314**:144–149.
3. Johnson RT. *Viral Infections of the Nervous System*, 2nd edn. Philadelphia: Lippincott-Raven, 1998.
4. Chua KB, Bellini WJ, Rota PA, *et al.* Nipah virus: a recently emergent deadly paramyxovirus. *Science* 2000; **288**:1432–1435.
5. Nash D, Mostashari F, Fine A, *et al.* The outbreak of West Nile virus infection in the New York City area in 1999. *N Engl J Med* 2001; **344**:1807–1814.
6. Gendelman HE. Neural immunity: Friend or foe. *J NeuroVirol* 2002; **8**:474–479.

7. Kennedy PGE. Overview of HIV in the nervous system. In: Scaravilli F, ed. *The Neuropathology of HIV Infection*, London: Springer-Verlag Ltd, 1993: 259–265.
8. Solomon T, Vaughn DW. Clinical features and pathophysiology of Japanese encephalitis and West Nile virus infections. In: Mackenzie JS, Barrett AD, Deubel V, eds. *Current Topics in Microbiology and Immunology: Japanese Encephalitis and West Nile Virus Infections.* Berlin:Springer-Verlag, 2002:171–194.
9. Kennedy PGE, Clements GB, Brown SM. Differential susceptibility of human neural cell types in culture to infection with herpes simplex virus. *Brain* 1983; **166**:101–119.
10. Kennedy PGE, Gairns J, MacLean AR. Replication of the herpes simplex virus type 1 RL1 mutant 1716 in primary neuronal cell cultures — possible relevance to use as a viral vector. *J Neurol Sci* 2000; **179**:108–114.
11. Blumberg BM, Gelbard HA, Epstein LG. HIV-1 infection of the developing nervous system: central role of astrocytes in pathogenesis. *Virus Res* 1994; **32**: 253–267.
12. Wilcox CL, Johnson Jr EM. Nerve growth factor deprivation results in the reactivation of latent Herpes Simplex Virus *in vitro*. *J Virol* 1987; **61**:2311–2315.
13. Zu Rhein GM, Chou SM. Particles resembling papovaviruses in human cerebral demyelinating disease. *Science* 1965; **148**:1477–1479.
14. Kennedy PGE. A retrospective analysis of forty-six cases of herpes simplex encephalitis seen in Glasgow between 1962 and 1985. *Q J Med* 1988; **68**: 533–540.
15. Kennedy PGE. The widening spectrum of infectious neurological disease. *J Neurol Neurosurg Psychiat* 1990; **53**:629–632.
16. Jacobson S. Association of human herpesvirus-6 and multiple sclerosis: here we go again? *J NeuroVirol* 1998; **4**:471–473.
17. Kennedy PGE, Steiner I. On the possible viral aetiology of multiple sclerosis. *Q J Med* 1993; **87**:523–528.
18. Kimberlin DW, Whitley RJ. Human herpesvirus-6: neurologic implications of a newly-described viral pathogen. *J NeuroVirol* 1998; **4**:474–485.
19. Baringer JR. Herpes simplex virus encephalitis. In: Davis LE, Kennedy PGE, eds. *Infectious Diseases of the Nervous System*. Oxford:Butterworth-Heinemann, 2000:139–164.
20. Kennedy PGE. Molecular studies of viral pathogenesis in the central nervous system. *J Roy Coll Phys Lond* 1992; **26**:204–214.
21. Steiner I, Kennedy PGE. Molecular biology of Herpes Simplex virus Type 1 latency in the nervous system. *Molec Neurobiol* 1993; **7**:137–159.
22. Morrison KE. Polymerase chain reaction. *Prac Neurol* 2002; **2**:288–293.

23. McArthur JC. HIV-associated dementia. In: Davis LE, Kennedy PGE, eds. *Infectious Diseases of the Nervous System*, Oxford: Butterworth-Heinemann, 2000:165–213.
24. Cohrs RJ, Randall J, Smith J, *et al.* Analysis of individual human trigeminal ganglia for latent Herpes Simplex Virus Type 1 and Varicella-Zoster Virus nucleic acids using real-time PCR. *J Virol* 2000; **74**:11464–11471.
25. Kennedy PGE, Chaudhuri A. Herpes simplex encephalitis. *J Neurol Neurosurg Psychiat* 2002; **73**:237–238.
26. Davis LE. Diagnosis and treatment of acute encephalitis. *Neurologist* 2000; **6**:145–159.
27. Kennedy PGE. The use of molecular techniques in studying viral pathogenesis in the nervous system. *TINS* 1990; **13**:424–431.
28. Kennedy PGE, Grinfeld E, Gow JW. Latent varicella-zoster virus is located predominantly in neurons in human trigeminal ganglia. *Proc Natl Acad Sci USA* 1998; **95**:4658–4662.
29. Cosby SL, McQuaid S, Taylor MJ, *et al.* Examination of eight cases of multiple sclerosis and 56 neurological and non-neurological controls for genomic sequences of measles virus, canine distemper virus, simian virus 5 and rubella virus. *J Gen Virol* 1989; **70**:2027–2036.
30. Kennedy PGE, Grinfeld E, Bell JE. Varicella-Zoster Virus gene expression in latently infected and explanted human ganglia. *J Virol* 2000; **74**: 11893–11898.
31. Lungu O, Annunziato P, Gershon A, *et al.* Reactivated and latent varicella-zoster virus in human dorsal root ganglia. *Proc Natl Acad Sci USA* 1995; **92**: 10980–10984.
32. Kennedy PGE, Grinfeld E, Bontems S, *et al.* Varicella-Zoster Virus gene expression in latently infected rat dorsal root ganglia. *Virology* 2001; **289**:218–223.
33. Kennedy PGE, Narayan O, Ghotbi Z, *et al.* Persistent expression of Ia antigen and viral genome in visna-maedi virus-induced inflammatory cells: possible role of lentivirus-induced interferon. *J Exp Med* 1985; **162**:1970–1982.
34. Aitman TJ. DNA microarrays in medical practice. *BMJ* 2001; **323**:611–615.
35. Kellam P. Post-genomic virology: the impact of bioinformatics, microarrays and proteomics on investigating host and pathogen interactions. *Rev Med Virol* 2001; **11**:313–329.
36. Kennedy PGE. Varicella-zoster virus latency in human ganglia. *Rev Med Virol* 2002; **12**:327–334.
37. Jones JO, Arvin AM. Microarray analysis of host cell gene transcription in response to Varicella-Zoster Virus infection of human T cells and fibroblasts *in vitro* and SCIDhu skin xenografts *in vivo*. *J Virol* 2003; **77**:1268–1280.

38. Whitley R, Lakeman AD, Nahmias A, *et al.* DNA restriction enzyme analysis of herpes simplex virus isolates obtained from patients with encephalitis. *N Engl J Med* 1982; **307**:1060–1062.
39. Evans DMA, Dunn G, Minor PD, *et al.* Increased neurovirulence associated with a single nucleotide change in a non-coding region of the Sabin type 3 polio vaccine genome. *Nature* 1985; **314**:548–550.
40. Quinlivan M, Hawrami K, Barrett-Muir W, *et al.* The molecular epidemiology of varicella-zoster virus: evidence for geographic segregation. *J Infect Dis* 2002; **186**:888–894.
41. Lonsdale DM, Brown SM, Subak-Sharpe JH, *et al.* The polypeptide and the DNA restriction enzyme profiles of spontaneous isolates of herpes simplex virus type 1 from explants of human trigeminal, superior cervical and vagus ganglia. *J Gen Virol* 1979; **43**:151–171.
42. Steiner I, Spivack JG, Lirette RP, *et al.* Herpes simplex virus type 1 latency-associated transcripts are evidently not essential for latent infection. *EMBO* 1989; **8**:505–511.
43. Brown SM, Subak-Sharpe JH, Warren KG, *et al.* Detection by complementation of defective or uninducible (herpes simplex type 1) virus genomes latent in human ganglia. *Proc Natl Acad Sci USA* 1979; **76**:2364–2368.
44. Iizuka N, Kohara M, Hagino-Yamagishi, *et al.* Construction of less neurovirulent polioviruses by introducing deletions the 5′ noncoding sequence of the genome. *J Virol* 1989; **63**:5354–5363.
45. Tuffereau C, Leblois H, Benejean J, *et al.* Arginine or lysine in position 333 of ERA and CVS glycoprotein is necessary for rabies virulence in adult mice. *Virology* 1989; **172:**206–212.
46. Sharpe AH, Fields BN. Pathogenesis of viral infections: basic concepts derived from the reovirus model. *N Engl J Med* 1985; **312**:486–497.
47. Tyler KL. Pathogenesis of reovirus infections of the central nervous system. *Semin Neurosci* 1991; **3**:117–124.
48. Kennedy PGE. Potential use of herpes simplex virus (HSV) vectors for gene therapy of neurological disorders. *Brain* 1997; **120**:1245–1259.
49. Glorioso JC, Goins WF, Meaney CA, *et al.* Gene transfer to brain using herpes simplex virus vectors. *Ann Neurol* 1994; **35**:Suppl:S28–S34.
50. During MJ, Neagele JR, O'Malley KL, Geller AI. Long-term behavioral recovery in parkinsonian rats by an HSV vector expressing tyrosine hydroxylase. *Science* 1994; **266**:1399–1403.
51. Geller AL, During MJ, Oh YJ, *et al.* An HSV-1 vector expressing tyrosine hydroxylase causes production and release of L-dopa from cultured rat striatal cells. *J Neurochem* 1995; **64**:487–496.

52. Brown SM, Harland J, MacLean AR, *et al.* Cell type and cell state determine differential in vitro growth of non-neurovirulent ICP34.5-negative herpes simplex virus types 1 and 2. *J Gen Virol* 1994; **75**:2367–2377.
53. Rampling R, Cruickshank G, Papanastassiou V, *et al.* Toxicity evaluation of replication-competent herpes simplex virus (ICP 34.5 null mutant 1716) in patients with recurrent malignant glioma. *Gene Ther* 2000; **7**:859–866.
54. Papanastassiou V, Rampling R, Fraser M, *et al.* The potential for efficacy of the modified (ICP 34.5^{-}) herpes simplex virus HSV1716 following intratumoural injection into human malignant glioma: a proof of principle study. *Gene Ther* 2002; **9**:398–406.

10

Molecular Neurology of Prion Disease

*John Collinge**

INTRODUCTION TO PRIONS AND HISTORICAL PERSPECTIVE

The prion diseases, or transmissible spongiform encephalopathies, are neurodegenerative conditions that affect both humans and animals. The prototypic disease is scrapie, a naturally occurring disease of sheep and goats, present in many countries worldwide and recognised in Europe for over 200 years. Other animal prion diseases, recognised over the last few decades, include transmissible mink encephalopathy and chronic wasting disease of mule deer and elk — principally in the United States — and, since the 1980's, bovine spongiform encephalopathy (BSE) — first described in the UK and now recognised in most European Union countries, Japan, Canada and the United States. The more recently described feline spongiform encephalopathy of domestic cats and spongiform encephalopathies of a number of species of zoo animals[1] are now also recognised as prion diseases.

*MRC Prion Unit and National Prion Clinic, Institute of Neurology and National Hospital for Neurology and Neurosurgery, Queen Square, London WC1N 3BG.

The human prion diseases have been traditionally classified into Creutzfeldt-Jakob disease (CJD), Gerstmann-Sträussler-Scheinker disease (GSS) and kuru. Although these are rare disorders, affecting about 1–2 people per million world-wide per annum, remarkable attention has been focused on them in recent years. This is because of the unique biology of the transmissible agent or prion, and also because of fears that the epizootic of BSE could pose a threat to public health through dietary exposure to infected tissues.

Scrapie was demonstrated to be transmissible by inoculation between sheep (and goats) following remarkably prolonged incubation periods in 1936. It was assumed that some type of virus must be the causative agent and Sigurdsson coined the term "slow virus infection" in 1954. There was considerable interest in the 1950's in an epidemic of a neurodegenerative disease, kuru, characterised principally by a progressive cerebellar ataxia, amongst the Fore linguistic group of the Eastern Highlands of Papua New Guinea. Subsequent field work suggested that kuru was transmitted during cannibalistic feasts. In 1959 Hadlow drew attention to the similarities between kuru and scrapie at the neuropathological, clinical and epidemiological levels leading to the suggestion that these diseases may also be transmissible.[2] A landmark in the field was the transmission, by intracerebral inoculation with brain homogenates into chimpanzees, of kuru and then CJD by Gajdusek and colleagues in 1966 and 1968, respectively.[3] Transmission of GSS followed in 1981. This work led to the concept of the "transmissible dementias". The term "Creutzfeldt-Jakob disease (CJD)" was introduced by Spielmeyer in 1922 drawing from the case reports of Creutzfeldt (1920) and Jakob (1921) and was used in subsequent years to describe a range of neurodegenerative conditions, many of which would not meet modern diagnostic criteria for CJD. Interestingly, Jakob suspected that the condition may be transmissible and experimentally inoculated rabbits in an attempt to demonstrate this in the 1920's. This was unsuccessful and we now know that rabbits are unusually resistant to prion infection.

The criterion of transmissibility allowed diagnostic criteria for CJD to be assessed and refined. All the animal and human conditions share common histopathological features. The classical diagnostic triad of spongiform vacuolation (affecting any part of the cerebral grey matter), neuronal loss and astrocytic proliferation may be accompanied by amyloid plaques.

The nature of the transmissible agent has been a subject of intense and heated debate for many years. The initial assumption that it must be viral was challenged, however, both by the failure to directly demonstrate a virus (or an immunological response) and because the transmissible agent was resistant to treatments which inactivate nucleic acids (such as ultraviolet radiation or treatment with nucleases). These remarkable findings led to suggestions in 1966 by Tikvar Alper and others that the transmissible agent may be devoid of nucleic acid[4,5] and led John Griffith to suggest in 1967 that the transmissible agent may in fact be composed entirely of protein.[6] In this remarkable letter to *Nature*, he proposed three hypothetical mechanisms for propagation of such an agent, one of which closely mirrors current thinking; indeed his model also presciently predicted the existence of distinct strains of agent. Needless to say, such a proposal met with great scepticism at the time, in what was the heyday of the "central dogma" of biology: that DNA encodes RNA that in turn encodes protein.

More than a decade later this remarkable proposal was lent biochemical credibility by intensive purification studies allied with laborious rodent bioassay. Progressive enrichment of brain homogenates for infectivity resulted in the isolation of a protease-resistant glycoprotein — designated the prion protein (PrP) by Prusiner and co-workers in 1982. This protein was the major constituent of infective fractions and was found to accumulate in affected brains and sometimes to form amyloid deposits. The term prion (from *pro*teinaceous *in*fectious particle) was proposed[7] to distinguish the infectious pathogen from viruses or viroids. Prions were defined as "small proteinaceous infectious particles that resist inactivation by procedures which modify nucleic acids".

The protease resistant PrP extracted from affected brains was of 27–30 kDa and became known as PrP.$^{27-30}$ Determination of the some of the amino acid sequence of PrP$^{27-30}$ enabled production of oligonucleotides that were used to screen cDNA libraries prepared from scrapie-infected hamsters. These studies led to the recovery of cognate cDNA clones by Weissmann and colleagues in 1985. While it had been previously assumed that PrP would be virally encoded, remarkably, PrP$^{27-30}$ was encoded by a single copy host chromosomal gene rather than by a putative viral nucleic acid in fractions enriched for scrapie infectivity. It then became clear that PrP$^{27-30}$ was derived from a larger molecule of 33–35 kDa designated PrP^{Sc} (denoting the *Sc*rapie isoform of the protein).[8] The normal product of the

PrP gene however is protease sensitive and was designated PrP^C (denoting the Cellular isoform of the protein). No differences in amino acid sequence between PrP^{Sc} and PrP^C or indeed any consistent covalent differences have been identified and it is clear that PrP^{Sc} is derived from PrP^C by a post-translational process.[9,10]

Clinical neurology and neurogenetics have played a major role in the evolution of our understanding of the pathobiology of prion disease. Study of the various forms of human prion disease has been crucial, notably the recognition that the familial forms of the human diseases — already known to be transmissible to laboratory animals by inoculation — are in fact autosomal dominant inherited conditions associated with coding mutations in the gene encoding PrP^C (designated *PRNP*),[11,12] The very strong genetic linkage and high or complete penetrance of some such mutations[13,14] left no need to propose any external infectious agent and argued strongly for what had become known as the "protein-only" hypothesis of prion propagation.

MOLECULAR BIOLOGY OF PRIONS

Nature of the Infectious Agent

A number of converging lines of experimental evidence now firmly support the idea that prions consist principally, or entirely, of abnormal isoforms of host encoded prion protein (for review see[15]). PrP^{Sc} is derived from PrP^C by a post-translational mechanism[9] and no covalent differences[10] between PrP^C and PrP^{Sc} have been demonstrated. It is proposed that PrP^{Sc} acts as a template which promotes the conversion of PrP^C to PrP^{Sc} and that the difference between these isoforms lies in their conformation and state of aggregation.

The three dimensional fold or conformation of the cellular isoform was first established by NMR spectroscopy of recombinant mouse PrP.[16] Subsequent studies on hamster, human and other mammalian PrP's show that they have essentially the same conformation. Following cleavage of signal sequences, the mature PrP^C consists of an N-terminal region of about 100 amino acids (which is unstructured in the isolated molecule in solution) and a C-terminal domain, also of around 100 amino acids. The C-terminal domain is composed largely of α-helical structure (with three α-helices and a short anti-parallel β-sheet) stabilised by a single disulphide bond linking helices 2 and 3. PrP is a glycoprotein with two asparagine-linked glycosylation sites and is attached to the external cell surface via a glycosylphosphatidylinositol (GPI) anchor (see Fig. 1).

The N-terminal region contains five repeats of an eight-amino acid sequence (the octapeptide-repeat region). Mutations in this region, resulting in addition of integral numbers of additional repeats, lead to forms of inherited prion disease. While unstructured in the isolated molecule, this region is very highly conserved in evolution and contains two tight binding sites for Cu^{2+} ions.[17] It is proposed that the unstructured N-terminal region may acquire structure following copper binding and a role for PrP in copper metabolism or transport is possible. Disturbance of this function by the conformational transitions between isoforms of PrP could be involved in prion-related neurotoxicity.

PrP^{Sc} is extracted from affected brains as highly aggregated, detergent insoluble, material that is not amenable to high-resolution structural techniques. However, Fourier transform infrared (FTIR) spectroscopic methods show that PrP^{Sc}, in sharp contrast to PrP^{C}, has a high β-sheet content.[18] The

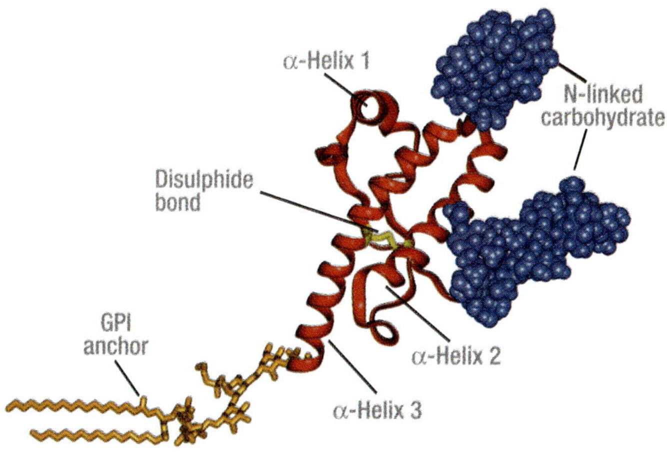

Fig. 1 Model of the C-terminal domain of human prion protein indicating positions of N-linked glycans (in blue), the single disulphide bond joining helixes 2 and 3 and the GPI anchor which attaches the protein to the outer surface of the cell membrane.

underlying molecular events during infection which lead to the conversion of PrP^{C} to the scrapie agent remain ill-defined. The most coherent and general model thus far proposed is that the protein, PrP, fluctuates between a dominant native state, PrP^{C}, and a series of minor conformations, one or a set of which can self-associate in an ordered manner to produce a stable supra-molecular structure, PrP^{Sc}, composed of misfolded PrP monomers. Once a stable "seed" structure is formed PrP can then be recruited leading to an explosive, auto-catalytic formation of PrP^{Sc}.

Little is known for certain about the molecular state of the protein that constitutes the self-propagating, infectious particle itself. There are examples of infectivity in the absence of detectable PrP^{Sc},[19–22] and different strains of prions (see below) are known to differ in their degree of protease resistance. A single infectious unit corresponds to around 10^5 PrP molecules.[23] It is unclear whether this indicates that a large aggregate is necessary for infectivity, or at the other extreme, whether only a single one of these PrP^{Sc} molecules is actually infectious. This relationship of PrP^{Sc} molecules to infectivity could simply, however, relate to the rapid clearance of prions from the brain known to occur following their intracerebral inoculation.

The difficulty in performing structural studies on native PrP^{Sc} has led to attempts to produce soluble β-sheet rich forms of PrP, which may be amenable to NMR or crystallographic structure determination. It is now recognised that the adage regarding protein folding: "one sequence, one conformation", is not strictly true. Depending on solvent conditions, probably any protein chain can adopt a variety of conformations in which there is a degree of periodic order (that is extensive regions of secondary structure). However such alternative states do not have precisely and tightly packed side chains which are the hallmark of the native state of orthodox globular proteins.

Studies on a large fragment of the human prion protein (PrP^{91-231}) have shown that at acidic pH PrP can fold to a soluble monomer comprised almost entirely of β-sheet in the absence of denaturants (Jackson *et al.* 1999). Reduction of the native disulphide bond was a prerequisite for β-sheet formation and these observations of alternative folding pathways dependent upon solvent pH and redox potential could have important implications for the mechanism of conversion to PrP^{Sc}. Indeed this monomeric β-sheet state was prone to aggregation into fibrils with partial resistance to proteinase K digestion, characteristic markers of PrP^{Sc}. Unusually for a protein with a predominantly helical fold, the majority of residues in PrP^{91-231} have a preference for β-conformation. In view of this property, it is possible that the PrP molecule is delicately balanced between radically different folds with a high energy barrier between them; one dictated by local structural propensity (the β-conformation) and one requiring the precise docking of side-chains (the native α-conformation). Such a balance would be influenced by mutations causing inherited human prion diseases.

The precise sub-cellular localisation of PrP^{Sc} propagation remains controversial. However, there is considerable evidence implicating either late-endosome-like organelles or lysosomes.[24–27] The environments of these organelles are evolved to facilitate protein unfolding at low pH prior to degradation by acid-activated proteases. It is possible that the α-PrP to β-PrP conversion, caused by reduction and mild acidification, is relevant to the conditions that PrP^{C} would encounter within the cell, following its internalisation during re-cycling. Such a mechanism could underlie prion propagation, and account for the transmitted, sporadic and inherited aetiologies of prion disease (see Table 1 and Fig. 2). Initiation of a pathogenic self-propagating conversion reaction, with accumulation of aggregated β-PrP, may be induced by exposure to a "seed" of aggregated β-PrP following prion inoculation, or as a rare stochastic conformational change, or as an inevitable consequence of expression of a pathogenic PrP^{C} mutant that is predisposed to form β-PrP. It remains to be demonstrated whether such alternative conformational states of the protein are sufficient to cause prion disease in an experimental host or whether other cellular co-factors are also required.

Prion Strains

In common with other pathogens, distinct naturally occurring isolates or strains are observed. However, it has been unclear how such strain properties could be encoded within an agent devoid of nucleic acids. Dickinson and colleagues isolated multiple distinct strains of naturally occurring sheep scrapie in mice. Such strains are distinguished by their biological properties: they produce distinct incubation periods and patterns of neuropathology in inbred lines of laboratory mice (for review see[28]). As they can be serially propagated in inbred mice with the same *Prnp* genotype

Table 1 Classification of human prion disease

Aetiology	Phenotype	Frequency
Sporadic Unknown: random distribution worldwide; incidence of 1–2 per million per annum	Sporadic CJD: sub-acute myoclonic form and range of atypical forms; multiple distinct prion strains associated with distinct clinicopathological phenotypes which includes sporadic fatal insomnia	~85%
Inherited Autosomal dominantly inherited conditions with high penetrance; all forms have germline *PRNP* coding mutations	Extremely variable: readily mimics familial Alzheimer's disease and other neurodegenerative conditions; over 30 mutations identified; includes Gerstmann-Sträussler-Scheinker disease (GSS), familial CJD and fatal familial insomnia	~10–15%
Acquired Iatrogenic infection with human prions via medical or surgical procedures: human cadaveric-derived pituitary hormones, tissue grafts and contaminated neurosurgical instruments	Iatrogenic CJD: typical CJD when direct CNS exposure; ataxic onset when peripheral infection	<5% (most from USA, UK, France and Japan)
Exposure to human prions via endocannibalism	Kuru	Unique to small area Papua New Guinea; major epidemic in 1950's with gradual decline since cessation of cannibalism
Environmental exposure (presumed dietary) to BSE prion strain; probable secondary transmission via blood transfusion	Variant CJD	Mainly UK (total to date ~150), 7 in France, individual patients in several other countries

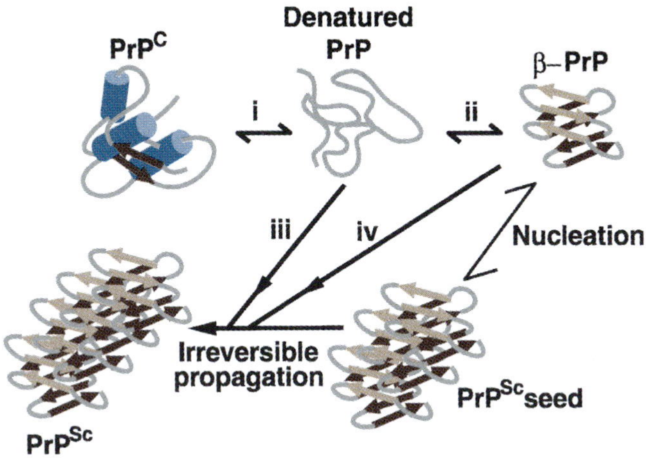

Fig. 2 Possible mechanism for prion propagation. Largely α-helical PrP^C proceeds via an unfolded state (i) to re-fold into a largely β-sheet form, β-PrP (ii). β-PrP is prone to aggregation in physiological salt concentrations. Prion replication may require a critical "seed" size. Further recruitment of unfolded PrP (iii) or β-PrP monomers (iv) then occurs as an essentially irreversible process driven thermodynamically by intermolecular interactions.

they cannot be encoded by differences in PrP primary structure. Furthermore, strains can be re-isolated in mice after passage in intermediate species with different PrP primary structures.[29] Conventionally, distinct strains of conventional pathogen are explained by differences in their nucleic acid genome. However, in the absence of such a scrapie genome, alternative possibilities must be considered. The protein-only hypothesis,[6] if correct, must be able to explain how a single polypeptide chain could encode multiple disease phenotypes. Clearly, understanding how a protein-only infectious agent could encode such phenotypic information is of wide biological interest.

Support for the idea that strain specificity may be encoded by PrP itself was provided by study of two distinct strains of transmissible mink encephalopathy prions which can be serially propagated in hamsters, designated hyper (HY) and drowsy (DY). These strains can be distinguished by differing physiochemical properties of the accumulated PrP^{Sc} in the brains of affected hamsters.[30] Following limited proteolysis, strain-specific migration patterns of PrP^{Sc} on polyacrylamide gels were seen which related to different N-terminal ends of HY and DY PrP^{Sc} following protease treatment and implying differing conformations of HY and DY PrP^{Sc}.[31]

Distinct human PrP^{Sc} types have been identified which are associated with different phenotypes of CJD[32,33] (Fig. 3). The different fragment sizes seen on Western blots following treatment with proteinase K suggests that there are several different human PrP^{Sc} conformations. However, while such biochemical modifications of PrP are clearly candidates for the molecular substrate of prion strain diversity, it is necessary to be able to demonstrate that these properties fulfil the biological properties of strains. In particular, that they are transmissible to the PrP in a host of both the same and different species. This has been demonstrated in studies with CJD isolates, with both PrP^{Sc} fragment sizes and the ratios of the three PrP glycoforms (diglycosylated, monoglycosylated and unglycosylated PrP) maintained on passage in transgenic mice expressing human PrP.[33] Furthermore, transmission of human prions and bovine prions to wild type mice results in murine PrP^{Sc} with fragment sizes and glycoform ratios which correspond to the original inoculum.[33] Variant CJD is associated with PrP^{Sc} glycoform ratios which are distinct from those seen in classical CJD. Similar ratios are seen in cattle BSE and BSE when transmitted to several other species.[33] These data strongly support the "protein only" hypothesis of infectivity and suggest that strain variation could be encoded by a combination of PrP conformation and glycosylation. Furthermore, polymorphism in PrP sequence can influence the generation of particular PrP^{Sc} conformers.[33] As

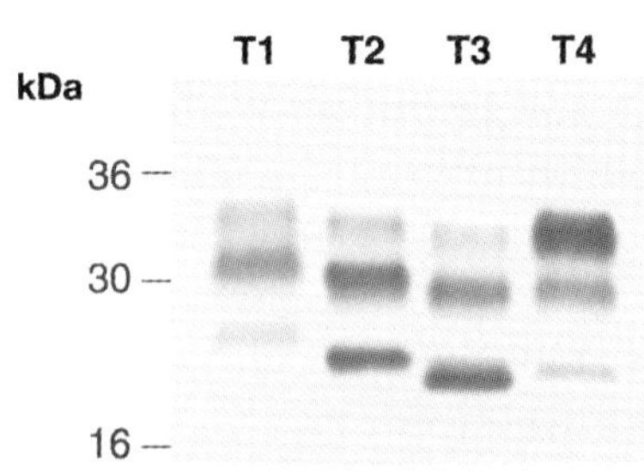

Fig. 3 Molecular strain typing of human prions. Western blot of brain homogenate after treatment with proteinase K shows different apparent molecular mass and glycoform ratios in patients with forms of sporadic or iatrogenic (T1-3) or variant CJD (T4).

PrP glycosylation occurs before conversion to PrP^{Sc}, the different glycoform ratios may represent selection of particular PrP^{C} glycoforms by PrP^{Sc} of different conformations. According to such a hypothesis, PrP conformation would be the primary determinant of strain type with glycosylation being involved as a secondary process. However, since it is known that different cell types may glycosylate proteins differently, PrP^{Sc} glycosylation patterns may provide a substrate for the neuropathological targeting that distinguishes different prion strains.[33] Particular PrP^{Sc} glycoforms may replicate most favourably in neuronal populations with a similar PrP glycoform expressed on the cell surface. Such targeting could also explain the different incubation periods which also discriminate strains, targeting of more critical brain regions, or regions with higher levels of PrP expression, producing shorter incubation periods. Recent work has shown strain-specific protein conformation can be influenced by the binding of copper and zinc to PrP^{Sc} providing a novel mechanism for post-translational modification of PrP, and for the generation of multiple prion strains.[34]

Molecular strain typing of prion isolates now allows molecular diagnosis of vCJD[33,35] and a new classification of human prion diseases with implications for epidemiological studies investigating the aetiology of sporadic CJD (Fig. 3). Such methods allow strain typing to performed in days rather than the 1–2 years required for classical biological strain typing. This technique may also be applicable to determining whether BSE has transmitted to other species,[33] and thereby pose a threat to human health, for instance to sheep.[36–38]

Such ability of a single polypeptide chain to encode information specifying distinct phenotypes of disease raises intriguing evolutionary questions. Do other proteins behave in this way? The novel pathogenic mechanisms involved in prion propagation may be of far wider significance and relevant to other neurological and non-neurological illnesses, indeed other prion-like mechanisms have now been described and the field of yeast and fungal prions has emerged.[39,40]

The "Species Barrier"

Transmission of prion diseases between different mammalian species is restricted by a "species barrier".[41] On primary passage of prions from species A to species B typically not all inoculated animals of species B develop disease, and those that do have much longer and more variable incubation periods than is seen with transmission of prions within the same species — where typically all inoculated animals would succumb with a relatively short, and remarkably consistent, incubation period. On second passage of infectivity to further animals of species B, transmission parameters resemble within-species transmissions, with most, if not all, animals developing the disease with short and consistent incubation periods. Species barriers can therefore be quantitated by measuring the fall in mean incubation period on primary and second passage, or, perhaps more rigorously, by a comparative titration study. The latter involves inoculating serial dilutions of an inoculum in both the donor and host species and comparing the doses required to kill 50% of inoculated animals (LD_{50}). The effect of a very substantial species barrier is that few, if any, animals succumb to disease at all on primary passage, and then at incubation periods approaching the natural lifespan of the species concerned.

Early studies of the molecular basis of the species barrier argued that it principally resided in differences in PrP primary structure between the species from which the inoculum was derived and the inoculated host. Transgenic mice expressing hamster PrP were, unlike wild type mice, highly susceptible to infection with Sc237 hamster prions (Prusiner *et al.* 1990). That most sporadic and acquired CJD occurred in individuals homozygous at *PRNP* polymorphic codon 129 supported the view that prion propagation proceeded most efficiently when the interacting PrP^{Sc} and PrP^{C} were of identical primary structure (Collinge *et al.* 1991; Palmer *et al.* 1991). However, it has been long recognised that prion strain type affects ease of transmission to another species. Interestingly, with BSE prions the strain component to the barrier seems to predominate, with BSE not only transmitting efficiently to a range of species, but maintaining its transmission characteristics even when passaged through an intermediate species with a distinct PrP gene.[29] For instance, transmission of CJD prions to conventional mice is difficult, with few if any inoculated mice succumbing after prolonged incubation periods, consistent with a substantial species barrier.[35,42] In sharp contrast, transgenic mice expressing only human PrP are highly susceptible to CJD prions, with 100% attack rate and consistent short incubation periods which are unaltered by second passage, consistent with a complete lack of species barrier.[42] However, vCJD prions (again comprising human PrP of identical primary structure) transmit much more readily to wild type mice than do classical CJD prions, while transmission to transgenic mice is relatively less efficient than with classical CJD.[35] The term "species barrier" does not seem appropriate to describe such effects and "transmission barrier" may be preferable.[43] Both PrP amino acid sequence and strain type affect the 3D-structure of glycosylated PrP which will presumably, in turn, affect the efficiency of the protein protein interactions thought to determine prion propagation.

Mammalian PrP genes are highly conserved. Presumably only a restricted number of different PrP^{Sc} conformations (that are highly stable and can therefore be serially propagated) will be permissible thermodynamically and will constitute the range of prion strains seen. PrP glycosylation may be important in stabilising particular PrP^{Sc} conformations. While a significant number of different such PrP^{Sc} conformations may be possible amongst the range of mammalian PrP's, only a sub-set of these would be allowable for a given single mammalian PrP. Substantial overlap between the favoured conformations for PrP^{Sc} derived from species A and species B might therefore result in relatively easy transmission of prion diseases between these two species, while two species with no preferred PrP^{Sc} conformations in common would have a large barrier to transmission (and indeed transmission would necessitate a change of strain type). According to such a *conformational selection model*[43] of a prion transmission barrier, BSE may represent a thermodynamically highly favoured PrP^{Sc} conformation that is permissive for PrP expressed in a wide range of different species, accounting for the remarkable promiscuity of this strain in mammals. Contribution of other components to the species barrier are possible and may involve interacting co-factors which mediate the efficiency of prion propagation, although no such factors have yet been identified.

Recent data has further challenged our understanding of transmission barriers.[44] The assessment of species barriers has relied on the development of a clinical disease in inoculated animals. However, it is now clear that *subclinical prion infections* are sometimes established on prion inoculation of a second species.[45] Such animals may harbour high levels of prion infectivity but do not develop clinical signs of disease during a normal lifespan. The barrier to primary passage may then be to the development of rapid neurodegeneration and the resulting clinical syndrome rather than a barrier to prion propagation itself. The existence of such subclinical carrier states of prion infection has important potential animal and public health implications and argues against direct neurotoxicity of prions.[45]

Pathogenesis

In some experimental rodent scrapie models, as in natural sheep scrapie, infectivity is first detectable in the spleen and other lymphoreticular tissues (for review see.[46]) Spleen titres rise to a plateau early in the incubation period, long before neuroinvasion is detectable. CNS prion replication then rises to high levels and the clinical phase then follows. The route of entry of prions following oral exposure may follow invasion of Peyer's patches and other gut lymphoid tissues, the relative protease–resistance of prions presumably allows a significant proportion of infectivity to survive the digestive tract. It is unclear how prions transit the intestinal mucosa, although M cells may be involved.[47] It has been suggested that myeloid dendritic cells mediate transport within the lymphoreticular system.[48] While mature B cells are required for peripheral prion propagation, this appears to be because they are required for maturation of follicular dendritic cells (FDC's). PrP^{Sc} accumulates in FDC's which are a long-lived cell type and it is thought that they are the site of prion propagation in the spleen.[49–51] However, neuroinvasion is possible without FDC's, indicating that other peripheral cell types can replicate prions.[52,53] Neuroinvasion involves the autonomic nervous system innervating lymphoid tissue with retrograde spread to the spinal cord or via the vagus to the brain stem.[54,55] Prions have been detected in the blood at low levels in some rodent models and experimental BSE-infected primates[56–58] and transmission of BSE prions between sheep by transfusion has been reported.[59] Several reports of infrequent transmission from human blood to rodents have been reported (for review see[60]) and there is now evidence for transmission of vCJD prion infection by blood transfusion.[61,62]

While prominent lymphoreticular involvement is seen in some experimental models or natural prion diseases, it is undetectable in others (see[46] for review). Both host and prion strain effects are relevant. For example infection of sheep with BSE prions results in a wide tissue distribution of infectivity while infection of cattle with this strain does not, infectivity being largely confined to the CNS. In humans infected with sporadic CJD prions, infectivity is largely confined to the CNS, while in variant CJD there is prominent involvement of lymphoreticular tissues.[63–65]

APPLICATION TO CLINICAL NEUROLOGY

Molecular Classification of Prion Disease

While neurologists have until recent years had to rely largely on clinical features to differentiate neurodegenerative disorders, the major advances in molecular genetics and in understanding molecular pathogenesis increasingly enable diagnosis using criteria higher in the diagnostic hierarchy of pathology. While prion diseases have traditionally been classified into CJD, GSS and kuru, we can now divide them aetiologically into inherited, sporadic, and acquired forms (Table 1) with sub-classification according to molecular criteria.

The terms CJD, GSS, and kuru simply represent clinical and/or pathological syndromes within a wider spectrum of disease and are also becoming problematic. For example the term CJD — now widely associated in the public mind with BSE — may be misleading. Variant "CJD" and familial "CJD" may be clinically very different to classical "CJD" and to each other, and of course have radically different aetiologies — both from each other and from classical sporadic CJD.

Around 15% of recognised prion disease is an *inherited* Mendelian disorder associated with one of the more than thirty recognised coding mutations in PRNP (for review see[66]). For a single gene inherited disorder of high penetrance such as inherited prion disease, the diagnostic supremacy of direct demonstration of causative mutation by DNA analysis is clear. Indeed, the availability of such definitive diagnostic markers has long allowed diagnosis of inherited prion disease in patients not only atypical on clinical grounds, but in whom classical neuropathological features are absent.[67,68] Indeed, the inherited prion diseases were the first neurodegenerative conditions for which direct gene tests were available and definitive pre-symptomatic testing was performed.[67,69] Such definite molecular genetic diagnosis has allowed recognition of a wider phenotypic range than hitherto appreciated.[70] Kindreds with inherited prion disease have been described with phenotypes of classical CJD, GSS, and also with other neurodegenerative syndromes including fatal familial insomnia.[71]

GSS is a pathological term referring to the presence of particular amyloid plaque morphologies now recognised to a variable extent in several different inherited prion diseases. Kindreds are documented in which some individuals have the classical syndromes of "CJD" and "Gerstmann-Sträussler-Sheinker disease" while others do not fit these rubrics at all.[13] Cases diagnosed by PrP gene analysis have been reported which are not only clinically atypical but which lack the classical histological features entirely.[68] Neuropathology in such patients is no longer the "Gold Standard": rather the recognised clinicopathological manifestation of a particular inherited condition simply widens. Significant clinical overlap exists with familial Alzheimer's disease, Pick's disease, Frontal lobe degeneration of non-Alzheimer type and amyotrophic lateral sclerosis with dementia. All the inherited forms can now be definitively diagnosed and sub-classified according to pathogenic mutation.

Sporadic CJD makes up around 85% of all recognised human prion disease. It occurs in all countries with an apparently random distribution and annual incidence of 1–2 per million; hypothesised causes include spontaneous production of PrP^{Sc} via rare stochastic events, somatic mutation of *PRNP* or unidentified environmental prion exposure. An association with sheep scrapie is not supported by epidemiological studies,[72] however the lack of such evidence does not exclude the possibility that a fraction of sporadic CJD is caused by environmental exposure to animal or human prions. There is marked genetic susceptibility in sporadic CJD in that most cases occur in homozygotes at codon 129 of *PRNP*, where either methionine (M) or valine (V) may be encoded. Heterozygotes (MV) appear significantly protected against developing sporadic CJD.[73–75] Additionally, a *PRNP* susceptibility haplotype has been identified indicating additional, as yet uncharacterised, genetic susceptibility to sporadic CJD at or near to the *PRNP* locus.[76]

The *acquired* prion diseases include iatrogenic CJD and kuru, and arise from accidental exposure to human prions through medical or surgical procedures or participation in cannibalistic feasts. The two most frequent causes of iatrogenic CJD occurring through medical procedures have arisen as a result of implantation of dura mater grafts and treatment with human growth hormone derived from the pituitary glands of human cadavers.[77,78] Other iatrogenic cases have resulted from corneal transplantation and use of contaminated neurosurgical instruments.[77,78] *PRNP* codon 129 genotype is also relevant to susceptibility and incubation period.[73,79] The occurrence of cases of apparently sporadic CJD in unusually young people in the UK in 1995[80–82] led to concerns that BSE transmission to humans may have occurred. Arrival of further cases in 1996 led to the recognition of a novel clinicopathological type of human prion disease, now known as variant CJD (vCJD),[83] indicating the arrival of a new risk factor for CJD in the UK.[84] A link with BSE seemed highly likely on epidemiological grounds and this was strongly supported by experimental data, firstly from molecular strain typing studies[33] and later by transmission studies into both transgenic and conventional mice.[35,85] *PRNP* mutations are absent in vCJD, and all cases studied to date have been 129MM[86,87] (and unpublished). That vCJD is caused by the same prion strain as that causing BSE in cattle, raised the possibility that a major epidemic of vCJD will occur in the UK and other countries as a result of dietary or other exposure to BSE prions and also[88] concerns of potential iatrogenic transmission of pre-clinical vCJD via medical and surgical procedures. That only *PRNP* 129MM individuals are susceptible to BSE infection is questionable, since the other acquired human prion diseases, iatrogenic CJD and kuru, occur in all codon 129 genotypes as the epidemic evolves, with MV having the longest mean incubation periods.[14,43,89] Human BSE infection of other *PRNP* genotypes may simply have a longer latency[43] (and may also have a different phenotype[35]).

The acquired prion diseases, such as vCJD, although not contagious in humans, are infectious diseases. In infectious disease, while again clinical and histopathological features may be key, confirmation of diagnosis, not least in life-threatening conditions, is by identification of the infectious pathogen itself or a specific immune response to it. Isolation and strain typing of the pathogen is at the apex of the diagnostic hierarchy. Strain typing in particular may allow the source of an outbreak to be identified and the best available prognostic and therapeutic advice to be provided. While it is essential to balance the potential risks and discomfort involved in an invasive diagnostic test against the improved diagnostic accuracy, it will only be by progressing steadily to greater use of molecular analysis of neurological disease that we will be able to deliver the diagnostic and ultimately therapeutic advances to patients with neurodegenerative diseases that are so desperately needed.

The marked phenotypic heterogeneity observed in human prion diseases has yet to be explained. However, it is likely that a significant proportion of this relates to the propagation of distinct human prion strains. The identification of strain-specific PrP^{Sc} biochemical properties now allows an aetiology-based classification by typing of the infectious agent itself. Four types of human PrP^{Sc} have now been reliably identified from patients with sporadic or iatrogenic prion disease using molecular strain typing[33–35,90] and a preliminary molecular classification proposed[90] (Fig. 3). Sporadic and iatrogenic CJD are associated with PrP^{Sc} types 1–3, while type 4 human PrP^{Sc} is uniquely associated with vCJD and is characterised by a fragment size and glycoform ratio that is distinct from PrP^{Sc} types 1–3 observed in classical CJD.[33–35,90] The M/V polymorphism at codon 129 of *PRNP* is associated with different PrP^{Sc} types. PrP^{Sc} types 1 and 4 have so far only been detected in MM individuals, type 3 cases are predominantly associated with at least one V allele (MV or VV), while type 2 is seen in any *PRNP* codon 129 genotype.[33,34,64,86,90] An earlier molecular classification,[32,91] based on subdivision into only two molecular sub-types by fragment size is also in use, and an international consensus has yet to be reached.

Molecular strain typing has major implications for epidemiological surveillance of sporadic CJD, whose aetiology remains obscure. While spontaneous conversion of PrP^C to PrP^{Sc} as a rare stochastic event, or somatic mutation of the PrP gene, resulting in expression of a pathogenic PrP mutant, are plausible explanations for sporadic CJD,[92] other causes for at least some cases, including environmental exposure to human or animal prions has not been ruled out by existing epidemiological studies.[93] Sub-classification of sporadic CJD based upon PrP^{Sc} type immediately allows a more precise molecular classification of human prion disease and re-analysis of epidemiological data using these molecular sub-types may reveal important risk factors obscured when sporadic CJD is analysed as a single entity. For example, it will be important to review the incidence of sporadic CJD associated with PrP^{Sc} type 2 and other molecular sub-types in both BSE-affected and unaffected countries in the light of recent findings suggesting that BSE prion infection of "humanised" transgenic mice of human prion disease may result in propagation of either type 4 PrP^{Sc} or type 2 PrP^{Sc}.[94] Individuals that propagate type 2 PrP^{Sc} as a result of BSE exposure may present with prion disease that would be indistinguishable on clinical, pathological and molecular criteria from that found in classical CJD.

Epidemiology of vCJD and other Possible Forms of Human BSE Infection

Estimates of the mean incubation period of human-to-human prion transmission come from study of growth hormone-related iatrogenic CJD and kuru: in both cases with estimates of around 12 years. In kuru, incubation periods can exceed 50 years (Whitfield, Alpers and Collinge, unpublished). The effect of a species barrier is to considerably increase both the mean and range of incubation periods seen — which may approach the usual lifespan of the species concerned.[43] For example, the cattle-to-mouse barrier for the BSE strain results typically in a 3–4 fold increase in mean incubation period. Mean incubation periods of human BSE infection of 30 years or more should be considered.[43] Furthermore, prion disease in mice follows a well-defined course with a highly distinctive and repeatable incubation time for a given prion strain in a defined inbred mouse line. In addition to the PrP gene, several additional genetic loci with a major effect on incubation period have now been mapped.[89] It can be anticipated that the human homologues of such loci may play a key role in human susceptibility to prion disease, both following accidental human prion exposure and exposure to the BSE agent. By definition, the patients identified to date with vCJD are those with the shortest incubation periods for BSE. These in turn, given that no unusual history of dietary, occupational or other exposure to BSE has been identified, would be expected to be predominantly those individuals with short incubation time alleles at these multiple genetic loci in addition to having the 129MM *PRNP* genotype. The vCJD cases reported to date may therefore represent a distinct genetic sub-population with unusually short incubation periods to BSE prions. It is possible therefore that recent estimates of the size of the vCJD epidemic based on uniform genetic susceptibility may substantially underestimate the eventual size.[95,96] In this context, it will be difficult to accurately predict a human epidemic until such loci are identified and their gene frequencies in the population can be determined.[97]

Long term transmission studies have been carried out using genetically modified mice to both characterise the distinct prion strains causing human disease and to model human susceptibility to infection with BSE and other prions.[70] These mice express human, but not mouse PrP^C. While these transgenic mouse models have been able to faithfully propagate human prion strains[33,35,42] and recapitulate the characteristic neuropathology of vCJD,[94] there are important caveats in extrapolating from such animal models to human susceptibility. However, these studies have found a much higher infection rate in transgenic mice expressing human PrP M129 than mice expressing human PrP V129 when challenged with either BSE or vCJD prions, and demonstrated that BSE prion infection can produce disease phenotypes resembling sporadic CJD infection of these mice and also novel prion strain phenotypes. Most recently, these studies have argued that the vCJD phenotype may only be expressed in the presence of the M form of human PrP.[98] While this would imply that only those humans expressing human PrP M129 may develop vCJD, this does not mean that VV individuals are completely resistant to BSE prion infection — but rather that if infected they would show a different phenotype.[98]

Early Diagnosis of Human Prion Disease

Despite the currently dire prognosis of established prion neurodegeneration, it is important to establish a firm early diagnosis. This is firstly because the differential diagnosis of prion disease includes potentially treatable conditions, for example cerebral vasculitis, and cortical biopsy, with full infection control precautions, may be considered in some patients. Secondly, although a devastating diagnosis, early confirmation removes uncertainty, which is itself distressing, obviates the need for further investigation and allows a care plan to be established including infection control measures and appropriate counselling of patient and their family. However, the advent of the first therapeutic trials in prion disease (see below) creates a new impetus to accelerate diagnosis. Drugs that retard, or even eliminate, prion infection will not reverse existing neuronal loss and there may be little benefit in arresting prion propagation at the advanced stage of neurodegeneration that diagnosis is often made currently (Table 2). Advances in early diagnosis must therefore go hand in hand with therapeutic advances to deliver real benefit to patients in the years ahead. While such effective drugs are not yet available, much research is ongoing in both diagnostics and therapeutics. However, given both the rarity and speed of evolution of prion disease, such research can only effectively progress with a high level of early referral — at a stage where the diagnosis may be very unclear — to research centres. For this reason, a UK-wide early referral system has been established in the UK with the support of the Department of Health and the Medical Research Council.

Table 2 Diagnosis of prion disease.

Sporadic (classical) CJD

- Rapidly progressive† dementia with two or more of myoclonus, cortical blindness, pyramidal signs, cerebellar signs, extrapyramidal signs, akinetic mutism
- Most cases age 45–75
- Serial EEG shows pseudoperiodic complexes in most cases
- CSF 14-3-3 protein usually positive
- CT and MRI normal, or atrophy, or abnormal signal basal ganglia
- *PRNP* analysis: no pathogenic mutations, most are 129MM (VV and MV may be longer duration, clinically atypical and EEG less often positive)
- Brain biopsy in highly selected cases (to exclude treatable alternative diagnoses): PrP immunocytochemistry or Western blot for PrP^{Sc} types 1–3

Iatrogenic CJD

- Progressive cerebellar syndrome and behavioral disturbance or classical CJD-like syndrome with history of iatrogenic exposure to human prions (pituitary derived hormones, tissue grafting or neurosurgery)
- May be young
- EEG, CSF and MRI generally less helpful than in sporadic cases
- *PRNP* analysis: no pathogenic mutations, most are 129 homozygotes
- Brain biopsy in highly selected cases (to exclude treatable alternative diagnoses): PrP immunocytochemistry or Western blot for PrP^{Sc} types 1–3

Variant CJD (Human BSE)

- Early features: depression, anxiety, social withdrawal, peripheral sensory symptoms
- Cerebellar ataxia, chorea or athetosis often precedes dementia, advanced disease as sporadic CJD
- Most in young adults; however, age at onset 12–74 years seen
- EEG non-specific slow waves, CSF 14-3-3 may be elevated or normal
- MRI: may be high T2-weighted signal in posterior thalamus bilaterally
- *PRNP* analysis: no mutations, all 129MM to date
- Tonsil biopsy: characteristic PrP immunostaining and PrP^{Sc} on Western blot (type 4t)

Inherited prion disease

- Varied clinical syndromes between and within kindreds: should consider in all pre-senile dementias and ataxias irrespective of family history
- *PRNP* analysis: diagnostic, codon 129 genotype may predict age at onset in pre-symptomatic testing

†Clinical duration typically 6 months or less but high variability: type 1 PrP^{Sc} associated with short duration (~8 weeks); ~10% have duration >2 years

While a confident clinical diagnosis of vCJD may be possible clinically when allied with a positive pulvinar sign on MRI, this sign is not always present, may not be an early feature, and false positives have been reported.[99] Further, while the initial patients with vCJD had a relatively uniform clinical presentation, not only are more atypical presentations now becoming apparent, but larger variability can be anticipated with involvement of additional genotypes in BSE prion infection (see above). It will be important to remain open minded about such phenotypes. The early clinical features of vCJD: depression, anxiety, behavioural change and sensory disturbances, are highly non-specific. Differentiation from much commoner psychiatric causes requires the arrival of overtly neurological features such as ataxia, chorea and cognitive decline although pre-existing use of neuroleptics and other psychotropic drugs may initially delay their diagnostic recognition. Current clinically based diagnostic criteria for vCJD, undoubtedly important for disease surveillance, require the evolution of disease over at least 6 months and the development of several signs indicative of extensive cerebral damage (http://www.doh.gov.uk/cjd/cjd_stat.htm). However, early diagnosis, before extensive irreversible brain damage has occurred, is crucial to exclude an alternative treatable disorder and brain biopsy may well be considered, particularly in younger patients. A clear tissue diagnosis of vCJD can now be made by tonsil biopsy with detection of characteristic PrP immunostaining and type 4t PrP^{Sc}[65,100] (Fig. 4). Tonsillar PrP^{Sc} is only detectable in vCJD, and not other forms of human prion disease. Early tonsil biopsy, if positive, obviates the need for further investigation. While the diagnostic accuracy provided by a tonsil biopsy has to be balanced against the fact that it is an invasive procedure, early referral for investigation should allow much earlier diagnosis and access to clinical trials before extensive functional loss has occurred.

While sporadic CJD can often, following the exclusion of other causes, be diagnosed with a high degree of confidence on the basis of clinical criteria, atypical forms, which present much greater diagnostic difficulty, are not uncommon.[101] Relatively slowly progressive forms can be difficult to differentiate from a rapidly progressive form of Alzheimer's disease with myoclonus for example — and CSF 14-3-3, a marker of neuronal damage, may be unhelpful or misleading in these situations. The large majority of

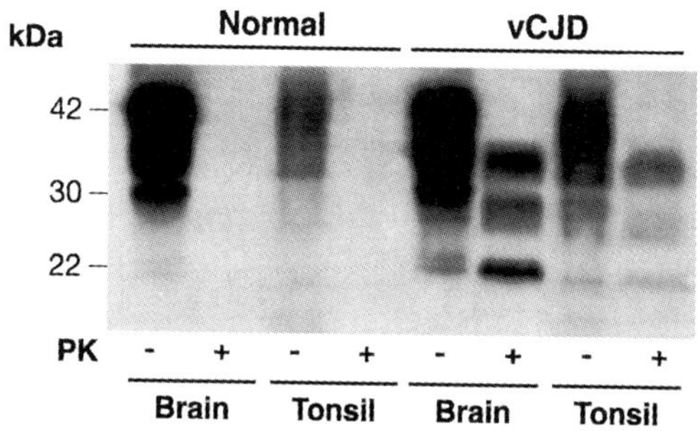

Fig. 4 Tonsil biopsy for diagnosis of vCJD prion infection. Western blot demonstrates PrP^{Sc} after proteinase K (PK) digestion. Characteristic PrP^{Sc} type is seen in tonsil from patients with vCJD allowing firm *ante mortem* tissue diagnosis.

patients with sporadic CJD will be of the *PRNP* codon 129 MM genotype. Patients with *PRNP* MV or VV genotypes are more likely to have an atypical course and negative EEG and this may also represent propagation of different prion strain types (see above). Other CSF markers have been proposed including NSE, S-100b and *Tau* but again none are specific for prion disease. There is a pressing need for a non-invasive diagnostic test, ideally of blood, which may allow screening of donated blood and tissue as well as identification of asymptomatically infected and affected individuals. Low levels of PrP^{Sc} have been detected in some peripheral tissues, including muscle, in a minority of patients with sporadic CJD at autopsy.[102] Such patients were often of long clinical duration and the diagnostic potential of muscle biopsy in the early clinical phase may be limited. Reports of infectivity of blood from patients with classical CJD are infrequent and have been questioned.[60] but there is now evidence that vCJD prions may be transmissible by blood transfusion. Such evidence for human prionaemia suggests that detection of PrP^{Sc} in human blood may be possible but is below the detection limit of current methods.[65] Much research is ongoing to develop more sensitive methods including promising developments with *in vitro* PrP^{Sc} amplification to increase diagnostic test sensitivity.[99] Other genetic markers may allow more precise identification of risk groups and both transcriptome and proteome analysis in sporadic and variant CJD may provide disease markers.

Presymptomatic and Ante-Natal Testing

Since a direct gene test has become available it has been possible to provide an unequivocal diagnosis in patients with inherited forms of the disease. This has also lead to the possibility of performing pre-symptomatic testing of unaffected but at-risk family members, as well as antenatal testing.[69] Because of the effect of *PRNP* codon 129 genotype on the age of onset of disease associated with some mutations it is possible to determine within a family whether a carrier of a mutation will have an early or late onset of disease. Most of the mutations appear to be fully penetrant, however experience with some is extremely limited. In families with the E200K mutation and in D178N (fatal familial insomnia) there are examples of elderly unaffected gene carriers who appear to have escaped the disease. Genetic counselling is essential prior to pre-symptomatic testing and follows a protocol similar to that established for Huntington's disease. A positive PrP gene analysis has important consequences for other family members, and it is preferable to have discussed these issues with others in the immediate family before testing. Following the identification of a mutation the wider family should be referred for genetic counselling. It is vital to counsel both those testing positive for mutations and those untested but at-risk that they should not be blood or organ donors and should inform surgeons, including dentists, of their risk status prior to significant procedures as precautions may be necessary to minimise risk of iatrogenic transmission.

Prevention and Public Health Management

While prion diseases can be transmitted to experimental animals by inoculation, it is important to appreciate that they are not contagious in humans. Documented case-to-case spread has only occurred by cannibalism (kuru) or following accidental inoculation with prions. Such iatrogenic routes include the use of inadequately sterilised intracerebral electrodes, dura mater, and corneal grafting, and from the use of human cadaveric pituitary-derived growth hormone or gonadotrophin. As discussed above, there is now evidence that vCJD prion infection is transmissible by blood transfusion. UK policy for some time has been to leucodeplete all whole blood and to source plasma for plasma products from outside the UK.

A further possible route of transmission of vCJD is via contaminated surgical and medical instruments. Prions resist conventional sterilisation methods and neurosurgical instruments are known to be able to act as a vector for prion transmission: several cases of iatrogenic transmission of sporadic CJD prions via neurosurgical instruments are documented.[103,104] Recent evidence suggests that classical CJD may also be transmitted by other surgical procedures.[93] The wider tissue distribution of prions in vCJD[65] together with the potential that significant numbers in the population may be silently infected has considerably increased these concerns. Prions adhere avidly to stainless steel and transmit the disease readily in experimental models.[105] Tonsillar PrP^{Sc} is readily detectable in all cases of vCJD studied at autopsy and lymphoreticular involvement is a very early feature of natural prion infection in sheep and in experimental scrapie models, where replication in the LRS is detectable early in the incubation period and rises to a plateau, which considerably precedes, and is maintained in, the clinical phase.[46] This suggests that tonsillar PrP^{Sc} has probably been present for a considerable period, perhaps years, before clinical presentation of vCJD in humans thereby providing the basis of prevalence screening of the general population for infection. Two anonymous screens of tonsil and appendix tissues, removed during routine surgery, have been performed. Tonsil appears a more sensitive reporter of vCJD prion infection than appendix,[106] and it is concerning that 3 positives were reported from a screen of around 12 000 largely appendix samples.[107] A pilot scale study of 2000 tonsils found no positives, despite use of high sensitivity methods.[108] A national scale study is now being organised by the UK Department of Health to obtain more accurate estimates of prion infection in the UK population.

Certain occupational groups are at risk of exposure to human prions, for instance neurosurgeons and other operating theatre staff, pathologists and morticians, histology technicians, as well as an increasing number of laboratory workers. Because of the prolonged incubation periods to prions following administration to sites other than the central nervous system (CNS), which is associated with clinically silent prion replication in the lymphoreticular tissue,[109] treatments inhibiting prion replication in lymphoid organs may represent a viable strategy for rational secondary prophylaxis after accidental exposure. A preliminary suggested regimen is a short course of immunosuppression with oral corticosteroids in individuals with significant accidental exposure to human prions.[101]

Therapeutic Strategies

All recognised prion diseases are invariably fatal following a relentlessly progressive course. However, there have been significant recent advances in understanding prion propagation and neurotoxicity and clear proof of principle studies of several therapeutic or secondary prophylactic approaches in animal models suggesting effective therapeutics for human disease is realistic.[110]

While recent attempts at therapeutics in experimental models have focussed on drugs that target PrP^{Sc} formation, the cause of cell death in prion neurodegeneration remains unclear. A key area of controversy is whether PrP^{Sc} is itself directly neurotoxic. There are prion diseases in which PrP^{Sc} levels in brain are very low and conversely subclinical prion infection occurs with high levels of PrP^{Sc} and no clinical signs.[44,45] Also, PrP^{Sc} itself is not directly toxic to neurons that do not express PrP^{C}.[111,112] However, the conversion of PrP^{C} to PrP^{Sc} is clearly central to prion pathogenesis as mice devoid of PrP (*Prnp*$^{0/0}$) are resistant to prion disease and do not propagate infectivity.[113,114] An alternative possibility is that neurodegeneration results, at least in part, from loss of normal PrP^{C} function as it is sequestered into PrP^{Sc} aggregates, or alternatively, that a toxic intermediate (designated PrP^{L}) may be formed during prion propagation.[45] Although PrP-null (*Prnp*$^{0/0}$) mice are essentially normal,[115,116] adaptive mechanisms during neurodevelopment might compensate for loss of PrP^{C} function, masking any phenotype, whilst loss of PrP^{C} function in the developed, adult CNS might be deleterious. However, this alternative has been effectively excluded by the demonstration that PrP^{C} depletion in the neurons of adult mice also results in healthy animals with no evidence of neuronal loss.[117] Importantly, this model validates PrP^{C} as a therapeutic target in prion disease. Also the availability of this model, in which PrP expression in neurones is lost at around 12 weeks post-natally, allows the effect of neuronal PrP ablation during established CNS prion infection to be studied. Remarkably, such mice were completely protected from the development of clinical disease and early spongiform neuropathology was reversed.[11] These animals continued to accumulate PrP^{Sc} in non-neuronal glial cells and indeed accumulated prions to reach infectious titres seen in end stage disease in conventional disease, again arguing against direct neurotoxicity of non-neuronal PrP^{Sc}. It is conceivable that reversal of early spongiosis in human disease may even allow some symptomatic recovery. Clearly, the transgenic approach used in this model to interfere with neuronal PrP^{C} expression is not applicable therapeutically in humans, but provides a clear proof of principle for agents or other methods to target PrP^{C} directly. One approach is the isolation of small molecule PrP ligands which bind and stabilise PrP^{C}[70] — rendering it less available for conversion, including antibodies that bind or sequester PrP^{C}, or methods to down regulate PrP transcription or translation.

While the precise molecular events involved in the conversion of PrP^{C} to PrP^{Sc} remain ill defined, it can be argued that any ligand which selectively stabilises PrP^{C} should block prion propagation. The use of hydrogen/deuterium exchange to measure the extent to which regions of PrP^{C} transiently unfold argue that the conversion of PrP^{C} to PrP^{Sc} must proceed through a highly unfolded state that retains little organised native structure and that compounds which can bind to any ordered region of PrP^{C} should therefore inhibit the conversion pathway.[118] The maintenance of effective brain levels of such drugs, to reduce prion propagation rates to below those of natural clearance mechanisms, could then plausibly cure prion infection. High throughput screening of large compound libraries can be applied to detect such ligands. This approach has been exploited therapeutically for at least two proteins where mutations or altered conformation result in disease, the central core of p53 and transthyretin (TTR).

Considerable progress has been made in immunotherapeutic approaches. Antibodies against several PrP epitopes inhibit PrP^{Sc} propagation in cell culture.[119–121] These antibodies had little or no affinity for native PrP^{Sc}, and might act by binding cell surface PrP^{C} and reducing its availability for incorporation into propagating prions. Transgenic mice expressing anti-PrP μ chains directed against similar epitopes are protected against peripheral (but not central) prion infection.[122] In a clear proof of principle, wild type mice that had been peripherally infected with prions were passively immunised with anti-PrP monoclonal antibodies. PrP^{Sc} levels and prion infectivity in the spleens of scrapie-infected mice were markedly reduced even when antibodies were administered at the point of near maximal PrP^{Sc} accumulation. Furthermore, treated animals remained healthy >300 days after untreated animals had succumbed to the disease.[123] Unsurprisingly, as antibodies do not readily cross the blood brain barrier, there was no protective effect in intracerebrally infected mice. Nevertheless, humanised anti-PrP monoclonal antibodies might be used for post-exposure prophylaxis of particular risk groups. For established clinical disease, such antibodies could in principle be given by intracerebroventricular infusion, although effective tissue penetration throughout the CNS would be a problem. An important caveat is that injection of large quantities of anti-PrP antibodies into the CNS caused massive neuronal apoptosis in one study,[124] and it will be important to characterise the epitope and dose dependency of this effect.

Active immunisation is limited by immune tolerance to PrP, which is a self protein widely expressed in the immune system. Approaches to overcome this tolerance are being actively investigated, and studies have reported modest protective effects.[125,126]

The rarity of human prion diseases and consequent difficulty in assessing safety and efficacy means that the development of immunisation methods for healthy individuals to protect against prion diseases are probably not realistic. However, such a programme might eradicate endemic animal prion diseases such as ovine scrapie. It should also be mentioned that clinical trial of active immunisation for Alzheimer's disease advise caution: anti-Aβ40–42 antibodies can have protective or noxious effects in cells[127] and a recent anti-Ab40–42 vaccination trial was halted owing to encephalitic illness in some recipients.[128]

Other possible future approaches to target PrP^C include the use of small duplex RNA molecules to silence gene expression in a sequence-specific manner — RNA interference (RNAi). This is a potenaully powerful method for therapeutic gene silencing which has been applied to suppress gene expression both *in vivo* and *in vitro* in mammalian cells,[129,130] including non-dividing cells such as neurons. The prospect of lentiviral-mediated RNA silencing of PrP^C as a therapeutic tool is conceivable and provides a potential means of extrinsically manipulating PrP gene expression. However, any successful therapeutic delivery system would require effective CNS penetration.

As discussed already, prion strains show considerable diversity and are associated with distinct PrP^{Sc} types, which differ in their conformation[31–33,131] and glycosylation.[33] Strain switching or "mutation" is well documented[28,33,35,132] and more than one identifiable strain may propagate in the same host. BSE prions induce replication of two distinct prion strains in transgenic mice expressing only human PrP.[94] The possibility of emergence of drug resistance can therefore be anticipated against agents targeting PrP^{Sc} by selection of a minor or sub-strain of prion. However, according to the "protein-only" hypothesis, targeting native PrP^{C} should block replication of all prion strains.

The Advent of Clinical Trials in Human Prion Disease

A variety of drugs have been tried in individual or small numbers of patients over many years. With a notable exception,[133] reports have been largely anecdotal. There is no clear evidence of efficacy of any agent, and there is no doubt that controlled clinical trials are urgently needed. Such trials will be highly challenging for several reasons. Prion diseases are rare, often rapidly progressive and always fatal which may make randomisation to placebo unacceptable. Patterns of disease overall are, however, extremely variable with clinical durations varying from weeks to >2 years in sporadic CJD, and >20 years in some inherited prion diseases.[70] As "first generation" treatments proposed for prion disease are likely, at best, to have only a modest effect on disease progression, even using survival duration as an outcome measure requires study of large numbers to reliably assess efficacy. There is a lack of systematic natural history studies of disease progression and an absence of biological markers of disease activity. The high profile of these diseases often leads to media reporting of putative prion therapeutics with optimistic headline coverage leading, understandably, to patient requests for immediate access and unrealistic expectations of efficacy — particularly in patients with advanced neurodegeneration. When such promise of dramatic effect is not rapidly realised, loss of interest in participation in controlled trials may then follow — which necessarily take considerable time to be planned, peer reviewed and ethically approved — to assess if the drug actually has a worthwhile therapeutic effect. In the UK, at the request of the Government's Chief Medical Officer, a clinical trial protocol (http://www.controlled-trials.com/isrctn/trial/PRION/0/06722585.html) and infrastructure has been developed to rigorously assess the drug quinacrine[134] and to provide a framework for assessment of novel therapeutics as these become available: the MRC PRION-1 trial. Importantly under these circumstances, a formal consultation with patient's representatives was organised to refine the protocol so that it would be acceptable to the majority of potential participants http://www.mrc.ac.uk/prn/pdf-cjd_workshop.pdf). Pentosan polyphosphate is another candidate anti-prion drug and has shown some efficacy in animal models.[135,136] Unlike quinacrine, it dues not enter the CNS readily and has been administered by intraventricular infusion in several patients. Major toxicity has been reported by this route in animal studies and such treatment was not supported by the UK's Committee of Safety on Medicines or CJD Therapy Advisory Group.[137,138]

Perspectives for the Decade Ahead

- Effective treatment of neurodegenerative disease is one of the major challenges facing biomedical research. The focus on prion disease following the BSE epidemic — added to the intrinsic interest in their unique pathobiology — has lead to these diseases being amongst the best understood causes of neurodegeneration.
- It is to be hoped prion diseases remain rare. If no epidemic of BSE-related human infection emerges, such advances in understanding neurodegenerative processes should have far wider value in commoner conditions — such as Alzheimer's and Parkinson's disease — also associated with aggregation of misfolded protein. Conversely, if extensive human infection with BSE prions emerges in exposed populations, and silent secondary transmission is efficient, the failure to have developed a treatment for what is, albeit inadvertently, a man-made disease, during the window of opportunity provided by the remarkable latency of these infections would not enhance public perception of biomedical research.
- Considerable uncertainty remains over the number of people pre- or sub-clinically *infected* with BSE prions and what proportion of these will develop *clinical disease* in the years ahead: the natural evolution of this infection in the UK population is likely to span decades. Identification of the key genes controlling incubation period will allow better estimates.

- Clinicians should remain alert to alternate phenotypes of infection with BSE prions, which could also manifest as sporadic CJD or novel phenotypes as well as vCJD. Molecular classification of prion disease, by both genetic typing of host and strain typing of infecting prion, is evolving and will allow more precise epidemiological studies.
- Blood-based diagnostic tests will probably become available — involving surrogate serum markers, high sensitivity detection of PrP^{Sc} or *in vitro* prion amplification. The arrival of such tests, particularly if associated with a significant false positive rate, will raise important ethical issues about anonymity of testing and notification which should be debated in advance.
- Effective means to decontaminate prions on surgical and medical instruments will become available over the next few years and should reduce the risk of iatrogenic prion transmission. Secondary transmission of vCJD will continue until a suitable screening test for the infected state arrives.
- Proof of principle studies in animal models suggest that humanised anti-PrP monoclonal antibodies could be used for passive immunisation in the early pathogenesis to block neuroinvasion. This treatment could be considered for known iatrogenically infected individuals. Active human immunisation may be feasible technically but is unlikely to be pursued commercially.
- While the precise molecular events in prion propagation are not clear and the atomic level structure of prions is not yet determined, it is clear that PrP^C is the essential substrate. It is also clear that interference with PrP^C expression in adult brain is without serious effect and blocks onset of neurological disease in animal models. It should be possible to identify small molecules which penetrate the CNS to bind to PrP^C and to prevent its recruitment into prions, or to use one of a number of emerging technologies to reduce PrP^C expression in brain. If such methods are able to reduce prion propagation rates to below those of natural clearance mechanisms it ought to be possible to cure prion infection. New methods for early diagnosis — and their timely use — will be crucial, as such methods will not reverse neuronal cell loss which is appreciable or severe by the time clinical diagnosis is typically reached.

BOX 1 GLOSSARY

- **TSE**: transmissible spongiform encephalopathy or prion disease
- **Prion**: the infectious agent causing prion diseases
- **Prion protein (PrP)**: a glycoprotein encoded by the host genome and expressed in many tissues but especially on the surface of neurons
- **PrP^C**: the normal Cellular isoform of PrP rich in α-helical structure
- **PrP^{Sc}**: the "*Sc*rapie" or disease associated isoform of PrP which differs from PrP^C in its conformation and is generally found as insoluble aggregated material rich in β-sheet structure
- **Protein-only hypothesis**: that prions lack a nucleic acid genome, are composed principally or solely of abnormal isomers of PrP (PrP^{Sc}) and replicate by recruitment of host PrP^C
- **Species barrier**: Better called the transmission barrier, this describes the observation that transmission of prions from one species to another is generally inefficient when compared to subsequent passage in the same host species
- **Prion incubation period**: the interval between exposure to prions and the development of neurological signs of prion disease; typically months even in rodent models and years to decades in humans
- **Prion strain**: distinct isolates of prions originally identified and defined by biological characteristics which breed true in inbred mouse lines
- **Molecular strain typing**: a means of rapidly differentiating prion strains by biochemical differences in PrP^{Sc}
- **Conformational selection model**: a hypothetical model which explains transmission barriers on the basis of overlap of permissible conformations of PrP^{Sc} (prion strains) between mammalian species
- **Subclinical infection**: a state where host prion propagation is occurring but which does not produce clinical disease during normal lifespan; essentially a carrier state of prion infection
- ***PRNP***: the human prion protein gene; mouse gene is designated *Prnp*
- **Codon 129 polymorphism**: unique to humans, there are two common forms of *PRNP* encoding either methionine or valine at codon 129; a major determinant of genetic susceptibility to and phenotypic expression of prion disease

BOX 2 Useful websites

UK National Prion Clinic, National Hospital for Neurology and Neurosurgery, London
http://www.nationalprionclinic.org
Medical Research Council Prion Unit, Institute of Neurology, London
http://www.prion.ucl.ac.uk/
UK CJD Surveillance Unit, Western General Hospital, Edinburgh
http://www.cjd.ed.ac.uk/
UK Department of Health
http://www.dh.gov.uk/PolicyAndGuidance/HealthAndSocialCareTopics/CJD/fs/en
Human BSE Foundation
http://www.hbsef.org/
CJD Support Network
http://www.cjdsupport.net/

Key Points for Clinical Practice

- Prion diseases are transmissible but not contagious: no risk to carers or partners. They are not sexually transmissible and there is no evidence to date for maternal transmission, including via breast feeding. Risk to medical professionals arises principally in the context of accidental inoculation with sharps used on infected tissues or fluids.
- Prions may resist conventional sterilisation and there is evidence for transmission of vCJD prion infection by blood transfusion. All patients with, or at risk[b] of, prion disease should be counselled not to be blood donors and to inform surgeons and dentists of their risk status prior to surgery. Hospitals should attempt to identify such individuals prior to surgery: a suggested questionnaire is available.[a]
- Inherited prion diseases are a significant cause of pre-senile dementia and clinically mimic other conditions; a family history is not always apparent: should analyse *PRNP* in all suspected cases of CJD, and consider *PRNP*

[a]Available via UK National Prion Clinic (www.nationalprionclinic.org).
[b]For example those carrying or at risk of having inherited *PRNP* mutations, those treated with human cadaveric pituitary-derived growth hormone or gonadotrophin (usage ceased in UK in 1985) or relevant tissue grafts (cornea, dura mater) or identified recipients of potentially vCJD contaminated blood and blood products.

analysis in all early onset dementia and ataxias.[a] Always exclude *PRNP* mutations prior to brain or tonsil biopsy.

- In variant CJD — in contrast to classical CJD — there is extensive infection of lymphoreticular (LRS) tissues. Firm tissue diagnosis of variant CJD can be made by identification of typical PrP^{Sc} at tonsil biopsy. LRS infection is thought to precede neuroinvasion and is likely to have been present long (years) before clinical onset allowing early diagnosis.
- Since 2004, the Department of Health established a National Referral System for prion disease in the UK. Neurologists are asked to complete a *referral form*[a] to notify both the National Prion Clinic in London and the National CJD Surveillance Unit in Edinburgh of any suspected cases. New diagnoses of prion disease should be notified to local consultant in communicable disease control.
- Specialist advice and care is available via the NHS National Prion Clinic and the National CJD Surveillance Unit.
- A clinical trial centre has been established to co-ordinate therapeutic trials in prion disease in the UK. The MRC PRION-1 trial,[a] a partially randomised patient preference study is underway and is recruiting patients into a study of quinacrine and also studying the natural history and disease markers in patients who do not elect treatment currently.

REFERENCES

1. Kirkwood JK, Wells GA, Wilesmith JW, Cunningham AA, Jackson SI. Spongiform encephalopathy in an arabian oryx (Oryx leucoryx) and a greater kudu (Tragelaphus strepsiceros). *Vet Rec* 1990; **127**:418–420.
2. Hadlow WJ. Scrapie and kuru. *Lancet* 1959; **ii**:289–290.
3. Gibbs CJ Jr, Gajdusek DC, Asher DM, Alpers MP, Beck E, Daniel PM, Matthews WB. Creutzfeldt-Jakob Disease (Spongiform Encephalopathy): Transmission to the Chimpanzee. *Science* 1968; **161**:388–389.
4. Alper T, Haig DA, Clarke MC. The exceptionally small size of the scrapie agent. *Biochem Biophys Res Commun* 1966; **22**:278–284.
5. Alper T, Cramp WA, Haig DA, Clarke MC. Does the agent of scrapie replicate without nucleic acid? *Nature* 1967; **214**:764–766.
6. Griffith JS. Self Replication and Scrapie. *Nature* 1967; **215**:1043–1044 .
7. Prusiner SB. Novel proteinaceous infectious particles cause scrapie. *Science* 1982; **216**:136–144.

8. Oesch B, Westaway D, Walchli M, McKinley MP, Kent SB, Aebersold R, Barry RA, Tempst P, Teplow DB, Hood LE, Raeber AJ. A cellular gene encodes scrapie PrP 27–30 protein. *Cell* 1985; **40**:735–746.
9. Borchelt DR, Scott M, Taraboulos A, Stahl N, Prusiner SB. Scrapie and cellular prion proteins differ in their kinetics of synthesis and topology in cultured cells. *J Cell Biol* 1990; **110**:743–752.
10. Caughey B, Raymond GJ. The scrapie-associated form of PrP is made from a cell surface precursor that is both protease- and phospholipase-sensitive. *J Biol Chem* 1991; **266 No 27**:18217–18223.
11. Owen F, Poulter M, Lofthouse R, Collinge J, Crow TJ, Risby D, Baker HF, Ridley RM, Hsiao K, Prusiner SB. Insertion in prion protein gene in familial Creutzfeldt-Jakob disease. *Lancet* 1989; **1**:51–52.
12. Hsiao K, Baker HF, Crow TJ, Poulter M, Owen F, Terwilliger JD, Westaway D, Ott J, Prusiner SB. Linkage of a prion protein missense variant to Gerstmann-Straussler syndrome. *Nature* 1989; **338**:342–345.
13. Collinge J, Brown J, Hardy J, Mullan M, Rossor MN, Baker H, Crow TJ, Lofthouse R, Poulter M, Ridley R, Owen F, Bennett C, Dunn G, Harding AE, Quinn N, Doshi B, Roberts GW, Honavar M, Janota I, Lantos PL. Inherited prion disease with 144 base pair gene insertion: II: Clinical and pathological features. *Brain* 1992; **115**:687–710.
14. Poulter M, Baker HF, Frith CD, Leach M, Lofthouse R, Ridley RM, Shah T, Owen F, Collinge J, Brown J, Hardy J, Mullan M, Harding AE, Bennett C, Doshi B, Crow TJ. Inherited prion disease with 144 base pair gene insertion: I: Genealogical and molecular studies. *Brain* 1992; **115**:675–685.
15. Prusiner SB. Molecular Biology of Prion Diseases. *Science* 1991; **252**:1515–1522.
16. Riek R, Hornemann S, Wider G, Billeter M, Glockshuber R, Wuthrich K. NMR structure of the mouse prion protein domain PrP (121–231). *Nature* 1996; **382**:180–182.
17. Jackson GS, Murray I, Hosszu LLP, Gibbs N, Waltho JP, Clarke AR, Collinge J. Location and properties of metal-binding sites on the human prion protein. *Proc Natl Acad Sci USA* 2001; **98**:8531–8535.
18. Pan K-M, Baldwin MA, Nguyen J, Gasset M, Serban A, Groth D, Mehlhorn I, Huang Z, Fletterick RJ, Cohen FE, Prusiner SB. Conversion of α-helices into β-sheets features in the formation of the scrapie prion proteins. *Proc Natl Acad Sci USA* 1993; **90**:10962–10966.
19. Collinge J, Palmer MS, Sidle KCL, Gowland I, Medori R, Ironside J, Lantos PL. Transmission of fatal familial insomnia to laboratory animals. *Lancet* 1995; **346**:569–570.
20. Wille H, Zhang GF, Baldwin MA, Cohen FE, Prusiner SB. Separation of scrapie prion infectivity from PrP amyloid polymers. *J Mol Biol* 1996; **259**:608–621.

21. Lasmézas CI, Deslys JP, Robain O, Jaegly A, Beringue V, Peyrin JM, Fournier JG, Hauw JJ, Rossier J, Dormont D. Transmission of the BSE agent to mice in the absence of detectable abnormal prion protein. *Science* 1997; **275**:402–405.
22. Shaked GM, Fridlander G, Meiner Z, Taraboulos A, Gabizon R. Protease-resistant and detergent-insoluble prion protein is not necessarily associated with prion infectivity. *J Biol Chem* 1999; **274**:17981–17986.
23. Bolton DC, McKinley MP, Prusiner SB. Identification of a protein that purifies with the scrapie prion. *Science* 1982; **218**:1309–1311.
24. Arnold JE, Tipler C, Laszlo L, Hope J, Landon M, Mayer RJ. The abnormal isoform of the prion protein accumulates in late-endosome-like organelles in scrapie-infected mouse brain. *J Pathol* 1995; **176**:403–411.
25. Mayer RJ, Landon M, Laszlo L, Lennox G, Lowe J. Protein processing in lysosomes: the new therapeutic target in neurodegenerative disease. *Lancet* 1992; **340**:156–159.
26. Taraboulos A, Raeber A, Borchelt DR, Serban D, Prusiner SB. Synthesis and trafficking of prion proteins in cultured cells. *Mol Biol of the Cell* 1992; **3**: 851–863.
27. Laszlo L, Lowe J, Self T, Kenward N, Landon M, McBride T, Farquhar C, McConnell I, Brown J, Hope J, Mayer RJ. Lysosomes as key organelles in the pathogenesis of prion encephalopathies. *J Pathol* 1992; **166**:333–341.
28. Bruce ME, Fraser H, McBride PA, Scott JR, Dickinson AG. The Basis of Strain Variation in Scrapie. In: Prusiner SB, Collinge J, Powell J, Anderton B, eds. *Prion Diseases in Human and Animals*. London: Ellis Horwood 1992.
29. Bruce M, Chree A, McConnell I, Foster J, Pearson G, Fraser H. Transmission of bovine spongiform encephalopathy and scrapie to mice: Strain variation and the species barrier. *Philos Trans R Soc Lond [Biol]* 1994; **343**:405–411.
30. Bessen RA, Marsh RF. Biochemical and physical properties of the prion protein from two strains of the transmissible mink encephalopathy agent. *J Virol* 1992; **66**:2096–2101.
31. Bessen RA, Marsh RF. Distinct PrP properties suggest the molecular basis of strain variation in transmissible mink encephalopathy. *J Virology* 1994; **68**:7859–7868.
32. Parchi P, Castellani R, Capellari S, Ghetti B, Young K, Chen SG, Farlow M, Dickson DW, Sims AAF, Trojanowski JQ, Petersen RB, Gambetti P. Molecular basis of Phenotypic Variability in Sporadic Creutzfeldt-Jakob Disease. *Annals of Neurology* 1996; **39**:669–680.
33. Collinge J, Sidle KCL, Meads J, Ironside J, Hill AF. Molecular analysis of prion strain variation and the aetiology of 'new variant' CJD. *Nature* 1996; **383**: 685–690.

34. Wadsworth JDF, Hill AF, Joiner S, Jackson GS, Clarke AR, Collinge J. Strain-specific prion-protein conformation determined by metal ions. *Nature Cell Biology* 1999; **1**:55–59.
35. Hill AF, Desbruslais M, Joiner S, Sidle KCL, Gowland I, Collinge J. The same prion strain causes vCJD and BSE. *Nature* 1997; **389**:448–450.
36. Hill A F, Sidle KCL, Joiner S, Keyes P, Martin TC, Dawson M, Collinge J. Molecular screening of sheep for bovine spongiform encephalopathy. *Neuroscience Letters* 1998; **255**: 159–162.
37. Kuczius T, Haist I, Groschup MH. Molecular analysis of bovine spongiform encephalopathy and scrapie strain variation. *J Infect Dis* 1998; **178**: 693–699.
38. Hope J, Wood SCER, Birkett CR, Chong A, Bruce ME, Cairns D, Goldmann W, Hunter N, Bostock CJ. Molecular analysis of ovine prion protein identifies similarities between BSE and an experimental isolate of natural scrapie, CH1641. *J Gen Virol* 1999; **80**:1–4.
39. Wickner RB, Masison DC. Evidence for two prions in yeast: [URE3] and [PSI]. *Curr Top Microbiol Immunol* 1996; **207**:147–160.
40. Wickner RB. A new prion controls fungal cell fusion incompatibility. *Proc Natl Acad Sci USA* 1997; **94**:10012–10014.
41. Pattison IH. Experiments with scrapie with special reference to the nature of the agent and the pathology of the disease. In: Gajdusek CJ, Gibbs CJ, Alpers MP, eds. *Slow, Latent and Temperate Virus Infections, NINDB Monograph 2*. Washington DC: US Government Printing 1965; 249–257.
42. Collinge J, Palmer MS, Sidle KCL, Hill AF, Gowland I, Meads J, Asante E, Bradley R, Doey LJ, Lantos PL. Unaltered susceptibility to BSE in transgenic mice expressing human prion protein. *Nature* 1995; **378**:779–783.
43. Collinge J. Variant Creutzfeldt-Jakob disease. *Lancet* 1999; **354**:317–323.
44. Hill AF, Joiner S, Linehan J, Desbruslais M, Lantos PL, Collinge J. Species barrier independent prion replication in apparently resistant species. *Proc Natl Acad Sci (USA)* 2000; **97**:10248–10253.
45. Hill AF, Collinge J. Subclinical prion infection. *Trends Microbiol* 2003; **11**: 578–584.
46. Fraser H, Bruce ME, Davies D, Farguhar CF, McBride PA. The Lymphoreticular system in the Pathogenesis of Scrapie. In: Prusiner SB, Collinge J, Powell J, Anderton B, eds. *Prion Diseases of Humans and Animals*. London: Ellis Horwood 1992.
47. Heppner FL, Christ AD, Klein MA, Prinz M, Fried M, Kraehenbuhl JP, Aguzzi A. Transepithelial prion transport by M cells. *Nat Med* 2001; **7**: 976–977.

48. Aucouturier P, Geissmann F, Damotte D, Saborio GP, Meeker HC, Kascsak R, Carp RI, Wisniewski T. Infected splenic dendritic cells are sufficient for prion transmission to the CNS in mouse scrapie. *J Clin Invest* 2001; **108**:703–708.
49. Montrasio F, Frigg R, Glatzel M, Klein MA, Mackay F, Aguzzi A, Weissmann C. Impaired prion replication in spleens of mice lacking functional follicular dendritic cells. *Science* 2000; **288**:1257–1259.
50. Mabbott NA, Mackay F, Minns F, Bruce ME. Temporary inactivation of follicular dendritic cells delays neuroinvasion of scrapie. *Nat Med* 2000; **6**:719–720.
51. Mabbott NA, Young J, McConnell I, Bruce ME. Follicular dendritic cell dedifferentiation by treatment with an inhibitor of the lymphotoxin pathway dramatically reduces scrapie susceptibility. *J Virol* 2003; **77**:6845–6854.
52. Prinz M, Montrasio F, Klein MA, Schwarz P, Priller J, Odermatt B, Pfeffer K, Aguzzi A. Lymph nodal prion replication and neuroinvasion in mice devoid of follicular dendritic cells. *Proc Natl Acad Sci USA* 2002; **99**:919–924.
53. Oldstone MB, Race R, Thomas D, Lewicki H, Homann D, Smelt S, Holz A, Koni P, Lo D, Chesebro B, Flavell R. Lymphotoxin-alpha- and lymphotoxin-beta-deficient mice differ in susceptibility to scrapie: evidence against dendritic cell involvement in neuroinvasion. *J Virol* 2002; **76**: 4357–4363.
54. Beekes M, McBride PA, Baldauf E. Cerebral targeting indicates vagal spread of infection in hamsters fed with scrapie. *J Gen Virol* 1998; **79**:601–607.
55. Bencsik A, Lezmi S, Baron T. Autonomous nervous system innervation of lymphoid territories in spleen: A possible involvement of noradrenergic neurons for prion neuroinvasion in natural scrapie. *J Neurovirol* 2001; **7**:447–453.
56. Holada K, Vostal JG, Theisen PW, MacAuley C, Gregori L, Rohwer RG. Scrapie infectivity in hamster blood is not associated with platelets. *J Virol* 2002; **76**:4649–4650.
57. Bons N, Lehmann S, Mestre-Frances N, Dormont D, Brown P. Brain and buffy coat transmission of bovine spongiform encephalopathy to the primate Microcebus murinus. *Transfusion* 2002; **42**:513–516.
58. Brown P, Cervenáková L, McShane LM, Barber P, Rubenstein R, Drohan WN. Further studies of blood infectivity in an experimental model of transmissible spongiform encephalopathy, with an explanation of why blood components do not transmit Creutzfeldt-Jakob disease in humans. *Transfusion* 1999; **39**:1169–1178.
59. Hunter N, Foster J, Chong A, McCutcheon S, Parnham D, Eaton S, MacKenzie C, Houston F. Transmission of prion diseases by blood transfusion. *J Gen Virol* 2002; **83**:2897–2905.

60. Brown P. Can Creutzfeldt-Jakob disease be transmitted by transfusion? *Curr op in Hematology* 1995; **2**:472–477.
61. Llewelyn CA, Hewitt PE, Knight RSG, Amar K, Cousens S, Mackenzie J, Will RG. Possible transmission of variant Creutzfeldt-Jakob disease by blood transfusion. *Lancet* 2004; **363**:417–421.
62. Peden AH, Head MW, Ritchie DL, Bell JE, Ironside JW. Preclinical vCJD after blood transfusion in a PRNP codon 129 heterozygous patient. *Lancet* 2004; **364**:527–529.
63. Hill AF, Zeidler M, Ironside J, Collinge J. Diagnosis of new variant Creutzfeldt-Jakob disease by tonsil biopsy. *Lancet* 1997; **349**:99–100.
64. Hill AF, Butterworth RJ, Joiner S, Jackson G, Rossor MN, Thomas DJ, Frosh A, Tolley N, Bell JE, Spencer M, King A, Al-Sarraj S, Ironside JW, Lantos PL, Collinge J. Investigation of variant Creutzfeldt-Jakob disease and other human prion diseases with tonsil biopsy samples. *Lancet* 1999; **353**:183–189.
65. Wadsworth JDF, Joiner S, Hill AF, Campbell TA, Desbruslais M, Luthert PJ, Collinge J. Tissue distribution of protease resistant prion protein in variant CJD using a highly sensitive immuno-blotting assay. *Lancet* 2001; **358**:171–180.
66. Collinge J. Creutzfeldt-Jakob Disease and Other Prion Diseases. In: O'Brien JT, Ames D, Burns A, eds. *Dementia*, 1999.
67. Collinge J, Harding AE, Owen F, Poulter M, Lofthouse R, Boughey AM, Shah T, Crow TJ. Diagnosis of Gerstmann-Straussler syndrome in familial dementia with prion protein gene analysis. *Lancet* 1989; **2**:15–17.
68. Collinge J, Owen F, Poulter M, Leach M, Crow TJ, Rossor MN, Hardy J, Mullan MJ, Janota I, Lantos PL. Prion dementia without characteristic pathology. *Lancet* 1990; **336**:7–9.
69. Collinge J, Poulter M, Davis MB, Baraitser M, Owen F, Crow TJ, Harding AE. Presymptomatic Detection or Exclusion of Prion Protein Gene Defects in Families with Inherited Prion Diseases. *Am J Hum Genet* 1991; **49**:1351–1354.
70. Collinge J. Prion diseases of humans and animals: their causes and molecular basis. *Annu Rev Neurosci* 2001; **24**:519–550.
71. Medori R, Tritschler HJ, LeBlanc A, Villare F, Manetto V, Chen HY, Xue R, Leal S, Montagna P, Cortelli P, Tinuper P, Avoni P, Mochi M, Baruzzi Q, Hauw JJ, Ott J, Lugaresi E, Autilio-Gambetti L, Gambetti P. Fatal familial insomnia, a prion disease with a mutation at codon 178 of the prion protein gene [see comments]. *N Engl J Med* 1992; **326**:444–449.
72. Brown P, Cathala F, Raubertas RF, Gajdusek DC, Castaigne P. The epidemiology of Creutzfeldt-Jakob disease: conclusion of a 15-year investigation in France and review of the world literature. *Neurology* 1987; **37**:895–904.

73. Collinge J, Palmer MS, Dryden AJ. Genetic predisposition to iatrogenic Creutzfeldt-Jakob disease. *Lancet* 1991; **337**:1441–1442.
74. Palmer MS, Dryden AJ, Hughes JT, Collinge J. Homozygous prion protein genotype predisposes to sporadic Creutzfeldt-Jakob disease. *Nature* 1991; **352**:340–342.
75. Windl O, Dempster M, Estibeiro JP, Lathe R, De Silva R, Esmonde T, Will R, Springbett A, Campbell TA, Sidle KCL, Palmer MS, Collinge J. Genetic basis of Creutzfeldt-Jakob disease in the United Kingom: a systematic analysis of predisposing mutations and allelic variation in the *PRNP* gene. *Hum Genet* 1996; **98**:259–264.
76. Mead S, Mahal SP, Beck J, Campbell T, Farrall M, Fisher E, Collinge J. Sporadic — but not variant — Creutzfeldt-Jakob disease is associated with polymorphisms upstream of *PRNP* Exon 1. *Am J Hum Genet* 2001; **69**:1225–1235.
77. Brown P, Preece MA, Will RG. "Friendly fire" in medicine: hormones, homografts, and Creutzfeldt-Jakob disease. *Lancet* 1992; **340**:24–27.
78. Brown P, Preece M, Brandel JP, Sato T, McShane L, Zerr I, Fletcher A, Will RG, Pocchiari M, Cashman NR, D'Aignaux JH, Cervenáková L, Fradkin J, Schonberger LB, Collins SJ. Iatrogenic Creutzfeldt-Jakob disease at the millennium. *Neurology* 2000; **55**:1075–1081.
79. Huillard d'Aignaux J, Costagliola D, Maccario J, Billette de Villemeur T, Brandel JP, Deslys JP, Hauw JJ, Chaussain JL, Agid Y, Dormont D, Alperovitch A. Incubation period of Creutzfeldt-Jakob disease in human growth hormone recipients in France. *Neurology* 1999; **53**:1197–1201.
80. Britton TC, Al-Sarraj S, Shaw C, Campbell T, Collinge J. Sporadic Creutzfeldt-Jakob disease in a 16-year-old in the UK. *Lancet* 1995; **346**:1155.
81. Bateman D, Hilton D, Love S, Zeidler M, Beck J, Collinge J. Sporadic Creutzfeldt-Jakob disease in a 18-year old in the UK. *Lancet* 1995; **346**: 1155–1156.
82. Tabrizi SJ, Scaravilli F, Howard RS, Collinge J, Rossor MN. GRAND ROUND. Creutzfeldt-Jakob disease in a young woman. Report of a Meeting of Physicians and Scientists, St. Thomas' Hospital, London. *Lancet* 1996; **347**:945–948.
83. Will RG, Ironside JW, Zeidler M, Cousens SN, Estibeiro K, Alperovitch A, Poser S, Pocchiari M, Hofman A, Smith PG. A new variant of Creutzfeldt-Jakob disease in the UK. *Lancet* 1996; **347**:921–925.
84. Collinge J, Rossor M. A new variant of prion disease. *Lancet* 1996; **347**:916-917.
85. Bruce ME, Will RG, Ironside JW, McConnell I, Drummond D, Suttie A, McCardle L, Chree A, Hope J, Birkett C, Cousens S, Fraser H, Bostock CJ. Transmissions to mice indicate that 'new variant' CJD is caused by the BSE agent. *Nature* 1997; **389**:498–501.

86. Collinge J, Beck J, Campbell T, Estibeiro K, Will RG. Prion protein gene analysis in new variant cases of Creutzfeldt-Jakob disease. *Lancet* 1996; **348**:56.
87. Zeidler M, Stewart GE, Barraclough CR, Bateman DE, Bates D, Burn DJ, Colchester AC, Durward W, Fletcher NA, Hawkins SA, Mackenzie JM, Will RG. New variant Creutzfeldt-Jakob disease: neurological features and diagnostic tests. *Lancet* 1997; **350**:903–907.
88. Ghani AC, Ferguson NM, Donnelly CA, Hagenaars TJ, Anderson RM Epidemiological determinants of the pattern and magnitude of the vCJD epidemic in Great Britain. *Proc R Soc Lond B* 1999; **265**:2443–2452.
89. Lloyd SE, Uphill JB, Targonski PV, Fisher EM, Collinge J. Identification of genetic loci affecting mouse-adapted bovine spongiform encephalopathy incubation time in mice. *Neurogenetics* 2002; **4**:77–81.
90. Hill AF, Joiner S, Wadsworth JD, Sidle KC, Bell JE, Budka H, Ironside JW, Collinge J. Molecular classification of sporadic Creutzfeldt-Jakob disease. *Brain* 2003; **126**:1333–1346.
91. Parchi P, Giese A, Capellari S, Brown P, Schulz-Schaeffer W, Windl O, Zerr I, Budka H, Kopp N, Piccardo P, Poser S, Rojiani A, Streichenberger N, Julien J, Vital C, Ghetti B, Gambetti P, Kretzschmar H. Classification of sporadic Creutzfeldt-Jakob Disease based on molecular and phenotypic analysis of 300 subjects. *Annals of Neurology* 1999; **46**:224–233.
92. Collinge J. Human prion diseases and bovine spongiform encephalopathy (BSE). *Hum Mol Genetics* 1997; **6**:1699–1705.
93. Collins S, Law MG, Fletcher A, Boyd A, Kaldor J, Masters CL. Surgical treatment and risk of sporadic Creutzfeldt-jakob disease: a case-control study. *Lancet* 1999; **353**:693–697.
94. Asante EA, Linehan JM, Desbruslais M, Joiner S, Gowland I, Wood A, Welch J, Hill AF, Lloyd SE, Wadsworth JDF, Collinge J. BSE prions propagate as either variant CJD-like or sporadic CJD-like prion strains in transgenic mice expressing human prion protein. *EMBO J* 2002; **21**(23):6358–6366.
95. Ghani AC, Ferguson NM, Donnelly CA, Anderson RM. Predicted vCJD mortality in Great Britain. *Nature* 2000; **406**:583–584.
96. D'Aignaux JNH, Cousens SN, Smith PG. Predictability of the UK variant Creutzfeldt-Jakob disease epidemic. *Science* 2001; **294**:1729–1731.
97. Lloyd SE, Onwuazor ON, Beck JA, Mallinson G, Farrall M, Targonski P, Collinge J, Fisher EMC. Identification of multiple quantitative trait loci linked to prion disease incubation period in mice. *Proc Natl Acad Sci USA* 2001; **98**:6279–6283.
98. Wadsworth JD, Asante EA, Desbruslais M, Linehan JM, Joiner S, Gowland I, Welch J, Stone L, Lloyd SE, Hill AF, Brandner S, Collinge J. Human Prion

Protein with Valine 129 Prevents Expression of Variant CJD Phenotype. *Science* 2004; **306**:1793–1796.

99. Mallucci G, Collinge J. Update on Creutzfeldt-Jakob disease. *Curr Opin Neurol* 2004; **17**:641–647.
100. Collinge J, Hill AF, Ironside J, Zeidler M. Diagnosis of new variant Creutzfeldt-Jakob disease by tonsil biopsy — authors' reply to Arya and Evans. *Lancet* 1997; **349**:1322–1323.
101. Collinge J. Human prion diseases: aetiology and clinical features. In: Growdon JH, Rossor M, eds. *The Dementias*. Newton, MA: Butterworth-Heinemann 1998: 113–148.
102. Glatzel M, Abela E, Maissen M, Aguzzi A. Extraneural pathologic prion protein in sporadic Creutzfeldt-Jakob disease. *N Engl J Med* 2003; **349**:1812–1820.
103. Bernoulli C, Siegfried J, Baumgartner G, Regli F, Rabinowicz T, Gajdusek DC, Gibbs CJ, Jr. Danger of accidental person-to-person transmission of Creutzfeldt-Jakob disease by surgery [letter]. *Lancet* 1977; **1**:478–479.
104. Blattler T. Implications of prion diseases for neurosurgery. *Neurosurg Rev* 2002; **25**:195–203.
105. Flechsig E, Hegyi I, Enari M, Schwarz P, Collinge J, Weissmann C. Transmission of scrapie by steel-surface-bound prions. *Mol Med* 2001; **7**:679–684.
106. Joiner S, Linehan J, Brandner S, Wadsworth JD, Collinge J. Irregular presence of abnormal prion protein in appendix in variant Creutzfeldt-Jakob disease. *J Neurol Neurosurg Psychiatry* 2002; **73**:597–598.
107. Hilton DA, Ghani AC, Conyers L, Edwards P, McCardle L, Ritchie D, Penney M, Hegazy D, Ironside JW. Prevalence of lymphoreticular prion protein accumulation in UK tissue samples. *J Pathol* 2004; **203**:733–739.
108. Frosh A, Smith LC, Jackson CJ, Linehan JM, Brandner S, Wadsworth JD, Collinge PJ. Analysis of 2000 consecutive UK tonsillectomy specimens for disease-related prion protein *Lancet* 2004; **364**:1260–1262.
109. Aguzzi A. Neuro-immune connection in spread of prions in the body. *Lancet* 1997; **349**:742–743.
110. Mallucci G, Collinge J. Rational targeting for prion therapeutics *Nat Rev Neurosci* 2005; **6**:23–34.
111. Brandner S, Isenmann S, Raeber A, Fischer M, Sailer A, Kobayashi Y, Marino S, Weissmann C, Aguzzi A. Normal host prion protein necessary for scrapie-induced neurotoxicity. *Nature* 1996; **379**:339–343.
112. Bueler H, Fischer M, Lang Y, Bluthmann H, Lipp H-P, DeArmond SJ, Prusiner SB, Aguet M, Weissmann C. PrP protein is not essential for normal development and behavior of the mouse. *Nature* 1993.
113. Bueler H, Aguzzi A, Sailer A, Greiner RA, Autenried P, Aguet M, Weissmann C. Mice devoid of PrP are resistant to scrapie. *Cell* 1993; **73**: 1339–1347.

114. Manson JC, Clarke AR, McBride PA, McConnell I, Hope J. PrP gene dosage determines the timing but not the final intensity or distribution of lesions in scrapie pathology. *Neurodegen* 1994; **3**:331–340.
115. Manson JC, Clarke AR, Hooper ML, Aitchison L, McConnell I, Hope J. 129/Ola mice carrying a null mutation in PrP that abolishes mRNA production are developmentally normal. *Mol Neurobiol* 1994; **8**:121–127.
116. Bueler H, Fischer M, Lang Y, Bluethmann H, Lipp H-P, DeArmond SJ, Prusiner SB, Aguet M, Weissmann C. Normal development and behaviour of mice lacking the neuronal cell-surface PrP protein. *Nature* 1992; **356**: 577–582.
117. Mallucci GR, Ratté S, Asante EA, Linehan J, Gowland I, Jefferys JGR, Collinge J. Post-natal knockout of prion protein alters hippocampal CA1 properties, but does not result in neurodegeneration. *EMBO J* 2002; **21**: 202–210.
118. Hosszu LLP, Baxter NJ, Jackson GS, Power A, Clarke AR, Waltho JP, Craven CJ, Collinge J. Structural mobility of the human prion protein probed by backbone hydrogen exchange. *Nature Struct Biol* 1999; **6**:740–743.
119. Enari M, Flechsig E, Weissmann C. Scrapie prion protein accumulation by scrapie-infected neuroblastoma cells abrogated by exposure to a prion protein antibody. *Proc Natl Acad Sci USA* 2001; **98**:9295–9299.
120. Peretz D, Williamson RA, Kaneko K, Vergara J, Leclerc E, Schmitt-Ulms G, Mehlhorn IR, Legname G, Wormald MR, Rudd PM, Dwek RA, Burton DR, Prusiner SB. Antibodies inhibit prion propagation and clear cell cultures of prion infectivity. *Nature* 2001; **412**:739–743.
121. Gilch S, Wopfner F, Renner-Muller I, Kremmer E, Bauer C, Wolf E, Brem G, Groschup MH, Schatzl HM. Polyclonal anti-PrP auto-antibodies induced with dimeric PrP interfere efficiently with PrPSc propagation in prion-infected cells. *J Biol Chem* 2003.
122. Heppner FL, Musahl C, Arrighi I, Klein MA, Rülicke T, Oesch B, Zinkernagel RM, Kalinke U, Aguzzi A. Prevention of scrapie pathogenesis by transgenic expression of anti-prion protein antibodies. *Science* 2001; **294**:178–182.
123. White AR, Enever P, Tayebi M, Mushens R, Linehan J, Brandner S, Anstee D, Collinge J, Hawke S. Monoclonal antibodies inhibit prion replication and delay the development of prion disease. *Nature* 2003; **422**:80–83.
124. Solforosi L, Criado JR, McGavern DB, Wirz S, Sanchez-Alavez M, Sugama S, DeGiorgio LA, Volpe BT, Wiseman E, Abalos G, Masliah E, Gilden D, Oldstone MB, Conti B, Williamson RA. Cross-Linking Cellular Prion Protein Triggers Neuronal Apoptosis in Vivo. *Science* 2004.
125. Sigurdsson EM, Brown DR, Daniels M, Kascsak RJ, Kascsak R, Carp R, Meeker HC, Frangione B, Wisniewski T. Immunization delays the onset of prion disease in mice. *Am J Pathol* 2002; **161**:13–17.

126. Schwarz A, Kratke O, Burwinkel M, Riemer C, Schultz J, Henklein P, Bamme T, Baier M. Immunisation with a synthetic prion protein-derived peptide prolongs survival times of mice orally exposed to the scrapie agent. *Neurosci Lett* 2003; **350**:187–189.
127. Mattson MP, Chan SL. Good and bad amyloid antibodies. *Science* 2003; **301**:1847–1849.
128. Nicoll JA, Wilkinson D, Holmes C, Steart P, Markham H, Weller RO. Neuropathology of human Alzheimer disease after immunization with amyloid-beta peptide: a case report. *Nat Med* 2003; **9**:448–452.
129. Hannon GJ. RNA interference. *Nature* 2002; **418**:244–251.
130. Dykxhoorn DM, Novina CD, Sharp PA. Killing the messenger: short RNAs that silence gene expression. *Nat Rev Mol Cell Biol* 2003; **4**:457–467.
131. Telling GC, Parchi P, DeArmond SJ, Cortelli P, Montagna P, Gabizon R, Mastrianni J, Lugaresi E, Gambetti P, Prusiner SB. Evidence for the conformation of the pathologic isoform of the prion protein enciphering and propagating prion diversity. *Science* 1996; **274**:2079–2082.
132. Lloyd SE, Linehan JM, Desbruslais M, Joiner S, Buckell J, Brandner S, Wadsworth JD, Collinge J. Characterization of two distinct prion strains derived from bovine spongiform encephalopathy transmissions to inbred mice. *J Gen Virol* 2004; **85**:2471–2478.
133. Otto M, Cepek L, Ratzka P, Doehlinger S, Boekhoff I, Wiltfang J, Irle E, Pergande G, Ellers-Lenz B, Windl O, Kretzschmar HA, Poser S, Prange H. Efficacy of flupirtine on cognitive function in patients with CJD: A double-blind study. *Neurology* 2004; **62**:714–718.
134. Korth C, May BC, Cohen FE, Prusiner SB. Acridine and phenothiazine derivatives as pharmacotherapeutics for prion disease. *Proc Natl Acad Sci USA* 2001; **98**:9836–9841.
135. Adjou KT, Simoneau S, Sales N, Lamoury F, Dormont D, Papy-Garcia D, Barritault D, Deslys JP, Lasmezas CI. A novel generation of heparan sulfate mimetics for the treatment of prion diseases. *J Gen Virol* 2003; **84**:2595–2603.
136. Doh-ura K, Ishikawa K, Murakami-Kubo I, Sasaki K, Mohri S, Race R, Iwaki T. Treatment of Transmissible Spongiform Encephalopathy by Intraventricular Drug Infusion in Animal Models. *J Virol* 2004; **78**:4999–5006.
137. Barrow PA, Mallucci GR, Collinge J, Jefferys JGR. Rescue of a physiological phenotype in mice with ablated PrP gene but expressing a transgene cassette for PrP. *J Physiology* 523P:203P. 2000.
138. Mayer V, Rajcani J, Mitrova E. Autoantibodies against neurofilaments in Creutzfeldt-Jakob disease: immunofluorescence in sections of human caudate nucleus. *Acta Virol (Praha)* 1981; **25**:390–394.

11

Monitoring and Interpretation of Intracranial Pressure

M Czosnyka and J D Pickard*

Intracranial pressure (ICP) is derived from cerebral blood and cerebrospinal fluid (CSF) circulatory dynamics and can be affected in the course of many diseases of the central nervous system. Monitoring of ICP requires an invasive transducer, although some attempts have been made to measure it non-invasively. Because of its dynamic nature, instant CSF pressure measurement using the height of a fluid column via lumbar puncture may be misleading. An averaging over 30 minutes should be the minimum, with a period of overnight monitoring in conscious patients providing the optimal standard. Computer-aided recording with online waveform analysis of ICP is very helpful.

Although there is no "Class I" evidence, ICP monitoring is useful, if not essential, in head injury, poor grade subarachnoid haemorrhage, stroke, intracerebral haematoma, meningitis, acute liver failure, hydrocephalus, benign intracranial hypertension, craniosynostosis etc. Information which

*Academic Neurosurgical Unit, University of Cambridge Clinical School, Cambridge, UK. Dr M Czosnyka, Academic Neurosurgery, Box 167, Addenbrooke's Hospital, Cambridge CB2 2QQ, UK; MC141@medschl.cam.ac.uk

can be derived from ICP and its waveforms includes cerebral perfusion pressure (CPP), regulation of cerebral blood flow and volume, CSF absorption capacity, brain compensatory reserve, and content of vasogenic events. Some of these parameters allow prediction of prognosis of survival following head injury and optimisation of "CPP-guided therapy". In hydrocephalus CSF dynamic tests aid diagnosis and subsequent monitoring of shunt function.

In most organs of the human body the environmental pressure for blood perfusion is either low, or coupled to atmospheric pressure. The environmental pressure for the brain differs in this respect as brain is surrounded and protected by a stiff skull. A rise in intracranial pressure (ICP) may impede blood flow and cause ischaemia.

Intracranial pressure is derived from the circulation of cerebral blood and cerebrospinal fluid (CSF) (however, it is not certain whether the operator in the formula should be represented by a simple addition): $ICP = ICP_{vascular} + ICP_{CSF}$.

The vascular component[1,2] is difficult to express quantitatively. It is probably derived from the pulsation of the cerebral blood volume detected and averaged by non-linear mechanisms of regulation of cerebral blood volume. More generally, multiple variables such as the arterial pressure, autoregulation, and cerebral venous outflow all contribute to the vascular component. The other circulatory CSF component may be expressed using Davson's equation[3]: ICP_{CSF} = (resistance to CSF outflow) × (CSF formation) + (pressure in sagittal sinus).

Any factor, which under physiological (for example, compression of jugular veins during a Queckenstedt test) or pathological conditions (brain swelling, space occupying lesion, obstruction of CSF pathway) disturbs this circulation, may provoke an increase in ICP.

ICP measurements are used to estimate cerebral perfusion pressure (CPP) as follows: mean CPP = mean arterial blood pressure (ABP) − mean ICP. CPP represents the pressure gradient acting across the cerebrovascular bed, and hence is an important factor in regulation of the cerebral blood flow (CBF).[4] Sufficient CPP is required to maintain a stable CBF. The autoregulatory reserve is interpreted as the difference between current mean CPP and lower limit of autoregulation. Low CPP (a threshold of 60–70 mm Hg is generally accepted in adults) may result in exhaustion of the autoregulatory reserve. However, policies to therapeutically maintain a high CPP are controversial. If the cerebral vessels are non-reactive, an increase in CPP may result in hyperaemia, increase in vasogenic oedema, and a secondary increase in ICP.[5,6] It is also probable that patient and time dependent differences in the optimal level of CPP may be considerable. Thus, the border between adequate and non-adequate CPP should be assessed individually and frequently.

METHODS OF MEASUREMENT

An intraventricular drain connected to an external pressure transducer is still considered to be "golden standard" method.[7,8] ICP can be controlled by CSF drainage and the transducer may be zeroed externally. However, after five days of monitoring the risk of infection starts to increase, with an overall risk estimated to be about 5%.[9] Insertion of the ventricular catheter may be difficult or impossible in cases of advanced brain swelling. Modern ventricular, subdural, or intraparenchymal microtransducers (most popular types: Camino ICP Bolt; Camino Laboratories, San Diego, California, USA; and Codman MicroSensor, Johnson and Johnson Professional Inc, Raynham, Massachusetts, USA) reduce infection rate and risk of haemorrhage[9] and have excellent metrological properties as revealed during bench tests — that is, bandwidth, linearity.[10] Generally, uniformly distributed ICP can be probably seen only when CSF circulates freely between all its natural pools, equilibrating pressure everywhere. When little or no CSF volume is left due to brain swelling, the assumption of one, uniform value of ICP is questionable. With the most common intraparenchymal probes, measured pressure may be compartmentalised and not necessarily representative of real ICP — that is, ventricular CSF pressure.[11] Microtransducers cannot be generally re-zeroed after insertion and considerable zero drift can sometimes occur in long term monitoring.[12] This problem has been addressed in the balloon-like Spiegelberg transducer which may zero itself every set time interval, although its limited bandwidth may make most of the methods used for the ICP waveform analysis impossible.[13] Contemporary epidural sensors are much more reliable than 10 years ago. But the question as to whether epidural pressure can express ICP with confidence and under all circumstances is still unanswered.

Lumbar CSF pressure is very seldom measured in neuro-intensive care. This form of assessment of craniospinal dynamics is more often used in hydrocephalus and benign intracranial hypertension. It is important to emphasise that decent monitoring over a period of at least half an hour with recording of pressure and pulse amplitude should be used as a "golden standard". Manometric assessment by measuring the height of the CSF column may be misleading as CSF pressure may vary considerably with time.[14] Attempts to monitor ICP non-invasively are still in a phase of technical evaluation.[15] The most promising methods are based on transcranial ultrasonography[16–18] — see below.

TYPICAL EVENTS AND TRENDS IN ICP MONITORING

It is difficult to establish a universal "normal value" for ICP as it depends on age, body posture, and clinical conditions. In the horizontal position, the normal ICP in healthy adult subjects was reported to be within the range of 7–15 mm Hg.[19] In the vertical position it is negative with a mean of around −10 mm Hg, but not falling below −15 mm Hg.[20]

The definition of raised ICP depends on the specific pathology. In hydrocephalus, a pressure above 15 mm Hg can be regarded as elevated. Following head injury, anything above 20 mm Hg is abnormal and aggressive treatment usually starts above 25 mm Hg. It is important to recognise that ICP in most cases varies with time. Decent averaging for at least 30 minutes is needed to calculate "mean ICP". The patient should rest in a horizontal position during the measurement and avoid movement, speaking, etc. Overnight monitoring during natural sleep, which provides a "grand average" with a good description of the dynamics of the pressure, should be regarded as the "gold standard" in conscious patients.

When monitored continuously in acute states (head injury, poor grade subarachnoid haemorrhage, intracerebral haematoma etc.), mean ICP may be classified into relatively few patterns (Fig. 1).

- Low and stable ICP (below 20 mm Hg) — for example, seen in patients following uncomplicated head injury (Fig. 1a). Such a pattern is also commonly seen in the initial period after brain trauma before the brain swelling evolves.
- High and stable ICP (above 20 mm Hg) — the most common picture to follow head injury (Fig. 1b).
- Vasogenic waves — "B" waves (Fig. 1c) and plateau waves (Fig. 1d).

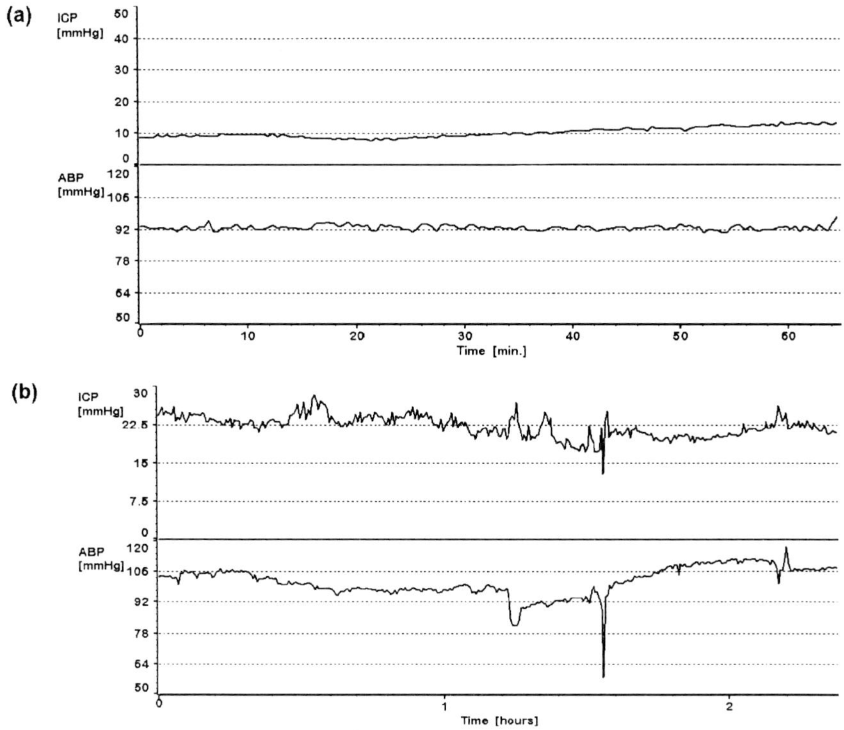

Fig. 1 (a) Low and stable intracranial pressure (ICP). Mean arterial blood pressure (ABP) is plotted along the lower panel. (b) Stable and elevated ICP — this can be seen most of the time in head injury patients. (c) "B" waves of ICP. They are seen both in mean ICP and spectrally resolved pulse amplitude of ICP (AMP, upper panel). They are also usually seen in plots of time averaged ABP, but not always. (d) Plateau waves of ICP. Cerebrospinal compensatory reserve is usually low when these waves are recorded (RAP (correlation coefficient (R) between AMP amplitude (A) and mean pressure (P)) close to +1; index of compensatory reserve). At the height of the waves, during maximal vasodilatation, integration between pulse amplitude and mean ICP fails as is indicated by fall in RAP. After the plateau wave, ICP usually falls below baseline level and cerebrospinal compensatory reserve improves. (e) High, spiky waves of ICP caused by sudden increases in ABP. (f) Increase in ICP caused by temporary decrease in ABP. (g) Increase in ICP of "hyperaemic nature". Both blood flow velocity and jugular venous oxygen saturation (Sjo_2) increased in parallel with ICP. (h) Refractory intracranial hypertension. ICP increased within a few hours to above 100 mm Hg. The vertical line denotes the moment when the ischaemic wave probably reached the vasomotor centres in the brain stem: heart rate increased and ABP (cerebral perfusion pressure) decreased abruptly. Note that the pulse amplitude of ICP (AMP) disappeared around 10 minutes before this terminal event.

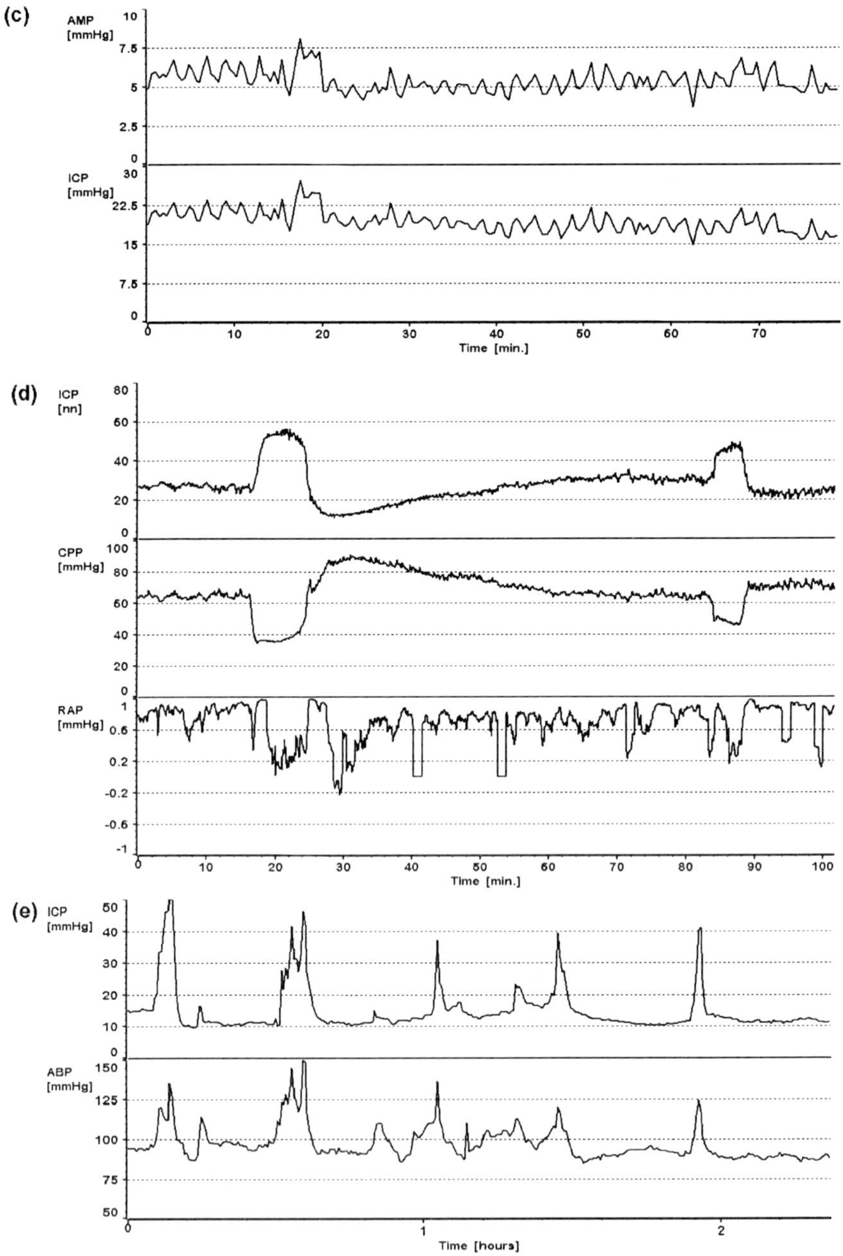

Fig. 1 *(Continued)*

(f)

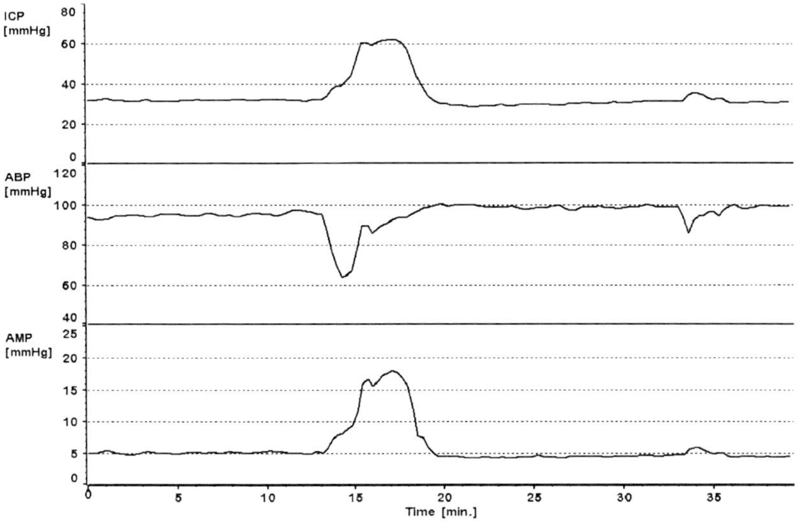

(g)

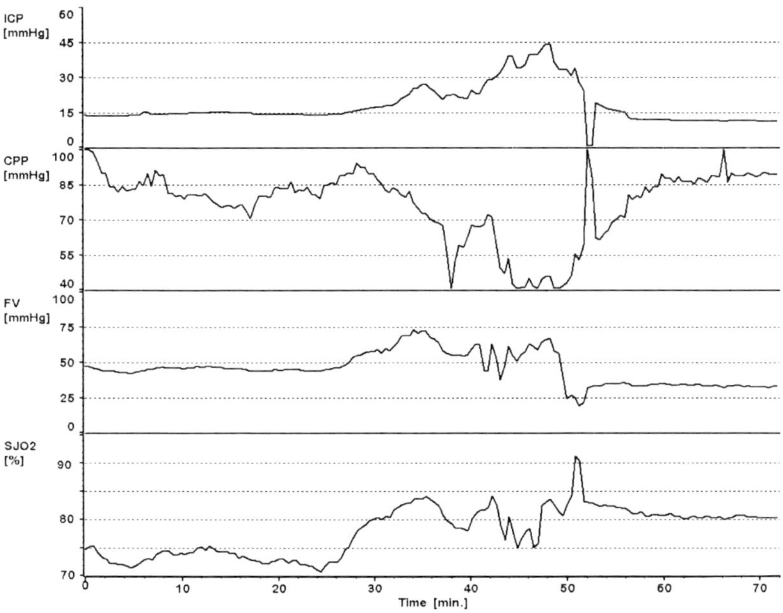

Fig. 1 *(Continued)*

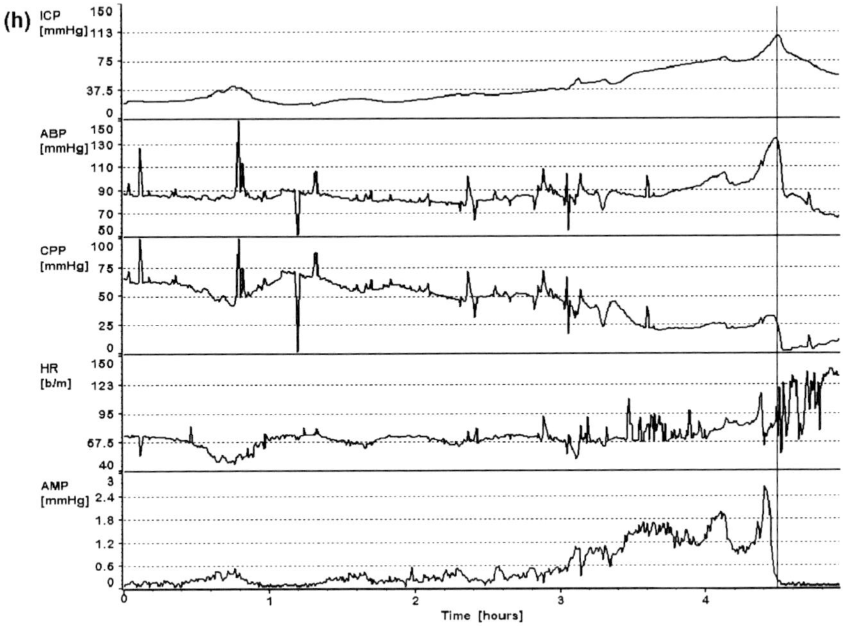

Fig. 1 *(Continued)*

- ICP waves related to changes in arterial pressure and hyperaemic events (Fig. 1 e–g).
- Refractory intracranial hypertension (Fig. 1h). This usually leads to death unless radical measures — for example, surgical decompression, are applied.

However, there is much more information in the ICP waveform than in the time averaged ICP mean value alone. It is important to bear in mind how little information is available in the fluid column in the manometer line connected to a lumbar puncture needle!

WAVEFORM ANALYSIS OF ICP

The ICP waveform consists of three components, which overlap in the time domain, but can be separated in the frequency domain (Fig. 2). The pulse waveform has several harmonic components; of these the fundamental component has a frequency equal to the heart rate. The amplitude of this component (AMP) is very useful for the evaluation of various indices.

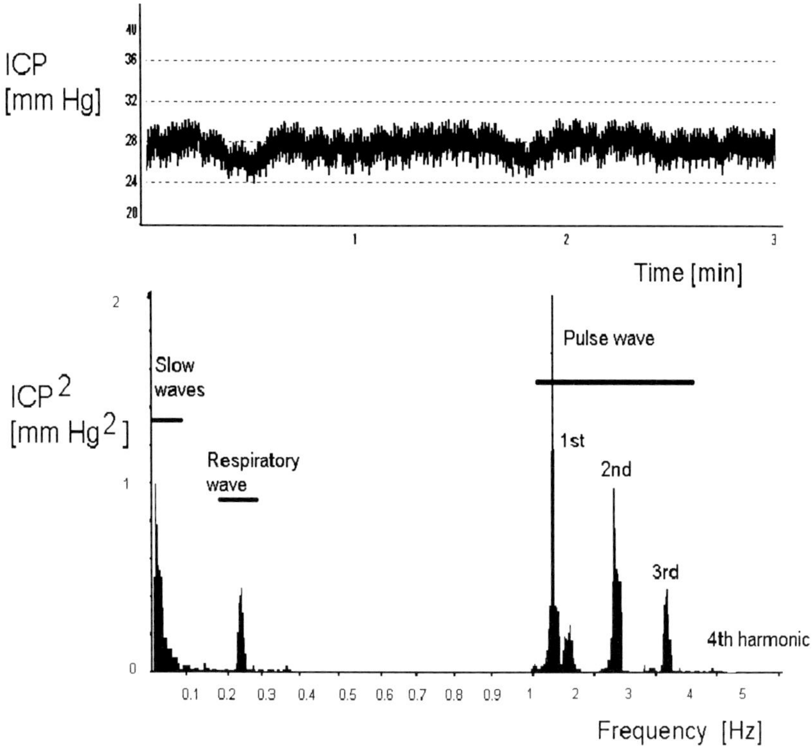

Fig. 2 Example of ICP recording showing pulse, respiratory, and "slow waves" overlapped in time domain (upper panel). Taken from Balestreri M, Czosnyka M. *Acta Neurochir* 2004; (in press)[21] with permission from the publishers.

The respiratory waveform is related to the frequency of the respiratory cycle (8–20 cycles per minute). "Slow waves" are usually not as precisely defined as in the original Lundberg thesis.[8] All components that have a spectral representation within the frequency limits of 0.05–0.0055 Hz (20 s to 3 min period) can be classified as slow waves. The magnitude of these waves can be calculated as the square root of the power of the signal of the passband of the equivalent frequency range at the output of the digital filter or any other equivalent detection technique.

PRESSURE–VOLUME COMPENSATORY RESERVE

Theoretically, the compensatory reserve can be studied through the relation between ICP and changes in volume of the intracerebral space, known as the "pressure–volume curve".[22,23] The index called RAP (correlation coefficient (R) between AMP amplitude (A) and mean pressure (P); index of compensatory reserve) can be derived by calculating the linear correlation between consecutive, time averaged data points of AMP and ICP (usually 40 of such samples are used) acquired over a reasonably long period to average over respiratory and pulse waves (usually 6–10 s periods). This index indicates the degree of correlation between AMP and mean ICP over short periods of time (~4 min).

A RAP coefficient close to 0 indicates lack of synchronisation between changes in AMP and mean ICP. This denotes a good pressure–volume compensatory reserve at low ICP (Fig. 3), where a change in volume produces no or very little change of the pressure. When RAP rises to +1, AMP varies directly with ICP and this indicates that the "working point" of the intracranial space shifts to the right towards the steep part of the pressure–volume curve. Here compensatory reserve is low; therefore any further rise in volume may produce a rapid increase in ICP. Following head injury and subsequent brain swelling, RAP is usually close to +1. With any further increase in ICP, AMP decreases and RAP values fall below zero. This occurs when the cerebral autoregulatory capacity is exhausted and the pressure–volume curve bends to the right as the capacity of cerebral arterioles to dilate in response to a CPP decrement is exhausted, and they tend to collapse passively. This indicates terminal cerebrovascular derangement with a decrease in pulse pressure transmission from the arterial bed to the intracranial compartment.

CEREBROVASCULAR PRESSURE REACTIVITY

Another ICP derived index is the pressure-reactivity index (PRx), which incorporates the philosophy of assessing cerebrovascular reactions by observing the response of ICP to slow spontaneous changes in ABP.[24] When the cerebrovascular bed is normally reactive, any change in ABP produces an inverse change in cerebral blood volume and hence ICP. When reactivity is disturbed, changes in ABP are passively transmitted to ICP. Using computational methods similar to those used for the calculation of the RAP index, PRx is determined by calculating the correlation coefficient between 40 consecutive data points of time averaged (over 6 to 10 seconds) ICP

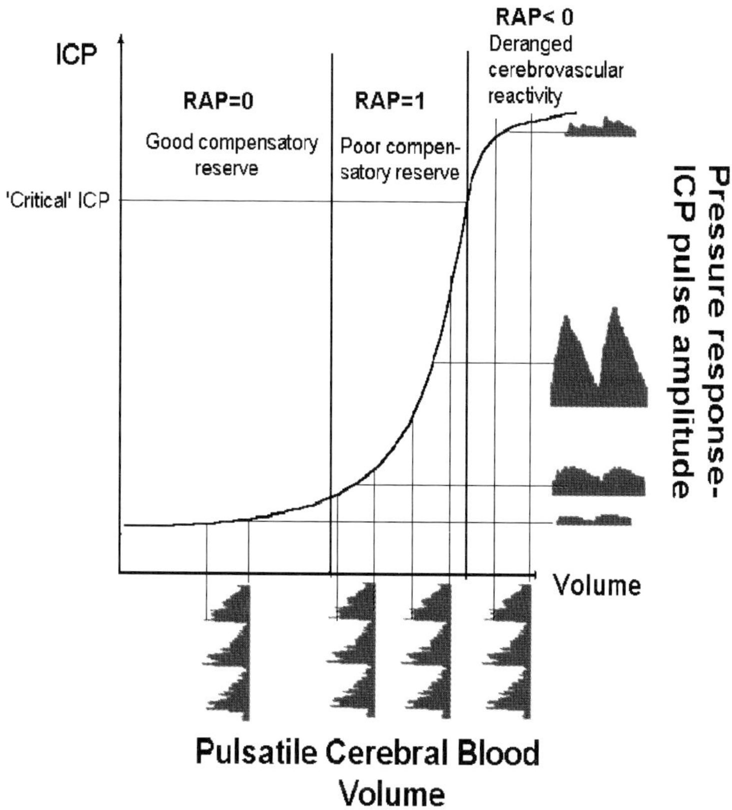

Fig. 3 In a simple model, pulse amplitude of intracranial pressure (ICP) (expressed along the y-axis on the right side of the panel) results from pulsatile changes in cerebral blood volume (expressed along the x-axis) transformed by the pressure–volume curve. This curve has three zones: a flat zone, expressing good compensatory reserve, an exponential zone, depicting poor compensatory reserve, and a flat zone again, seen at very high ICP (above the "critical" ICP) depicting derangement of normal cerebrovascular responses. The pulse amplitude of ICP is low and does not depend on mean ICP in the first zone. The pulse amplitude increases linearly with mean ICP in the zone of poor compensatory reserve. In the third zone, the pulse amplitude starts to decrease with rising ICP. RAP, index of compensatory reserve. Taken from Balestreri M, Czosnyka M. *Acta Neurochir* 2004; (in press)[21] with permission from the publishers.

and ABP. A positive PRx signifies a positive gradient of the regression line between the slow components of ABP and ICP, which we hypothesise to be associated with passive behaviour of a non-reactive vascular bed (Fig. 4a). A negative value of PRx reflects a normally reactive vascular bed, as ABP

(a)

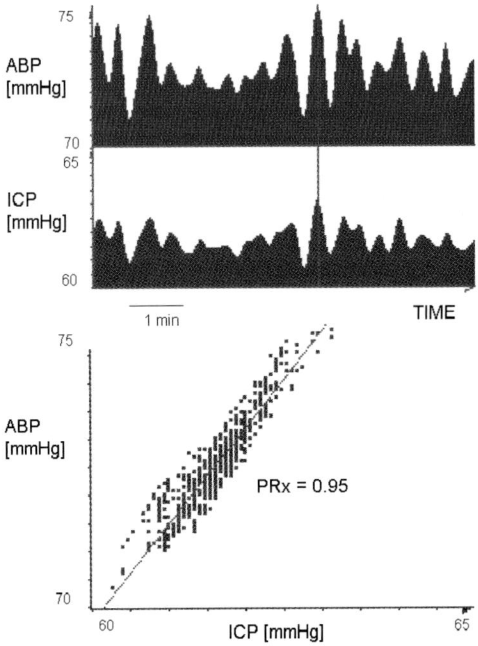

Fig. 4 Relation between slow waves of arterial blood pressure (ABP) and intracranial pressure (ICP). (a) Slow waves in ICP and ABP produce obviously positive correlation (lower panel), giving a positive value of PRx. This indicated loss of cerebrovascular reserve. (b) Coherent waves both in ABP and ICP (upper panel) produced a negative correlation coefficient, when plotted on the regression graph (lower panel), giving clearly negative values of PRx. Taken from Czosnyka M, Smielewski P, Kirkpatrick P, *et al.* Continuous assessment of the cerebral vasomotor reactivity in head injury. *Neurosurgery* 1997; **41**:11–17[25] with permission from the publisher.

waves provoke inversely correlated waves in ICP (Fig. 4b). This index correlates well with indices of autoregulation based on transcranial Doppler ultrasonography.[25] Furthermore, abnormal values of both PRx and RAP, indicative respectively of poor autoregulation or deranged cerebrospinal compensatory reserve, have been demonstrated to be predictive of a poor outcome following head injury.[5]

OTHER METHODS OF ICP ANALYSIS

One of the priorities in brain monitoring is to develop a technique which helps in predicting decompensation or herniation. Early works focused on

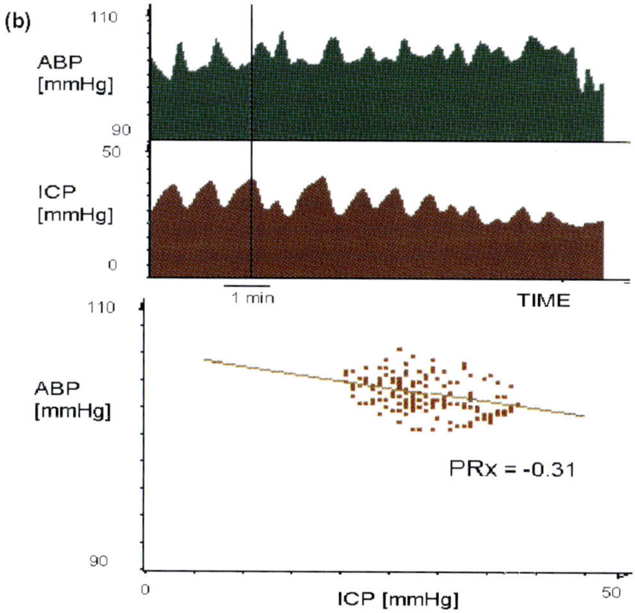

Fig. 4 (*Continued*)

intracranial volume–pressure response[26] which evolved in time into continuous monitoring of brain compliance.[27] Such a method relies on the evaluation of the pressure response to known small volume additions by inflating and deflating a balloon inserted within the cerebrospinal space. The method has been implemented in the Spiegelberg Brain Compliance monitor (Spiegelberg GmbH, Hamburg, Germany) and initial trials indicated it might be useful in various scenarios.[28,29] Its correlation with outcome remains to be demonstrated.

Analysis of the pulse waveform of ICP, known as high-frequency-centroid, was based on evaluation of the power spectrum of a single pulse ICP waveform and calculation of its "power-weighted average frequency" within the frequency range of 5–15 Hz. High-frequency centroid was demonstrated to decrease with increasing ICP and then increase in the state of refractory intracranial hypertension where the blood flow regulation mechanism failed.[30]

Pulse transmission between arterial pressure and mean ICP has been investigated by various groups.[31–33] It depends upon assumptions about the linearity of the transmission model. Such assumptions are probably unrealistic particularly in pathological circumstances.

The ratio of respiratory wave to pulse amplitude of ICP was believed to be predictive of a worse outcome after head injury. Modulation of pulse waveform by the respiratory cycle has been demonstrated to correlate with brain compliance. Attempts have been made to use this for diagnosis in patients with hydrocephalus.[34]

Recently, the power of slow waves of ICP was reported to be predictive of outcome in patients with intracranial hypertension following head injury. A low content of slow waves in the overall ICP dynamics was associated with a fatal outcome[21] highlighting a possible link of these events with cerebral autoregulation.[35]

PRACTICAL USE OF THE ICP DERIVED PARAMETERS: "OPTIMAL CPP" AND MULTIPLE TREND ANALYSIS

Both PRx and RAP can be used to evaluate secondary variables that combine the value of absolute ICP and CPP with information about the state of autoregulatory and compensatory reserves. PRx plotted against CPP gives a "U shape" curve.[5] This indicates that for the majority of patients there is a value of the CPP in which pressure-reactivity is optimal. This optimal pressure can be estimated by plotting and analysing the PRx–CPP curve in a sequential six hour wide time-moving window online. It has been demonstrated in a group of retrospectively evaluated patients that the greater the distance between the current and the "optimal" CPP the worse the outcome. This potentially useful methodology attempts to refine CPP oriented therapy.[5] Both, too low (ischaemia), and, too high, CPP (hyperaemia and secondary increase in ICP) are detrimental. Hence, we have suggested that CPP should be optimised to maintain cerebral perfusion in the globally most favourable state.[5]

No matter how sophisticated new variables or outcome-predicting models become, the most useful tool at the bedside is a computer screen, which presents the trends of multiple parameters with time. This gives an opportunity to react to a crisis situation, understand the cerebrospinal dynamics in multiple dimensions and predict an optimal strategy for the individual patient's care (Fig. 5).

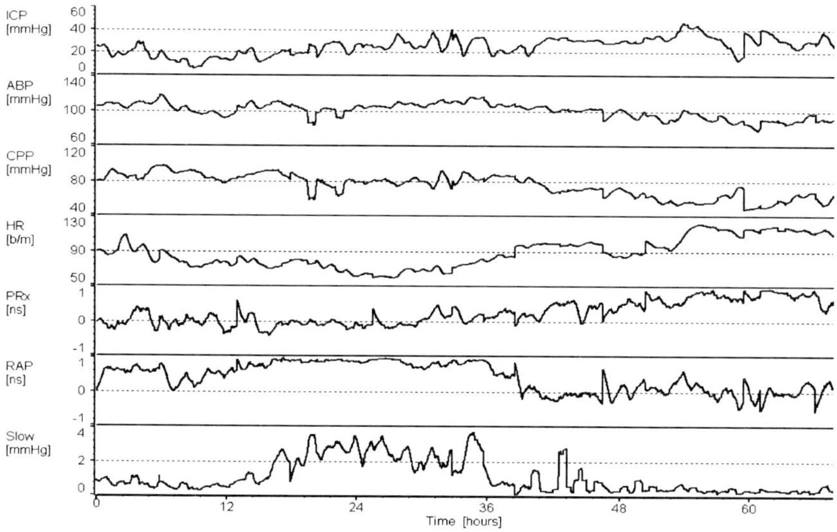

Fig. 5 A typical example of intracranial hypertension in a patient who died. In this case the progressive increase in intracranial pressure (ICP) caused a severe fall in cerebral perfusion pressure (CPP) (time 48 h), although normal values of arterial blood pressure (ABP) were maintained. The pressure-reactivity index (PRx), at first slightly above the normal range, started to climb in the second part of the monitoring (time > 40 h) when the progressive increase in ICP was associated with CPP reduction; similarly RAP (index of compensatory reserve), initially above 0.6, started to fall suggesting a loss in vasomotor reserve (time > 40 h). Slow waves were low at the beginning, increased around 24 hours after injury and fell 36 hours after injury. Taken from Balestreri M, Czosnyka M. *Acta Neurochir* 2004. With permission from the publishers.

ASSOCIATION WITH OUTCOME FOLLOWING SEVERE HEAD INJURY

In severe head injury an averaged ICP above 25 mm Hg over the whole period of monitoring increases risk of death twofold. Averaged values of the RAP and PRx indices are also strong predictors of fatal outcome. Both these indices suggest that good vascular reactivity is an important element of brain homoeostasis, enabling the brain to protect itself against an uncontrollable rise of the intracerebral volume. A low value of slow waves of ICP is also indicative of a fatal outcome following head injury.

As ICP, PRx, and the power of ICP slow waves are independent predictors of outcome, these three variables, although mutually correlated, should be considered jointly in any outcome analysis.

Mean CPP has become an actively controlled variable and hence has lost its predictive power for outcome. It does not mean that traditional short term decreases in CPP ("CPP insults") have become more benign, but they are probably better managed nowadays, and with baseline CPP above 65–70 mm Hg they do not frequently produce ischaemia. This is, probably, one of the most spectacular "success stories" of CPP oriented protocols. However, there is still lack of "Class I" evidence indicating that CPP oriented therapy is clearly beneficial. Robertson *et al.*[36] compared CPP and ICP oriented therapies and showed a decrease in ischaemic insults in CPP oriented therapy but an increase in respiratory complications with no overall difference in outcome.

ICP IN HYDROCEPHALUS AND BENIGN INTRACRANIAL HYPERTENSION

ICP in chronic diseases should be interpreted slightly differently. The problems are related more to disturbance of the CSF circulation. Increased ICP signifies increased resistance to CSF outflow[37] or increased cerebral venous outflow pressure,[38] rather than exhausted cerebrospinal compensatory reserve created by a decreased volume in either one or both of the two main buffering components — CSF and venous blood volume, as in head injury.

In more chronic conditions of ventricular dilatation, where ICP is not greatly raised, obstruction to CSF absorption may be confirmed by CSF infusion tests (ventricular or lumbar) taking care to adapt the technique to the site of any obstruction.[39] The infusion study can be performed via the lumbar CSF space or via a preimplanted ventricular access device. In both cases, two needles are inserted (22 G spinal needles for lumbar tests; 25 G butterfly needles for ventricular studies). One needle is connected to a pressure transducer via a stiff saline-filled tube and the other to an infusion pump mounted on a purpose-built trolley containing a pressure amplifier and personal computer running software written inhouse. After 10 minutes of baseline measurement, the infusion of normal saline at a rate of 1.5 ml/min or 1 ml/min (if the baseline pressure was higher than

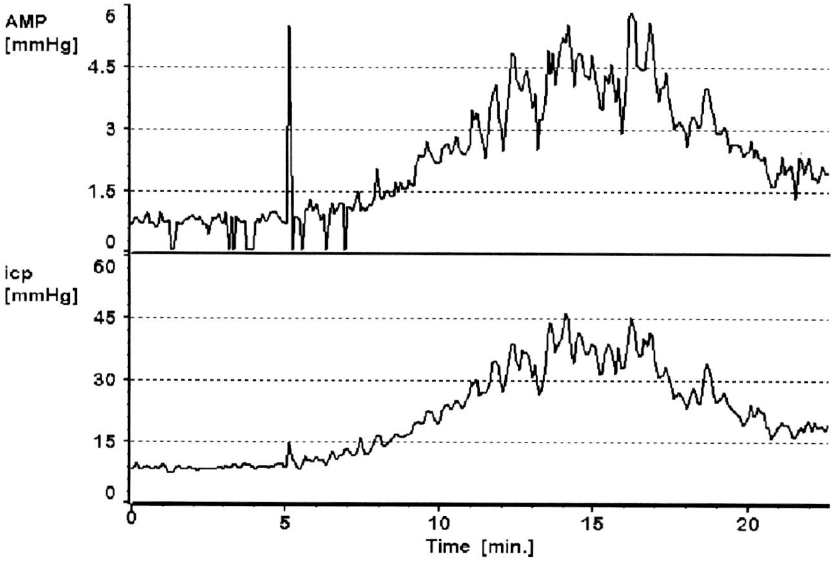

Fig. 6 Infusion test. Pulse amplitude of intracranial pressure (ICP) (AMP) on upper graph and mean ICP (lower graph) plotted versus time. Normal saline infusion was started at 6 minutes of monitoring. ICP reached a level of 37 mm Hg. Infusion was discontinued at 17 minutes after start of recording. Resistance to CSF outflow was elevated: 20 mm Hg/(ml/min). AMP always increases proportionally to an increase of mean ICP.

15 mm Hg) is started and continued until a steady state ICP plateau is achieved (Fig. 6). If the ICP increases to 40 mm Hg, the infusion is interrupted. Following cessation of the saline infusion, ICP is recorded until it decreases to steady baseline levels. All compensatory parameters are calculated using computer supported methods based on physiological models of the CSF circulation.[40] Baseline ICP and R_{CSF} characterise the static properties of the CSF circulation. R_{CSF} is calculated as the pressure increase during the infusion, divided by the infusion rate. A value below 13 mm Hg/(ml/min) characterises normal CSF circulation[41]; above 18 mm Hg/(ml/min) the CSF circulation is clearly disturbed[42]; between 13 mm Hg/(ml/min) and 18 mm Hg/(ml/min) there is a grey zone, when other compensatory parameters and other clinical investigations should be considered to make a decision about shunting. Because the resistance to CSF outflow both in normal individuals[43] and in patients with normal pressure hydrocephalus (NPH)[44] increases with age, it is very likely that

the "critical threshold" of normal and abnormal R_{CSF} should be also age adjusted.

The cerebrospinal elasticity coefficient (E1) and AMP waveform express the dynamic components of CSF pressure–volume compensation.[45–47] E1 describes the compliance of the CSF compartment according to the formula: compliance of CSF space $= C_i = 1/[E1 \times (ICP - p_0)]$, where p_0 is the unknown reference pressure level, representing the hydrostatic difference between the site of ICP measurement and the pressure indifferent point of the cerebrospinal axis.[48] Cerebrospinal compliance is inversely proportional to ICP, therefore comparison between different subjects can be made only at the same level of the difference: $ICP - p_0$. E1 is independent of ICP, thus this coefficient is a much more convenient parameter when comparing individual patients. A low value of E1 (less than 0.2/ml) is specific for a compliant system, whereas a higher value indicates decreased pressure–volume compensatory reserve.

The AMP increases proportionally when the mean ICP rises. The proportionality ratio (the AMP/P index) characterises both the elastance of the cerebrospinal space and the transmission of arterial pulsations to the CSF compartment.[41] Finally, the production of CSF fluid can be estimated using Davson's equation. However, the sagittal sinus pressure (P_{SS}) is unknown and cannot be easily measured without increasing the invasiveness of the whole procedure. Consequently, the P_{SS} and CSF formation are estimated jointly using a non-linear model using the least square distance method during the computerised infusion test.[40] It is important to mention that such an "estimate" of CSF production rate approximates CSF absorption, rather than the actual production rate. It is based upon the assumption that all circulating CSF is reabsorbed via the arachnoid granulations. In cases where significant CSF suffusion into brain parenchyma occurs, CSF production may be grossly underestimated.

An infusion study may be useful for the assessment of shunt function in vivo. The end-equilibrium pressure achieved during the test should not significantly exceed the shunt's operating pressure increased by the hydrodynamic resistance of the opened shunt multiplied by the infusion rate. These values are readily available from the UK shunt evaluation laboratory (Academic Neurosurgical Unit, Addenbrooke's Hospital, Cambridge, UK) and can be used in clinical measurements to confirm shunt malfunction objectively.[49] This is particularly important as an increasing number of shunt revisions obviously impedes the chance for uneventful management of hydrocephalus.[50]

Overnight ICP monitoring in patients with NHP may reveal a high incidence of slow waves during sleep which is a very helpful prognostic sign for the outcome following shunting[51,52] (Fig. 7a). Benign intracranial hypertension seldom requires more than CSF pressure monitoring through a lumbar catheter or needle for an hour. When overnight ICP monitoring is performed in a patient with benign intracranial hypertension, baseline ICP is usually increased, the amplitude and frequency of "slow

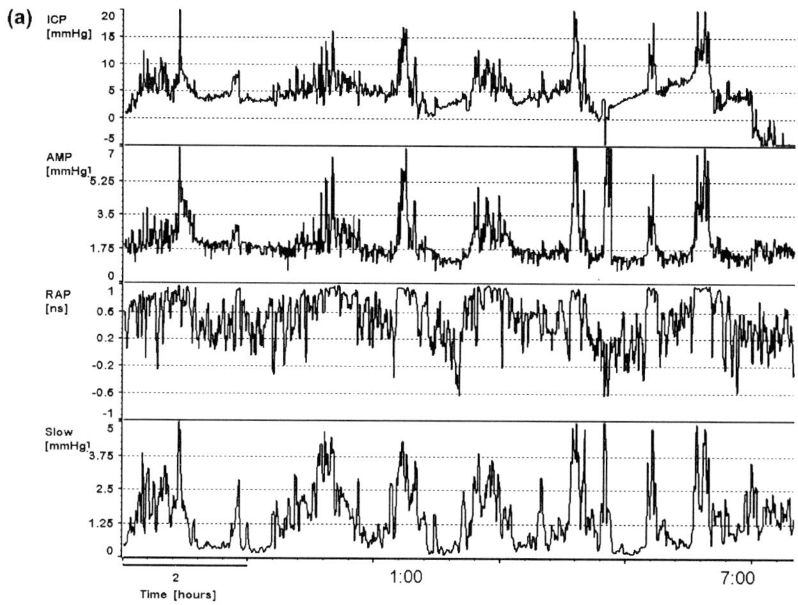

Fig. 7 Overnight monitoring of intracranial pressure. (a) Normal pressure hydrocephalus: baseline pressure normal (5 mm Hg) with periodical vasogenic increases reaching 20 mm Hg (ever hour), with associated decrease of compensatory reserve (RAP increasing towards +1) and an increase in magnitude of slow waves. These are vasogenic events, most frequently triggered by the REM phase of sleep. (b) Benign intracranial hypertension: baseline ICP elevated to 20 mm Hg with limited dynamics (although vasogenic waves clearly present around 5–6 am) and with permanently reduced compensatory reserve (RAP close to +1). (c) Intermittent hypertension due to sleep apnoea — case reported in Ref. 53. ICP, mean ICP level (1 minute averaged); AMP, pulse amplitude of ICP; RAP, index of compensatory reserve; Slow, magnitude of slow waves of ICP; FV, blood flow velocity; Sao_2, arterial blood saturation, REM, rapid eye movement.

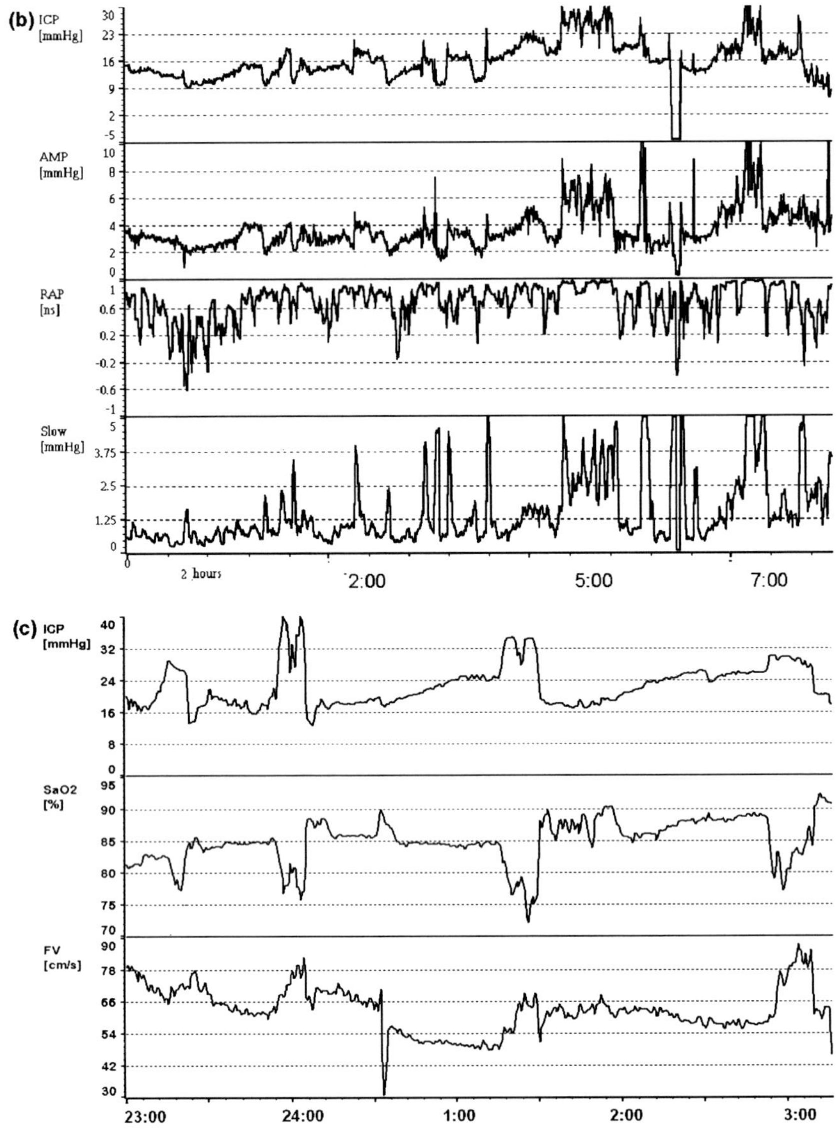

Fig. 7 *(Continued)*

waves" is moderate, but the increased RAP index indicates reduced cerebrospinal compensatory reserve (Fig. 7b). Intermittent intracranial hypertension caused by sleep apnoea syndrome may cause symptoms similar to benignintracranial hypertension. Multimodal monitoring including ICP,

Doppler blood flow velocity, and arterial blood oxygen saturation is useful (Fig. 7c).[53]

Although a number of centres employ ICP measurement and infusion studies, the selection of patients has not been shown to be more precise than when using the CSF tap test.[54] Positive predictive power of the increased resistance to CSF outflow is generally high, but the negative predictive power is low — that is, patients with normal resistance to CSF outflow may sometimes improve following shunting. However, the same problem concerns CSF tap tests.[55] Extending the period of CSF drainage to 72 hours seems to be an efficient solution but increases hospital stay and risk of complications.

ATTEMPTS TO MEASURE ICP AND CPP NON-INVASIVELY

It would be very helpful to measure ICP or CPP without invasive transducers. Transcranial Doppler examination,[16] tympanic membrane displacement,[15] and ultrasound "time of flight" techniques have been advocated.[17]

The description of transcranial Doppler sonography by Aaslid *et al.* in 1982 permitted bedside monitoring of one index of cerebral blood flow, non-invasively, repeatedly, and even continuously.[56] The problem has been that it is a "big tube technique", which measures flow velocity in branches of the circle of Willis, most commonly the middle cerebral artery (MCA). Compliant branches of the MCA can be compared with two physiological pressure transducers. The pattern of blood flow within these tubes is certainly modulated by transmural pressure — that is, CPP, and the distal vascular resistance (also modulated by the CPP). But what is the calibration factor and how should we compensate for unknown non-linear distortion?

There is a reasonable correlation between the pulsatility index of MCA velocity and CPP after head injury but absolute measurements of CPP cannot be extrapolated.[57] Others have suggested that "critical closing pressure" derived from flow velocity and arterial pressure waveform approximated the ICP.[58] The accuracy of this method has, however, never been satisfactory.[59]

Aaslid *et al.*[60] suggested that an index of CPP could be derived from the ratio of the amplitudes of the first harmonics of the ABP and the MCA velocity (detected by transcranial Doppler sonography) multiplied by mean flow velocity. Recently, a method for the non-invasive assessment of CPP has been reported, derived from mean arterial pressure multiplied by the ratio of diastolic to mean flow velocity.[61,62] This estimator can predict real CPP with an error of less than 10 mm Hg for more than 80% of measurements. This is of potential benefit for the continuous monitoring of changes in real CPP over time in situations where the direct measurement of CPP is not readily available.

A more complex method aimed at the non-invasive assessment of ICP has been introduced and tested by Schmidt *et al.*[18] The method is based on the presumed linear transformation between arterial pressure and ICP waveforms. Coefficients of this transformation are derived from the database of real ABP and ICP recordings. Similar linear transformation is built, using the same database between flow velocity and arterial pressure. Then the model assumes linear relationship between arterial pressure and flow velocity and arterial pressure to ICP transformations. Multiple regression coefficients are calculated. Finally, for each prospective study, ICP is calculated using ABP to ICP transformation, formed from ABP to flow velocity transformation transposed using precalculated regression coefficients.

IS ICP MONITORING USEFUL

The continuous measurement of ICP is an essential modality in most brain monitoring systems. After a decade of enthusiastic attempts to introduce new modalities for brain monitoring (tissue oxygenation, microdialysis, cortical blood flow, transcranial Doppler ultrasonography, jugular bulb oxygen saturation) it is increasingly obvious that ICP is robust, only moderately invasive, and can be realistically conducted in regional hospitals.

Although there has been no randomised controlled trial about influence of ICP monitoring on overall outcome after following head injury, recent audit[63] shows almost twofold lower mortality in neurosurgical centres, where ICP is usually monitored, versus general intensive care units, where it is not monitored. However, the availability of ICP monitoring is not the only difference between neurosurgical and general intensive care units that might explain the difference in mortality after head injury.

ICP waveform contains valuable information about the nature of cerebrospinal pathophysiology. Autoregulation of cerebral blood flow and compliance of cerebrospinal system are both expressed in ICP. Methods of waveform analysis are useful both to derive this information and to guide the management of patients.

The value of ICP in acute states such as head injury, poor grade subarachnoid haemorrhage, and intracerebral haematoma depends on a close link between monitoring and therapy. CPP oriented protocols,[64,65] osmotherapy[66] and the "Lund protocol"[6] cannot be conducted correctly without ICP guidance. A decision about decompressive craniectomy should be supported by the close inspection of the trend of ICP and, preferably, by information derived from its waveform.[67] In encephalitis,[68] acute liver failure,[69] and cerebral infarction after stroke,[70] ICP monitoring is used less commonly, however, an increasing number of reports highlight its importance.

A slightly different methodology for CSF pressure interpretation is applied in chronic states such as hydrocephalus or benign intracranial hypertension. In the first case assessment of CSF pressure–volume compensation and circulation are essential to optimise patient management.[42,71] Volume-adding tests with parallel measurement of ICP and/or overnight ICP monitoring with waveform analysis have a special role. In patients with a shunt in situ, who present with persistent or recurring clinical symptoms, it helps to avoid unnecessary shunt revisions. This is particularly important as patients with a history of multiple shunt revision have a lower chance to achieve good outcome in the future. In benign intracranial hypertension[72] or craniostenosis[73] ICP monitoring has been documented as useful both for diagnosis and to document response to therapy.

In summary, ICP is a complex modality, which contains combined information about cerebral compensatory and CBF regulation mechanisms. Control of ICP requires continuous monitoring. ICP-derived indices help to understand the pathophysiology of developing events and facilitate patient care.

ACKNOWLEDGEMENTS

Many thanks to my colleagues participating in the head injury monitoring programme in Cambridge, UK, 1991–2002, whose experience, knowledge, and clinical material helped us to write this review: P Al-Rawi, D Chattfield, C Turner, M Balestreri, M Hiler, P Smielewski, SK Piechnik, LA Steiner, E Guazzo, PC Whetfield, P Minhas, M Soehle, R Kett-White, P Hutchinson, E Schmidt, B Schmidt, A Kumar, A Raabe, M Schuhmann, PJ Kirkpatrick, FK Matta, A Gupta, DK Menon, and nursing and research staff from Neuro Critical Care Unit and Wolfson Brain Imaging Centre.

REFERENCES

1. Marmarou A, Maset AL, Ward JD, *et al.* Contribution of CSF and vascular factors to elevation of ICP in severely head-injured patients. *J Neurosurg* 1987; **66**:883–890.
2. Czosnyka M, Richards HK, Czosnyka Z, *et al.* Vascular components of cerebrospinal fluid compensation. *J Neurosurg* 1999; **90**:752–759.
3. Davson H, Hollingsworth G, Segal MB. The mechanism of drainage of the cerebrospinal fluid. *Brain* 1970; **93**:665–678.
4. Miller JD, Stanek A, Langfitt TW. Concepts of cerebral perfusion pressure and vascular compression during intracranial hypertension. *Prog Brain Res* 1972; **35**:411–432.
5. Steiner LA, Czosnyka M, Piechnik SK, *et al.* Continuous monitoring of cerebrovascular pressure reactivity allows determination of optimal cerebral perfusion pressure in patients with traumatic brain injury. *Crit Care Med* 2002; **30**:733–738.
6. Asgeirsson B, Grande PO, Nordstrom CH. A new therapy of post-trauma brain oedema based on haemodynamic principles for brain volume regulation. *Intensive Care Med* 1994; **20**:260–267.
7. Guillaume J, Janny P. Manometrie intracranienne continué interest de la methode et premiers resultants. *Rev Neurol (Paris)* 1951; **84**: 131–142.
8. Lundberg N. Continuous recording and control of ventricular fluid pressure in neurosurgical practice. *Acta Psychiatr Neurol Scand* 1960; **36**(suppl 149): 1–193.
9. Ghajar J. Intracranial pressure monitoring techniques. *New Horizions* 1995; **3**:395–399.

10. Czosnyka M, Czosnyka Z, Pickard JD. Laboratory testing of three intracranial pressure microtransducers: Technical report. *Neurosurgery* 1996; **38**: 219–224.
11. Wolfla CE, Luerssen TG, Bowman RM, *et al.* Brain tissue pressure gradients created by expanding frontal epidural mass lesion. *J Neurosurg* 1996; **84**: 642–647.
12. Piper I, Barnes A, Smith D, *et al.* The Camino intracranial pressure sensor: is it optimal technology? An internal audit with a review of current intracranial pressure monitoring technologies. *Neurosurgery* 2001; **49**:1158–1164.
13. Chambers IR, Siddique MS, Banister K, *et al.* Clinical comparison of the Spiegelberg parenchymal transducer and ventricular fluid pressure. *J Neurol Neurosurg Psychiatry* 2001; **71**:383–385.
14. Owler BK, Fong KC, Czosnyka Z. Importance of ICP monitoring in the investigation of CSF circulation disorders. *Br J Neurosurg* 2001; **15**:439–440.
15. Reid A, Marchbanks RJ, Martin R, *et al.* Mean intracranial pressure monitoring by an audiological technique — a pilot study. *J Neurol Neurosurg Psychiatry* 1989; **52**:610–612.
16. Schoser BG, Riemenschneider N, Hansen HC. The impact of raised intracranial pressure on cerebral venous hemodynamics: a prospective venous transcranial Doppler ultrasonography study. *J Neurosurg* 1999; **91**:744–749.
17. Petkus V, Ragauskas A, Jurkonis R. Investigation of intracranial media ultrasonic monitoring model. *Ultrasonics* 2002; **40**:829–833.
18. Schmidt B, Klingelhofer J, Schwarze JJ, *et al.* Noninvasive prediction of intracranial pressure curves using transcranial Doppler ultrasonography and blood pressure curves. *Stroke* 1997; **28**:2465–2472.
19. Albeck MJ, Borgesen SE, Gjerris F, *et al.* Intracranial pressure and cerebrospinal fluid outflow conductance in healthy subjects. *J Neurosurg* 1991; **74**: 597–600.
20. Chapman PH, Cosman ER, Arnold MA. The relationship between ventricular fluid pressure and body position in normal subjects and subjects with shunts: a telemetric study. *Neurosurgery* 1990; **26**:181–189.
21. Balestreri M, Czosnyka M. Intracranial hypertension: what additional information can be derived form ICP waveform after head injury? *Acta Neurochir* 2004; (in press).
22. Lofgren J, von Essen C, Zwetnow NN. The pressure-volume curve of the cerebrospinal fluid space in dogs. *Acta Neurol Scand* 1973; **49**:557–574.
23. Avezaat CJ, van Eijndhoven JH, Wyper DJ. Cerebrospinal fluid pulse pressure and intracranial volume-pressure relationships. *J Neurol Neurosurg Psychiatry* 1979; **42**:687–700.

24. Muizelaar JP, Ward JD, Marmarou A, *et al.* Cerebral blood flow and metabolism in severely head-injured children. Autoregulation. *Neurosurg* 1989; **71**:72–76.
25. Czosnyka M, Smielewski P, Kirkpatrick P, *et al.* Continuous assessment of the cerebral vasomotor reactivity in head injury. *Neurosurgery* 1997; **41**:11–17.
26. Miller JD, Pickard JD. Intracranial volume pressure studies in patients with head injury. *Injury* 1974; **5**:265–268.
27. Piper IR, Miller JD, Whittle IR, *et al.* Automated time-averaged analysis of craniospinal compliance (short pulse response). *Acta Neurochir Suppl (Wien)* 1990; **51**:387–390.
28. Piper I, Spiegelberg A, Whittle I, *et al.* A comparative study of the Spiegelberg compliance device with a manual volume-injection method: a clinical evaluation in patients with hydrocephalus. *Br J Neurosurg* 1999; **13**:581–586.
29. Yau YH, Piper IR, Contant CF, *et al.* Clinical experience in the use of the Spiegelberg automated compliance device in the assessment of patients with hydrocephalus. *Acta Neurochir Suppl* 2002; **81**:171–172.
30. Robertson C, Narayan R, Contant C, *et al.* Clinical experience with continuous monitoring of intracranial compliance. *J Neurosurg* 1989; **71**:673–680.
31. Portnoy HD, Chopp M, Branch C, *et al.* Cerebrospinal fluid pulse waveform as an indicator of cerebral autoregulation. *J Neurosurg* 1982; **56**:666–678.
32. Takizawa H, Gabra-Sanders T, Miller JD. Changes in the cerebrospinal fluid pulse wave spectrum associated with raised intracranial pressure. *Neurosurgery* 1987; **20**:355–361.
33. Piper I, Miller JD, Dearden M, *et al.* System analysis of cerebrovascular pressure transmission: an observational study in head injured patients. *J Neurosurg* 1990; **73**:871–880.
34. Foltz EL, Blanks JP, Yonemura K. CSF pulsatility in hydrocephalus: respiratory effect on pulse wave slope as an indicator of intracranial compliance. *Neurol Res* 1990; **12**:67–74.
35. Lemaire JJ, Khalil T, Cervenansky F, *et al.* Slow pressure waves in the cranial enclosure. *Acta Neurochir (Wien)* 2002; **144**:243–254.
36. Robertson CS, Valadka AB, Hannay HJ, *et al.* Prevention of secondary ischemic insults after severe head injury. *Crit Care Med* 1999; **27**:2086–2095.
37. Borgesen SE, Gjerris F. Relationships between intracranial pressure, ventricular size, and resistance to CSF outflow. *J Neurosurg* 1987; **67**:535–539.
38. Higgins JN, Owler BK, Cousins C, *et al.* Venous sinus stenting for refractory benign intracranial hypertension. *Lancet* 2002; **359**:228–230.
39. Katzman R, Hussey F. A simple constant-infusion manometric test for measurement of CSF absorption. I Rationale and method. *Neurology* 1970; **20**: 534–544.

40. Czosnyka M, Whitehouse H, Smielewski P, *et al.* Testing of cerebrospinal compensatory reserve in shunted and non-shunted patients: a guide to interpretation based on observational study. *J Neurol Neurosurg Psychiatry* 1996; **60**: 549–558.
41. Borgesen SE, Gjerris F. The predictive value of conductance to outflow of CSF in normal pressure hydrocephalus. *Brain* 1982; **105**:65–86.
42. Boon AJ, Tans JT, Delwel EJ, *et al.* Dutch normal-pressure hydrocephalus study: prediction of outcome after shunting by resistance to outflow of cerebrospinal fluid. *J Neurosurg* 1997; **87**:687–693.
43. Albeck MJ, Skak C, Nielsen PR, *et al.* Age dependency of resistance to cerebrospinal fluid outflow. *J Neurosurg* 1998; **89**:275–278.
44. Czosnyka M, Czosnyka ZH, Whitfield PC, *et al.* Age dependence of cerebrospinal pressure-volume compensation in patients with hydrocephalus. *J Neurosurg* 2001; **94**:482–486.
45. Avezaat CJJ, Eijndhoven JHM. Cerebrospinal fluid pulse pressure and craniospinal dynamics. A theoretical, clinical and experimental study [Thesis]. The Hague: Jongbloed A, 1984.
46. Ekstedt J. CSF hydrodynamic studies in man. Method of constant pressure CSF infusion. *J Neurol Neurosurg Psychiatry* 1977; **40**:105–119.
47. Marmarou A, Shulman K, Rosende RM. A non-linear analysis of CSF system and intracranial pressure dynamics. *J Neurosurg* 1978; **48**:332–344.
48. Raabe A, Czosnyka M, Piper I, *et al.* Monitoring of intracranial compliance: correction for a change in body position. *Acta Neurochir (Wien)* 1999; **141**: 31–36.
49. Czosnyka ZH, Czosnyka M, Pickard JD. Shunt testing in-vivo: a method based on the data from the UK shunt evaluation laboratory. *Acta Neurochir Suppl* 2002; **81**:27–30.
50. Tuli S, Drake J, Lawless J, *et al.* Risk factors for repeated cerebrospinal shunt failures in pediatric patients with hydrocephalus. *J Neurosurg* 2000; **92**: 31–38.
51. Pickard JD, Teasdale G, Matheson M, *et al.* Intraventricular pressure waves — the best predictive test for shunting in normal pressure hydrocephalus. In: Shulman K, Marmarou A, Miller JD, Becker DP, Hochwald GM, Brock M, eds. *Intracranial Pressure IV*. Berlin: Springer-Verlag, 1980:498–500.
52. Symon L, Dorsch NWC, Stephens RJ. Pressure waves in so-called low-pressure hydrocephalus. *Lancet* 1972; **2**:1291–1292.
53. Kirkpatrick PJ, Meyer T, Sarkies N, *et al.* Papilloedema and visual failure in a patient with nocturnal hypoventilation. *J Neurol Neurosurg Psychiatry* 1994; **57**:1546–1547.

54. Kahlon B, Sundbarg G, Rehncrona S. Comparison between the lumbar infusion and CSF tap tests to predict outcome after shunt surgery in suspected normal pressure hydrocephalus. *J Neurol Neurosurg Psychiatry* 2002; **73**:721–726.
55. Walchenbach R, Geiger E, Thomeer RT, *et al.* The value of temporary external lumbar CSF drainage in predicting the outcome of shunting on normal pressure hydrocephalus. *J Neurol Neurosurg Psychiatry* 2002; **72**:503–506.
56. Aaslid R, Markwalder TM, Nornes H. Noninvasive transcranial Doppler ultrasound recording of flow velocity in basal cerebral arteries. *J Neurosurg* 1982; **57**:769–774.
57. Chan KH, Miller JD, Dearden NM, *et al.* The effect of changes in cerebral perfusion pressure upon middle cerebral artery blood flow velocity and jugular bulb venous oxygen saturation after severe brain injury. *J Neurosurg* 1992; **77**: 55–61.
58. Dewey RC, Pieper HP, Hunt WE. Experimental cerebral hemodynamics. Vasomotor tone, critical closing pressure, and vascular bed resistance. *Neurosurgery* 1974; **41**:597–606.
59. Czosnyka M, Smielewski P, Piechnik S, *et al.* Critical closing pressure in cerebrovascular circulation. *J Neurol Neurosurg Psychiatry* 1999; **66**:606–611.
60. Aaslid R, Lundar T, Lindegaard K-F, *et al.* Estimation of cerebral perfusion pressure from arterial blood pressure and transcranial Doppler recordings. In: Miller JD, Teasdale GM, Rowan JO, Galbraith SL, Mendelow AD, eds. *Intracranial Pressure VI*. Berlin: Springer-Verlag, 1986:229–231.
61. Czosnyka M, Matta BF, Smielewski P, *et al.* Cerebral perfusion pressure in head-injured patients: a noninvasive assessment using transcranial Doppler ultrasonography. *J Neurosurg* 1998; **88**:802–808.
62. Schmidt EA, Czosnyka M, Matta BF, *et al.* Non-invasive cerebral perfusion pressure (nCPP): evaluation of the monitoring methodology in head injured patients. *Acta Neurochir Suppl* 2000; **76**:451–452.
63. Patel HC, Bouamra O, King A, *et al.* Mortality after head trauma. Effect of neurosurgical care. *Euroacademia Multidisciplinaria Neurotraumatologica*. Graz, 2003. [Abstract] *Acra Neurochir (Wein)* 2003; **145**:1135–1148.
64. Rosner MJ, Rosner SD, Johnson AH. Cerebral perfusion pressure: Management protocol and clinical results. *J Neurosurg* 1995; **83**:949–962.
65. Patel HC, Menon DK, Tebbs S, *et al.* Specialist neurocritical care and outcome from head injury. *Intensive Care Med* 2002; **28**:547–553.
66. Bullock R. Mannitol and other diuretics in severe neurotrauma. *New Horizons* 1995; **3**:448–452.
67. Whitfield PC, Patel H, Hutchinson PJ, *et al.* Bifrontal decompressive craniectomy in the management of posttraumatic intracranial hypertension. *Br J Neurosurg* 2001; **15**:500–507.

68. Rebaud P, Berthier JC, Hartemann E, *et al.* Intracranial pressure in childhood central nervous system infections. *Intensive Care Med* 1988; **14**:522–525.
69. Keays RT, Alexander GJ, Williams R. The safety and value of extradural intracranial pressure monitors in fulminant hepatic failure. *J Hepatol* 1993; **18**:205–209.
70. Schwab S, Aschoff A, Spranger M, *et al.* The value of intracranial pressure monitoring in acute hemispheric stroke. *Neurology* 1996; **47**:393–398.
71. Tisell M, Edsbagge M, Stephensen H, *et al.* Elastance correlates with outcome after endoscopic third ventriculostomy in adults with hydrocephalus caused by primary aqueductal stenosis. *Neurosurgery* 2002; **50**:70–77.
72. Sussman JD, Sarkies N, Pickard JD. Benign intracranial hypertension. Pseudotumour cerebri: idiopathic intracranial hypertension. *Adv Tech Stand Neurosurg* 1998; **24**:261–305.
73. Tuite GF, Evanson J, Chong WK, *et al.* The beaten copper cranium: a correlation between intracranial pressure, cranial radiographs, and computed tomographic scans in children with craniosynostosis. *Neurosurgery* 1996; **39**:691–699.

12

Cerebral Perfusion and Stroke

*H S Markus**

Stroke is a heterogeneous syndrome caused by multiple disease mechanisms, but all result in a disruption of cerebral blood flow with subsequent tissue damage. This review covers the mechanisms responsible for regulation of the normal cerebral circulation, and how they are disrupted in disease states. A central concept in treating patients with acute ischaemic stroke is the existence of an ischaemic penumbra of potentially salvageable tissue, and the evidence for its existence in humans is reviewed.

The cerebral circulation has developed several specialised features to cope with the high metabolic demands of the brain and the devastating consequences of cerebral ischaemia. Unlike other organs, the brain uses glucose almost exclusively as its sole substrate for energy metabolism. As it is unable to store energy it requires a constant supply of oxygenated blood containing an adequate glucose concentration to maintain its function and structural integrity. The brain oxygen requirement in the adult human accounts for about 20% of that of the whole body, for a relative size of only 2%. Brain oxygen uptake is even higher in children in the first decade of life, where it may be up to 50% of the total body oxygen supply.[1] This high

*Centre for Clinical Neuroscience St George's, University of London, Cranmer Terrace, London SW17 0RE, UK; h.markus@sgul.ac.uk

metabolic demand is matched by a disproportionately large amount of the total cardiac output passing to the brain. Total cerebral blood flow (CBF) at rest is about 800 ml/min, which is 15–20% of total cardiac output.[2] Cerebral perfusion is a high flow, low pressure system with relatively preserved diastolic flow. This can be appreciated by comparing Doppler waveforms from cerebral vessels with systemic vessels; the relative ratio of diastolic to systolic flow is much higher for the cerebral circulation. Anatomical features and functional responses, including cerebral autoregulation, allow this high flow and provide protection against ischaemia.

CEREBROVASCULAR ANATOMY

The cerebral circulation has a well developed collateral circulation which plays an important protective role. Of crucial importance are communications between the basal cerebral arteries at the circle of Willis.[3] Natural variation in its integrity will determine whether carotid occlusion results in massive hemispheric stroke or is asymptomatic. Collateral supply may also occur through anastomoses between the external carotid artery branches and the intracerebral circulation, and there is a well developed plexus of anastomoses between cerebral vessels on the surface of the brain.[4] Despite this collateral supply, there are regions of the brain that are particularly vulnerable to a fall in perfusion pressure. Such border zone areas include those at the boundaries between the anterior and middle and the middle and posterior cerebral arteries, and the "internal" border zone areas in the corona radiata and centrum semiovale. Territories supplied by the penetrating or perforating arteries that supply the white matter and subcortical grey matter nuclei are particularly vulnerable to ischaemia. These arteries are end arteries and therefore ischaemia in their territory results in more rapid cell death.

METHODS FOR MEASUREMENT OF CEREBRAL BLOOD FLOW

Various different techniques (Table 1) allow measurement of regional cerebral perfusion, although many are only semiquantitative. Such techniques require a tracer or contrast agent and an imaging technique allowing its concentration to be measured. Commonly used techniques include positron

Table 1 Methods of measurement of cerebral blood flow.

Table width = E

Technology	Specific methods	Important points
Perfusion (regional CBF)		
Positron emission tomography	^{15}O labelled water	Allows oxygen consumption measurement Quantitative
Xenon based techniques	Stable xenon methods Intravenous xenon methods Xenon SPECT Xenon CT	Potentially quantitative
MR based techniques	Exogenous perfusion Endogenous perfusion	Using gadolinium bolus Non-invasive
CT based techniques	Contrast perfusion imaging Xenon CT	
SPECT	^{99m}TcHMPAO ^{123}I labelled IMP	Semiquantitative
Transcranial Duplex ultrasound	Ultrasound perfusion	Quantification not yet possible
Perfusion (regional CBF) Volume flow (in major arteries)		
Carotid and vertebral ultrasound	Doppler based techniques Non-Doppler based techniques	Flow in major arteries and not tissue perfusion
Transcranial Doppler ultrasound		Velocity rather than flow measurement

CBF, cerebral blood flow; CT, computed tomography; IMP, inosine 5′-monophosphate; MR, magnetic resonance; SPECT, single photon emission computed tomography; ^{99m}TcHMPAO, technetium-99m hexamethylpropylene amine oxime.

emission tomography (PET), single photon emission computed tomography (SPECT), xenon computed tomography (CT) and contrast CT perfusion, and magnetic resonance imaging (MRI) perfusion studies.

Nuclear Medicine Methods

Nuclear medicine methods use radionucleides as tracers and the tomographic approach in image reconstruction. These can be distinguished on the basis of the physical characteristics of the radionucleides. SPECT uses low energy, photon emitting radionucleides. PET uses coincidence radiation employing positron emitting radionucleides which, after annihilation, produce two 511 keV τ rays. PET allows quantification, while with SPECT only semiquantification is possible. Both involve exposure to significant radiation doses. The tracer most commonly used for PET CBF measurements is ^{15}O labelled water. Only PET allows measurement of oxygen utilisation and therefore the oxygen extraction fraction (OEF). SPECT is more widely available and uses lipophilic tracers that easily cross the blood–brain barrier and ideally are fully extracted during the first pass through the cerebral circulation. ^{99m}TcHMPAO is most widely used. It is stable for about six hours after being reconstituted, making it convenient for clinical use. A good correlation with CBF values has been obtained.[5] Xenon is commonly used in cerebral perfusion measurement both with CT and SPECT. Xenon gas can be inhaled in significant quantities with little risk, although at high concentrations it is an anaesthetic. It dissolves into blood and readily diffuses through the blood–brain barrier. Xenon has a high atomic number relative to air and soft tissues and as a result its presence can be measured by CT. A washout method can be used; when an initial equilibrium of substance has been reached in the tissue and the supply is acutely discontinued, the substance concentration decays in an exponential fashion and the decay of the exponential contains information on perfusion.[6]

Computed Tomography

Perfusion CT involves sequential acquisition of stationary cerebral CT sections during intravenous administration of a bolus of iodinated contrast material.[7] Analysis of the resultant contrast enhancement curves according to the central volume principle allows calculation of cerebral blood volume (CBV), mean transit time (MTT), and CBF.

Magnetic Resonance Imaging

Increasingly, MRI methods are used to measure perfusion. These fall into two categories.

Exogenous contrast methods are most widely used and rely on an intravenous bolus injection of a paramagnetic tracer and acquisition of rapid images, using echo-planar imaging, as the tracer passes through the cerebral vasculature.[8] The effect of the tracer is to reduce signal from tissue in its immediate vicinity owing to dephasing effects caused by the magnetic field gradients induced. This signal loss is closely related to the flow of the tracer through the brain and thus to cerebral perfusion. This method also offers the ability to measure CBV and MTT. Its advantage is that paramagnetic substances have a large effect on signal intensity, and therefore the signal to noise ratio is high. Recent studies have suggested that a degree of quantification is possible if an arterial input function is obtained by imaging through a supplying vessel,[8] although whether absolute quantification is reliable is debated.

Endogenous methods such as arterial spin labelling use blood water as an endogenous contrast agent for the measurement of perfusion. This is completely non-invasive and can be repeated as often as desired, but the signal to noise ratio is low.

Ultrasound

Ultrasound methods are non-invasive. Carotid and vertebral neck ultrasound uses either Doppler based[9] or non-Doppler based methods[10] to quantify volume flow in the major supplying arteries, rather than regional perfusion. In contrast, transcranial Doppler, most commonly of the middle cerebral artery, measures velocity and not flow and therefore only gives an accurate estimate of changes in flow if the diameter of the vessel being insonated does not change. Tissue perfusion can now be estimated using transcranial imaging ultrasound, but quantification using this technique is still being developed.[11]

REGULATION OF THE NORMAL CEREBRAL CIRCULATION

Basal Cerebral Blood Flow Regulation

Under normal conditions CBF is determined by both cerebral perfusion pressure (CPP) and cerebrovascular resistance (CVR). CVR is determined by the diameter of the intracranial arteries and also blood viscosity. CPP can be calculated from the difference between systemic arterial pressure and venous back pressure. In conditions were CPP remains constant any change in CBF must result from a change in CVR, usually as a result of alteration in the diameter of the small intracranial arteries, which are the predominant resistance vessels. Thus under normal circumstances there is a direct correlation between CBF and cerebral blood volume, which will both increase as vessels dilate and decrease as vessels constrict. This relation is disturbed in pathological states such as cerebral ischaemia.

Various physiological variables influence CBF, including arterial blood gases, cerebral autoregulation, and metabolic rate through vasoneuronal coupling. One important mediator controlling basal CBF is nitric oxide. In 1980[12] it was shown that an intact endothelial cell layer is required if isolated arterial segments were to relax in response to perfusion with acetylcholine. It was subsequently found that this "endothelium derived relaxing factor" is nitric oxide (NO), synthesised by endothelial cells from the amino acid L-arginine by the enzyme NO synthase (NOS).[13] There are three isoforms of this enzyme: endothelial NOS (eNOS), neuronal NOS (nNOS), and an inducible NOS (iNOS). Endothelium derived NO plays a crucial role in maintenance of blood vessel calibre and therefore blood flow throughout the vasculature. It is also important in preventing thrombosis through inhibition of platelet adhesion, activation, and aggregation, and in preventing atherosclerosis by inhibition of vascular smooth muscle cell proliferation. Under normal conditions there appears to be a functional balance between the endothelium dependent vasodilator effects of NO and the endothelium derived constrictor substances such as endothelin, thromboxane, and angiotensin II. Extensive animal data in a variety of species have shown that tonic release of NO plays an important role in maintaining resting CBF.[14] More recently the role of NO in maintaining basal CBF has been shown in the human: inhibition of NO by the NO synthase inhibitor N-monomethyl-L-arginine (L-NMMA) resulted in a 30% fall in CBF.[15,16]

Cerebral Autoregulation

CBF remains relatively constant despite moderate variations in perfusion pressure. This phenomenon, described as "cerebral autoregulation," plays an important protective role against the danger of hypoxia at low perfusion pressure, and the risk of brain oedema at higher arterial pressure.[17] As a rough approximation the lower and upper limits of autoregulation occur at mean arterial pressures of 60 and 150 mm Hg in the normotensive human.

Between these limits CBF is relatively but not absolutely constant. Once the limits of autoregulation are reached, CBF increases or decreases passively with increases or reductions in perfusion pressure.

Reductions in CBF below the limit of autoregulation result in brain hypoperfusion. In an attempt to compensate for this, the extraction coefficient of oxygen from the blood increases. Clinical symptoms of functional disruption are not observed until the reduction in CPP exceeds the ability of the increase in oxygen extraction to satisfy the metabolic demands of cerebral tissues. If mean arterial pressure increases above the upper limit of autoregulation, resistance arteries in the brain cannot sustain vasoconstriction. An early sign is the appearance of "sausage stringing," characterised by an alternating pattern of dilated arterial segments with focal regions of constriction. The dilated segments represent regions of passive dilatation, and the constricted segments regions of sustained autoregulation. Further increases in CPP result in dilatation along the entire length of the arterioles, and CBF increases passively. This is accompanied by damage to the cerebrovascular endothelium and disruption of the blood–brain barrier. The latter results in extravasation of plasma proteins through the vessel wall and subsequent oedema. These events are important in acute hypertensive encephalopathy.

The upper and lower limits of autoregulation are not fixed but vary under both physiological stimuli and disease states. Activation of the sympathetic nerves results in upward shift of both the lower and upper limits — a potentially protective response because acute elevations in arterial pressure are usually accompanied by sympathetic activation. The autoregulatory plateau is shifted to higher values in patients with chronic hypertension.[18] This protective response can have deleterious effects if blood pressure is excessively reduced, when symptoms of ischaemia may occur at a relatively higher blood pressure.

Cerebral autoregulation is impaired in various disease states including head injury,[19] ischaemic stroke,[20] and subarachnoid haemorrhage.[21] This may result in an already damaged brain being excessively sensitive to fluctuations in perfusion pressure.

Assessment of Cerebral Autoregulation in the Human

Traditionally, studies in the human have estimated "static" autoregulation. Steady state values of CBF are determined across a range of blood pressures.

This is achieved by the use of drugs or shifts in blood volume — for example, using tilt tables or negative pressure.[22] Determination of the full range of autoregulation requires blood pressure to be manipulated across a wide range, and the long time interval between measurements made at different pressures can make interpretation of studies difficult. More recently methods for estimating "dynamic" autoregulation in humans have been developed. These determine the rate of response of CBF or flow velocity after a sudden change in arterial blood pressure. A technique allowing measurement of CBF with very high temporal resolution is required. The only technique that can provide this is transcranial Doppler ultrasound. In the most widely used technique a sudden fall in blood pressure is induced by inflating leg cuffs above venous pressure and then rapidly deflating them.[23] The rate of rise of CBF velocity in the middle cerebral artery is then compared with the rate of rise of blood pressure continuously monitored using a non-invasive device such as a Finapres. Using this technique, autoregulation has been shown to be impaired distal to carotid artery stenosis,[24] following head injury,[25] and in ischaemic stroke.[26] Although the measurement of dynamic autoregulation gives a useful estimate of how the cerebral circulation can compensate for fluctuations in perfusion pressure, it may not necessarily be measuring the same mechanisms that determine static autoregulation.

Physiological Mechanisms of Autoregulation

The mechanisms responsible for CBF autoregulation in the human are not fully understood. Traditional hypotheses are that neurogenic, myogenic, and metabolic factors play a role. More recently NO has been implicated. Autoregulation is preserved in animals that have undergone sympathetic and parasympathetic denervation, suggesting that neurogenic factors are not of primary importance.[27]

The myogenic hypothesis states that smooth muscle in the resistance arteries responds directly to alterations in perfusion pressure by contracting during increases in pressure, and relaxing during reductions in pressure. This would provide a mechanism whereby responses in cerebral resistance arteries could occur within seconds of an autoregulatory stimulus. *In vitro* animal experiments support this hypothesis, with constriction of isolated arteries occurring following a rapid increase in intravascular pressure.

In vivo experiments have not, in general, provided such strong support.[27] The metabolic hypothesis states that reductions in CBF stimulate the release of vasoactive substances from the brain, which in turn stimulates the dilatation of cerebral resistance arteries. Several candidates for this role have been proposed — including carbon dioxide, hydrogen ions, oxygen, adenosine, potassium, and calcium — but no definite role has been demonstrated for any of these.[28]

Endothelial factors, particularly NO, have been suggested as mediators of cerebral autoregulation. In some animal studies, impaired autoregulation has been reported following inhibition of NO, but others have found no effect.[29] In eNOS knockout mice there was a greater fall in CBF during haemorrhagic hypotension than in wild type mice, consistent with an alteration in the lower limit of autoregulation.[30] In humans the NOS inhibitor L-NMMA resulted in a significant impairment of dynamic autoregulation.[29]

Arterial Blood Gases, the Hypercapnic Response, and CBF

CBF is very sensitive to changes in blood CO_2 concentrations, largely mediated by concomitant changes in the pH of brain tissue.[31] Apart from this direct effect of pH on cerebral vessels, secondary mechanisms may mediate vasoactive effects through a pH dependent alteration in the release of other vasoactive factors. Prostaglandins have been suggested as mediators of CO_2 dependent changes in CBF, based on the effects of the inhibitor of cyclo-oxygenase, indomethacin. However, further studies in animals[31] and humans[32] cast doubt on this mechanism in adults, although it does appear to be important in newborn animals.[31]

Studies in rats suggested NO dependence of the hypercapnic response, although this NO production is primarily through nNOS rather than eNOS. However, results in higher species have been less consistent, and studies in primates have shown conflicting results.[33,34] A complicating factor in animal studies is that anaesthesia may alter cerebrovascular responses. A study in humans showed that inhibition of eNOS by intravenous N^G-monomethyl-L-arginine (L-NNMA) did not alter the hypercapnic response.[15]

Reducing arterial Po_2 results in an increase in CBF, whereas an increase in arterial Po_2 above normal values has a much lesser effect. Hypoxia appears to act both through a direct effect on cerebral resistance vessels, and through indirect effects including the release of vasoactive factors such as hydrogen ions, adenosine, and potassium ions.[31] It has also been suggested that pH dependent mechanisms — mainly the production of lactate — participate in hypoxic vasodilatation.[35]

Clinical Relevance of the Hypercapnic Response

The response of CBF to hypercapnia is widely used in humans to estimate cerebral perfusion reserve, most often in patients with carotid stenosis or occlusion. The haemodynamic consequences of a carotid stenosis — for example, during a fall in blood pressure — will depend on distal collateral supply, particularly through the circle of Willis. An estimate of this cerebral perfusion reserve can be obtained by measuring CBF during both normocapnia and hypercapnia (with concentrations of 5–8% CO_2 in air). In the presence of impaired reserve, resistance vessels are already vasodilated, and the capacity to increase CBF further is reduced (Fig. 1). A wide range of imaging methods has been used to measure the hypercapnic response, including PET, SPECT, MRI, and xenon based techniques. For techniques

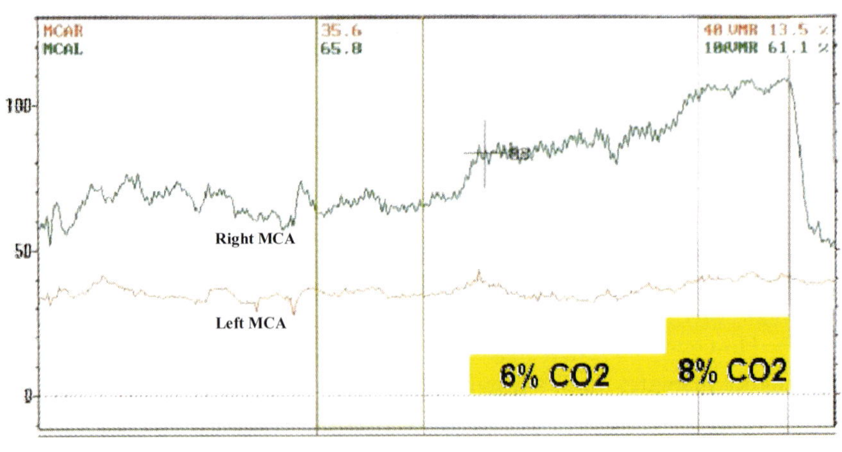

Fig. 1 Measurement of cerebral perfusion reserve or cerebral reactivity in a patient with unilateral left internal carotid artery occlusion. Blood flow velocity is recorded continuously from both middle cerebral arteries. As the patient breaths a 6% and then an 8% mixture of carbon dioxide in air, flow velocity increases on the right but not on the left. This indicates severely impaired perfusion reserve ipsilateral to the occlusion. MCA, middle cerebral artery.

that do not allow absolute quantification (such as SPECT), side to side differences are calculated; this makes interpretation difficult in bilateral carotid stenosis. Transcranial Doppler monitoring of middle cerebral artery flow velocity has become widely adopted in this context.[36] Its validity is suggested by angiographic studies showing that the diameter of the middle cerebral artery does not change during hypercapnia.[37] Acetazolamide, a carbonic anhydrase inhibitor, can be used as the vasodilator stimulus instead of CO_2.

Prospective studies have shown that a reduced hypercapnic response ipsilateral to a carotid occlusion is associated with an increased risk of stroke or transient ischaemic attacks (TIA).[38–40] It has been suggested this test may identify a subgroup of patients with carotid occlusion who benefit from extracranial–intracranial bypass, although this hypothesis has yet to be tested. There is some evidence from prospective studies that patients with tight carotid stenosis, as opposed to occlusion, and who have impaired perfusion reserve are at increased risk of stroke, but the association appears less strong than for carotid occlusion.[39]

Local Coupling of Cerebral Blood Flow and Metabolism

Local CBF is tightly coupled to changes in neuronal metabolism, and this forms the basis of functional brain imaging using BOLD MRI. Over a century ago Roy and Sherrington suggested that "...(the brain) is well fitted to provide for a local variation of the blood supply in accordance with local variations of the functional activity".[41] Numerous studies have shown that local increases in neuronal activity result in a local increase in glucose utilisation, accompanied by local increases in CBF. The precise mechanisms linking neuronal activation, metabolism, and flow are not fully understood. Various factors are likely to contribute to this coupling process, including potassium release with neuronal depolarisation, and H^+ and adenosine release when there is a mismatch between oxygen delivery and utilisation. NO may play a key mediator role, probably through production via nNOS rather than eNOS. In animal models, cerebral vasodilatation associated with simple somatosensory stimulation appears to be mediated by nNOS derived NO.[42] In the human, L-NMMA given intravenously did not alter the local CBF response to learned sequential movements, but, because of its poor blood–brain barrier penetration, it probably only inhibits eNOS.[16]

CEREBRAL CIRCULATION RESPONSES TO FOCAL ISCHAEMIA

Protective Responses to a Progressive Fall in Cerebral Perfusion Pressure

CPP can fall because of systemic arterial hypotension, or severe stenosis in an extracranial or intracranial supplying artery, or a combination of the two. As CPP falls, intracranial resistance vessels dilate to maintain CBF; this results in an increase in CBV. When vasodilatation is maximal, further falls in CPP result in a fall in CBF. Because oxygen delivery to the brain normally greatly exceeds demand, metabolic activity is maintained initially by increasing the OEF from blood. When oxygen extraction becomes maximal, flow is inadequate to meet metabolic demands, cellular metabolism is impaired, and the cerebral metabolic rate of oxygen ($CMRo_2$) begins to fall. This sequence of events is illustrated in Fig. 2.

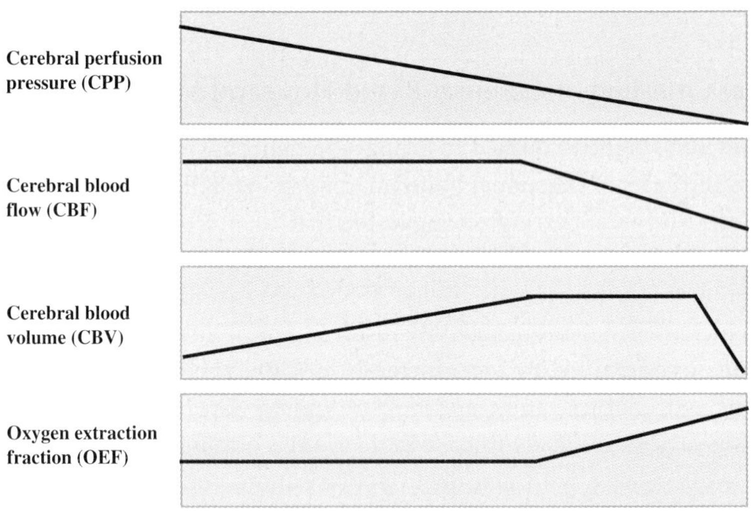

Fig. 2 A schematic illustration of the consequences of a progressive fall in cerebral perfusion pressure. This can occur because of a progressive fall in arterial blood pressure, or because of a stenosis or occlusion of a major supplying artery to the brain. As CPP falls, CBF is maintained by vasodilatation. This results in a progressive rise in CBV. When vasodilatation is maximal, cerebral autoregulation is exhausted, and a further reduction in CPP results in a decrease in CBF. As CBF falls, oxygen extraction is increased and functional consequences of reduced CBF only occur when OEF is maximal.

Critical Flow Thresholds

As cerebral autoregulation is impaired or lost in moderate to severe ischaemia, CBF varies passively with CPP. This relation has allowed investigators to gradually reduce CBF and assess critical flow thresholds at which certain functions are lost.[43] Experimental studies in primates[44] and cats,[45] and clinical studies in humans during carotid endarterectomy,[46] have shown that spontaneous and evoked electrical activity ceases when CBF falls below 16–18 ml/100 g/min. This level of ischaemia therefore represents a threshold for loss of neuronal electrical function (that is, electrical failure). It was subsequently shown that there is a lower threshold (10–12 ml/100 g/min) for loss of cellular ion haemostasis (that is, membrane failure). At this lower threshold, K^+ is released from and Ca^{2+} taken up by the cells.[47] Rapid efflux of K^+ and uptake of Ca^{2+} represents a generalised collapse of membrane function and at this point cells also take up Na^+ and Cl^- with osmotically obligated water.[43] The threshold for infarction appears similar to that for energy failure/loss of membrane haemostasis, but it varies with the duration of the insult.

More recent studies suggest that the pattern of thresholds may be more complex, although the general principle of two major critical flow thresholds (loss of electrical function and then loss of cellular ion haemostasis) still applies.[48] Protein synthesis is inhibited first (at a CBF of about 50 ml/100 g/min), and is completely suppressed below 35 ml/100 g/min. This is above the level at which glucose utilisation and energy metabolism are disrupted. Glucose utilisation transiently increases at flow rates below 35 ml/100 g/min, before it sharply declines below 25 ml/100 g/min. This corresponds to anaerobic glycolysis with the beginning of acidosis and the accumulation of lactate. At flow rates below 26 ml/100 g/min, tissue acidosis becomes pronounced and both phosphocreatine and ATP begin to decline.[48] Anoxic depolarisation, as assessed by recording extracellular potassium and calcium activities, occurs at even lower values (<15 ml/100 g/min) (Fig. 3).

THE ISCHAEMIC PENUMBRA

The concept of an ischaemic penumbra is crucial to the treatment of acute stroke. It follows on from the finding of separate thresholds for

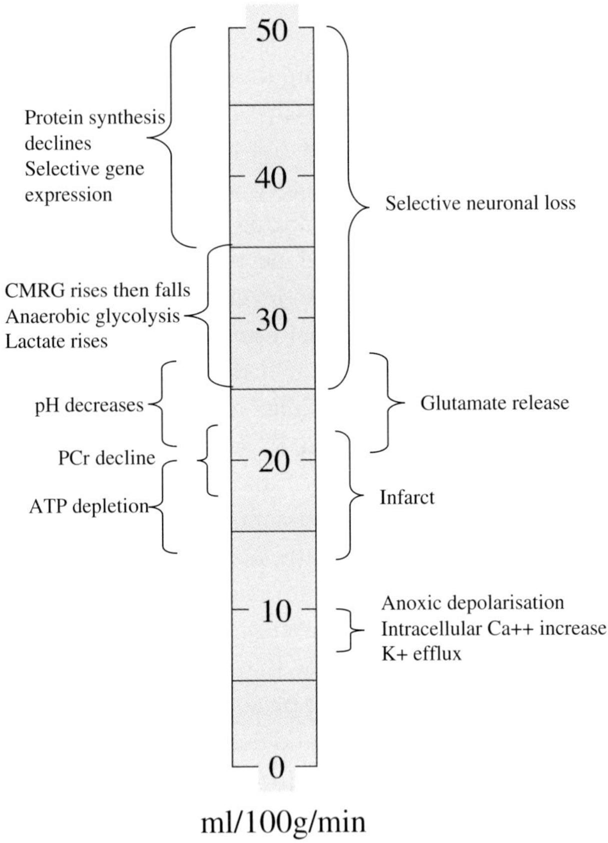

Fig. 3 Thresholds of ischaemia for the induction of functional, metabolic, and histological lesions. Exact levels vary slightly in different animal models and also with the duration of ischaemia for certain variables. CMRG, cerebral metabolic rate of glucose; PCr, phosphocreatine. Modified from Hossman,[45] Fig. 1.

electrical failure and loss of ion haemostasis. The concept is that following a focal ischaemic insult a penumbral region exists around a core of densely ischaemic and irreversibly damaged tissue. This penumbral region contains electrically inexcitable but viable cells. It follows that if perfusion is restored to this region, tissue salvage may be possible. The concept originated in the mid 1970s from experiments done on baboons.[49] After middle cerebral artery occlusion, brain tissue could be identified which had lost its electrical function but did not show the release of potassium that characterises cell death. This functionally silent tissue was nevertheless viable because it regained its function and did not undergo necrosis if flow was restored early enough. This tissue was named the ischaemic penumbra. Topographically the penumbra was shown to be mainly cortical surrounding the ischaemic core, which in the baboon, as in the human, predominantly involves the subcortical structures supplied by the perforating arteries. In later studies it was found that the duration of ischaemia, as well as the absolute flow, plays a crucial role in determining the fate of ischaemic tissue. For example, in the macaque monkey, tissue with CBF of around 15 ml/100 g/min could withstand about three hours of occlusion, while tissue with a perfusion of around 5 ml/100 g/min would stand only two hours.[50] This time dependence has crucial significance when considering acute treatment in humans, and implies that treatment will be most successful when it is given as early as possible. A crucial question when assessing the potential for acute stroke therapy is how extensive the ischaemic penumbra is in the human.

Is There an Ischaemic Penumbra in the Human?

PET studies have suggested that there is a significant ischaemic penumbra in humans, and that reversibly ischaemic tissue may persist for much longer than initial experiments in smaller animals suggested.[51] These studies used oxygen-PET and quantitatively measured the main physiological variables involved in tissue ischaemia — namely CBF, $CMRo_2$, OEF, and CBV. Using this technique, a pattern of changes termed "misery perfusion" can be identified. This is characterised by reduced CBF with relatively preserved or even normal $CMRo_2$, thus fitting the concept of penumbra with tissue of reduced perfusion but relatively maintained neuronal function. The hallmark of misery perfusion is an increased OEF, ranging from the normal value of about 30–40% up to a theoretical maximum of 100%, where the tissue extracts all the oxygen supplied by the blood flow reaching it.

In the late 1980s and early 1990s several groups applied PET to investigate the pathophysiology of ischaemic stroke. In the acute phase in patients with anterior circulation stroke, misery perfusion was found to affect mainly the cortex, while the capsule–basal ganglia region often showed markedly reduced $CMRo_2$, and only moderately increased OEF.[52,53] Deterioration of $CMRo_2$ from the acute to subacute stage of stroke was reported, and this was interpreted as the transition from ischaemia to necrosis. However, no study directly documented the presence of tissue that was still metabolically active within a region that went on to infarction, as determined by follow up CT or MRI.[51]

Baron and colleagues studied 30 patients with first ever middle cerebral artery territory ischaemia within 18 hours of stroke onset and co-registered these early PET scans with late CT scans done to assess final infarct size.[54] Those voxels or regions within the finally infarcted tissue with preserved $CMRo_2$ on acute stage PET were identified. Of the 30 patients, there were eight who survived until the late CT, had follow up PET, had a middle cerebral artery territory infarct of >16 ml as measured on late CT scanning, and had technically adequate PET studies. In seven of these, voxels were found which progressed to infarction but initially had $CMRO_2$ values above the threshold for irreversible damage. The tissue identified represented substantial volumes of the final infarct, being a mean of 35% and as high as 52% in one patient. Furthermore "penumbral" tissue was observed in two cases at 16 hours and in another case at 13 hours after stroke onset.

Proving that this is indeed penumbral tissue requires intervention studies in which the penumbra is salvaged by treatment that restores flow. However, the concept is supported by the results of clinical trials showing that thrombolysis improves outcome,[55] and also by more recent MRI studies (see below). Further support is provided by PET studies in the baboon which showed that tissue with acute misery perfusion went on to infarction if the middle cerebral artery was permanently occluded, but escaped infarction if the artery was reopened at six hours.[56] Interestingly, the time window for reperfusion in the baboon was much longer than in previous macaque experiments,[48] and considerably longer than PET studies in the cat.[57] These significant species differences emphasise the importance of undertaking studies in humans.

Can MRI Identify an Ischaemic Penumbra?

Although PET studies give quantitative information on CBF and oxygen metabolism they are not widely available, difficult to do in large numbers of patients, and involve the administration of radioisotopes. The ability to identify potentially salvageable tissue using more widely available techniques has great attraction. There is considerable interest in the potential of

MRI to identify an "ischaemic penumbra," based on a combination of perfusion and diffusion weighted imaging techniques. Most studies have used exogenous perfusion techniques to measure the perfusion deficit. The most promising technique for identifying ischaemic tissue is diffusion weighted imaging (DWI). Diffusion imaging is dependent on the apparent diffusion coefficient (ADC) of water in tissue. Following ischaemia, ADC values rapidly decline, corresponding to a region of high signal on DWI. This ADC decline is likely to reflect the accumulation of intracellular water, cytotoxic oedema, disruption of high energy metabolism, and loss of ion homeostasis. These ADC changes do not occur uniformly in the ischaemic region, and in experimental stroke models ischaemic tissue with the most severe perfusion deficit has the earliest and most severe fall in ADC. Serial DWI studies in animal models have demonstrated the evolution of ADC changes over time, and shown that mild declines in ADC early after stroke are potentially reversible.

The concept of diffusion–perfusion mismatch has been developed as a possible way of identifying potentially salvageable tissue in the human. The rationale is that tissue that is abnormal on DWI and has low perfusion is destined for infarction. In contrast, tissue that is normal on DWI but has perfusion reduced to penumbral levels may either recover, particularly if flow is restored, or progress to infarction. There is considerable support for this hypothesis. Within six hours of stroke onset approximately 70% of acute patients will have a perfusion volume greater than the diffusion volume, 20% will have equal volumes, and in 10% the perfusion volume will be smaller, representing spontaneous reperfusion.[58] Baird *et al.* studied 13 patients with ischaemic stroke in whom both DWI and perfusion imaging were measured at an initial time point (2 to 53 hours) and at follow up time points (7 to 725 days).[59] Lesion volume increased by 230% when the perfusion volume exceeded the initial DWI volume, but decreased by 47% when perfusion volume was smaller than or equivalent to the DWI volume. Barber *et al.* reported an increase of 62% in the DWI lesion volume between acute (<24 hours) and subacute (3–5 days) time points, when the perfusion volume was greater than the DWI lesion volume (11 of 17 patients).[60] The remaining six patients in whom the perfusion volume was less than the DWI volume had a stable DWI lesion volume.

Studies in patients receiving thrombolysis provide further evidence that this mismatch does represent reversibly ischaemic tissue. Patients undergoing intravenous thrombolysis, who underwent successful reperfusion, had reduced lesion volume as defined by DWI and later by T2 weighted MRI.[61] Similar results have been found in patients receiving intra-arterial thrombolytic treatment.[62] MRI studies also emphasised the heterogeneity of human stroke. In some patients mismatch is still present many hours after stroke, consistent with the PET studies,[51] while in others there is no mismatch early after stroke. This may partly explain why acute stroke trials have been disappointing and it has led to the suggestion that treatment needs to be tailored to the patient more effectively. Current therapeutic trials are targeting patients with persisting mismatch for treatment with thrombolytic and other agents.

Recent data have suggested that the diffusion–perfusion concept is more complex. Initially it was thought that tissue appearing abnormal on DWI in humans was almost always destined to infarction. Case reports were published showing reversibility of DWI changes, and a more recent series has confirmed that DWI abnormalities may recover,[63] although probably in a minority of patients. Furthermore on repeat scanning at seven days such recovered areas may remain normal, or DWI and T2 abnormalities may reappear. This is consistent with animal studies showing secondary delayed ADC declines after temporary focal occlusion. The clinical significance of these changes is not fully understood, but if they predict a worse clinical outcome they may represent a therapeutic target for post-reperfusion neuroprotective targets. Despite the MR diffusion–perfusion concept being more complicated than initially appreciated, there is considerable evidence that it does allow "penumbral" tissue to be identified. Its clinical use will depend upon current prospective randomised trials showing that it can identify a subgroup of patients who benefit particularly from therapeutic interventions such as thrombolysis.

MRI is much more widely available than PET, but nevertheless few units can offer it to all patients with acute stroke. This has led to attempts to use the more available technology of CT perfusion to identify the extent of reversibly ischaemic tissue. Compared with DWI, non-contrast CT imaging is poor at showing ischaemic tissue in the first few hours. However, with the use of a rapid intravenous injection of iodinated contrast in combination with newer multislice CT scanners, maps can be obtained of CBF, CBV, and MTT, in addition to conventional non-contrast CT images, with acquisition times of less than 10 minutes.[7] CT perfusion measurements have been found to correlate well with stable xenon-CT estimates.[64] Using this method, an ischaemic cerebral area (penumbra plus irreversibly damaged tissue) has been defined as a CBF reduction of more than 34% compared with the contralateral hemisphere.[65] Within this area, regions with a CBV below a predefined cut off (2.5 ml/100 g) were selected as irreversibly damaged tissue, with the remainder defined as penumbra. In a validation study against MRI in 13 subjects, a highly significant correlation was found between the CT defined irreversibly damaged tissue and the region of MR DWI abnormality.[65] The ischaemic cerebral area correlated strongly with the MR defined MTT defect, but less well with the MRI perfusion defect obtained from CBF maps. The authors suggested that CT perfusion provides a similar degree of information to MRI on penumbral tissue. Other small studies have also shown correlations with MRI estimates,[66] but much more data from larger series are required, ideally including comparisons with PET as well as with MRI.[67] This should both confirm these findings, and show that the CT derived information allows prediction of outcome and selection of a group who may benefit from thrombolytic and other treatments. A further limitation of CT perfusion in comparison with MRI is that slice coverage is limited.

MECHANISMS OF INFARCTION IN PENUMBRAL TISSUE

A complex cascade of mechanisms is responsible for the progression of penumbral tissue to infarction. Understanding these processes is of crucial importance for developing potentially effective treatment strategies. Mechanisms determining both flow and the cellular and metabolic consequences of hypoperfusion are important.

Key factors determining outcome are the presence and extent of collateral flow, and the time at which recanalisation occurs. Cerebral autoregulation is disrupted within the ischaemic penumbra, which may make the tissue particularly vulnerable to alterations in blood pressure. Continued embolisation and thrombus propagation may also play a role.[68] Experimental evidence suggests that progressive microvascular obstruction contributes to the progression of ischaemic damage following stroke.[69] Various processes may contribute to the activation of cerebral microvessels following ischaemia, including alterations in integrin–matrix interactions, leucocyte endothelial cell adhesion, blood–brain barrier permeability changes, and microvascular occlusion owing to adhesion of leucocytes, activated platelets, and fibrin deposition.[69]

A complex cascade of cellular and metabolic consequences follows focal ischaemic injury. This is illustrated in Fig. 4 and has been reviewed in detail elsewhere.[58,70] Only some of the more important aspects are covered below.

Excitotoxicity appears to play a crucial role.[71] Glutamate release activates postsynaptic N-methyl-D-aspartate (NMDA) and α-amino-3-hydroxy-5-methyl-4-isoxazole (AMPA) receptors. This results in calcium and sodium influx into cells, leading to cellular oedema and activation of the catabolic processes that destroy cellular integrity. These initial events are amplified by further increases in intracellular calcium through activation of voltage mediated calcium channels.[72] The intracellular calcium

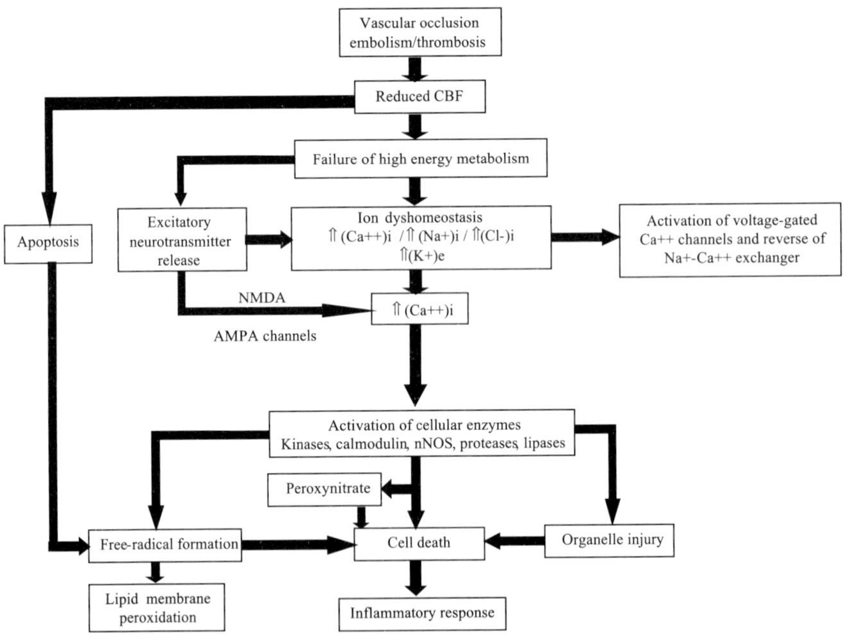

Fig. 4 Mechanisms mediating tissue injury in focal cerebral ischaemia. Modified from Fig. 4.5 in Wise *et al.*, 1983.[53] Multiple mechanisms which interact with each other are involved. This may partly explain why neuroprotective approaches inhibiting one pathway have failed in the treatment of human ischaemic stroke. AMPA, α-amino-3-hydroxy-5-methyl-4-isoxazole; NMDA, N-methyl-D-aspartate; NNOS, neuronal nitric oxide synthase; (…)i, intracellular; (…)e, extracellular.

increase induces protein kinases and calmodulin regulated enzymes. Glutamate release appears to be a primary process in a cascade of molecular reactions mediated by NO and free radicals. NO generated in neurones and microglia causes cell death by inhibiting mitochondrial functions, including apoptosis, and promoting the formation of free radicals such as the highly toxic peroxynitrite radicals. Knockout mice which lack nNOS develop infarcts of smaller size.[73] Glutamate diffusion from areas with high concentration and more severe ischaemia to less severely ischaemic regions may also lead to progression of injury. Cerebrospinal fluid and plasma concentrations of glutamate are higher in patients with progressing ischaemic stroke than in those with non-progressing stroke.[74]

Peri-infarct depolarisations may play an important role in progression of damage.[75] These depolarisations are similar to the phenomena of spreading depression observed after mechanical or chemical injury to normal cortical tissue in animal models. DWI studies of rat stroke models have shown that peri-infarct depolarisations lead to an increase in the size of the ischaemic lesion.[76] The underlying mechanism may be an increase in energy demands upon already compromised tissue by this energy consuming process.[58] In normal tissue, this increased energy demand is met by an increase in perfusion, but this is not possible when CBF is reduced. The relevance of peri-infarct depolarisations in human stroke has not been determined, although with DWI monitoring it may be possible to show an imaging correlate.

Inflammation is probably important in extending ischaemic injury. Most inflammatory reactions are mediated by cytokines. These have been implicated in several mechanisms that may potentiate ischaemic brain injury,[77] including release of NO from inducible NOS by astrocytes, recruitment, activation, and adhesion to the endothelium of infiltrating leucocytes, the promotion of a local procoagulant state, and the regulation of apoptotic processes.

SUMMARY

Many anatomical and physiological responses ensure that the brain receives adequate blood supply, and protect it against the devastating consequences of cerebral ischaemia. The disruption of these by focal ischaemia can now be investigated by an ever increasing range of methods to estimate perfusion, many of which can be implemented on routine clinical neuroimaging equipment. PET studies in humans have demonstrated reversibly ischaemic penumbral tissue following focal ischaemia; its extent varies markedly between patients but in a subgroup it may persist for hours. It is hoped that imaging techniques will allow patients with salvageable tissue to be identified for specific therapeutic interventions.

REFERENCES

1. Nehlig A. Metabolism of the central nervous system. In: Mraovitch S, Sercombe R, eds. *Neurophysiological Basis of Cerebral Blood Flow Control: An Introduction.* London: John Libby, 1996:177–196.
2. Kety SS. Circulation and metabolism of the human brain in health and disease. *Am J Med* 1950; **8**:205–217.
3. Riggs HE, Rupp C. Variation in form of the circle of Willis. *Arch Neurol* 1963; **8**:8–14.
4. Vander Ecken HM, Adams RD. The anatomy and functional significance of the meningeal arterial anastamoses of the human brain. *J Neuropathol Exp Neurol* 1953; **12**:132–157.
5. Lassen NA, Andersen AR, Fribert L, *et al.* The retention of ^{99m}Tc-d,l-HMPAO in the human brain after intra-arterial bolus injection. A kinetic analysis. *J Cereb Blood Flow Metab* 1988; **8**:S13–S22.
6. Yonas H, Pindzola RP, Johnson DW. Xenon/computed tomography cerebral blood flow and its use in clinical management. *Neurosurg Clin North Am* 1996; **7**:605–616.
7. Lev MH, Sagal AZ, Farkas J, *et al.* Utility of perfusion-weighted CT imaging in acute middle cerebral artery stroke treated with thrombolysis. Prediction of final infarct volume and clinical outcome. *Stroke* 2001; **32**:2021–2028.
8. Ostergaard L, Johannsen P, Host-Poulsen P, *et al.* Cerebral blood flow measurements by magnetic resonance imaging bolus tracking: comparison with [(15)O]H_2O positron emission tomography in humans. *J Cereb Blood Flow Metab* 1998; **18**:935–940.
9. Eicke BM, Tegeler CH. Ultrasonic quantification of blood flow volume. In: Tegeler CH, Babikian VL, Gomez CR, eds. *Neurosonology*. St Louis: CV Mosby, 1995:100–109.
10. Deane CR, Markus HS. Colour velocity flow measurement. In vitro validation and application to the human carotid arteries. *Ultrasound Med Biol* 1997; **23**: 447–452.

11. Eyding J, Wilkening W, Postert T. Brain perfusion and ultrasonic imaging techniques. *Eur J Ultrasound* 2002; **16**:91–104.
12. Furchcott RF, Zawadski JV. The obligatory role of endothelial cells in the relaxation of arterial smooth muscle by acetylcholine. *Nature* 1980; **288**: 373–376.
13. Palmer RM, Ferrige AG, Moncada S. Nitric oxide release accounts for the biological activity of endothelium-derived relaxing factor. *Nature* 1987; **327**: 524–526.
14. Iadecola C, Pelligrino DA, Moskowitz MA, *et al.* Nitric oxide synthase inhibition and cerebrovascular regulation. *J Cereb Blood flow Metab* 1994; **14**:175–192.
15. White RP, Deane C, Vallance P, *et al.* Nitric oxide synthase inhibition in humans reduces cerebral blood flow but not the hyperaemic response to hypercapnia. *Stroke* 1998; **29**:467–472.
16. White RP, Hindley C, Bloomfield PM, *et al.* The effect of the nitric oxide synthase inhibitor L-NMMA on basal CBF and vasoneuronal coupling in man: a PET study. *J Cereb Blood Flow Metab* 1999; **19**:673–678.
17. Lassen NA. Cerebral blood flow and oxygen consumption in man. *Physiol Rev* 1959; **39**:183–238.
18. Strandgaard S, Olsen J, Skinhoj E, *et al.* Autoregulation of brain circulation in severe arterial hypertension. *BMJ* 1973; **1**:507–510.
19. Enevoldsen EM, Jensen FT. Autoregulation and CO_2 responses of cerebral blood flow in patients with acute severe head injury. *J Neurosurg* 1978; **48**: 689–703.
20. Symon L, Branston NM, Strong AJ. Autoregulation in acute focal ischaemia. An experimental study. *Stroke* 1976; **7**: 547–554.
21. Voldby B. Pathophysiology of subarachnoid haemorrhage. Experimental and clinical data. *Acta Neurochir Suppl (Wien)* 1988; **45**: 1–6.
22. Panerai RB. Assessment of cerebral pressure autoregulation in humans — a review of measurement methods. *Physiol Meas* 1998; **19**:305–338.
23. Tiecks FP, Lam AM, Aaslid R, *et al.* Comparison of static and dynamic cerebral autoregulation measurements. *Stroke* 1995; **26**:1014–1019.
24. White RP, Markus HS. Non-invasive determination of impaired dynamic cerebral autoregulation in carotid artery stenosis. *Stroke* 1997; **28**:1340–1344.
25. Hlatky R, Furuya Y, Valadka AB, *et al.* Dynamic autoregulatory response after severe head injury. *J Neurosurg* 2002; **97**:1054–1061.
26. Dawson SL, Panerai RB, Potter JF. Serial changes in static and dynamic cerebral autoregulation after acute ischaemic stroke. *Cerebrovasc Dis* 2003; **16**: 69–75.
27. Busija DW, Heistad DD. Factors involved in the physiological regulation of the cerebral circulation. *Rev Physiol Biochem Pharmacol* 1984; **101**:161–211.

28. Chillon J-M, Baumbach GL. Autoregulation of cerebral blood flow. In: Welch KMA, Caplan LR, Reis DJ, *et al*, eds. Primer on Cerebrovascular Diseases. San Diego: Academic Press, 1997: 51–54.
29. White R, Vallance P, Markus HS. The effect of nitric oxide synthase inhibition on dynamic cerebral autoregulation in man. *Clin Sci (Colch)* 2000; **99**:555–560.
30. Huang Z, Huang PL, Ma J, *et al*. Enlarged infarcts in endothelial nitric synthase knockout mice are attenuated by nitro-l-arginine. *J Cereb Blood Flow Metab* 1996; **16**:981–987.
31. Kuschinsky W. Regulation of cerebral blood flow: an overview. In: Mraovitch S, Sercombe R, eds. *Neurophysiological Basis of Cerebral Blood Flow Control: An Introduction*. London: John Libby, 1996:245–262.
32. Markus HS, Vallance P, Brown MM. Differential effect of three cyclo-oxygenase inhibitors on human cerebral blood flow velocity and carbon dioxide reactivity. *Stroke* 1994; **25**: 1760–1764.
33. Thompson BG, Pluta RM, Girton ME, *et al*. Nitric oxide mediation of chemoregulation but not autoregulation of cerebral blood flow in primates. *J Neurosurg* 1996; **84**:71–78.
34. McPherson RW, Kirsch JR, Ghaly RF, *et al*. Effect of nitric oxide synthase inhibition on the cerebral vascular response to hypercapnia in primates. *Stroke* 1995; **26**:682–687.
35. Shinozuka T, Nemoto EM, Winter PM. Mechanisms of cerebrovascular O_2 sensitivity from hyperoxia in the rat. *J Cereb Blood Flow Metab* 1989; **9**:187–195.
36. Ringlestein EB, Sievers C, Ecker S, *et al*. Noninvasive assessment of CO_2-induced cerebral vasomotor response in normal individuals and patients with internal carotid artery occlusions. *Stroke* 1988; **19**:963–969.
37. Huber P, Handa J. Effect of contrast material, hypercapnia, hyperventilation, hypertonic glucose and papaverine on the diameter of cerebral arteries. *Invest Radiol* 1967; **2**: 17–32.
38. Kleiser B, Widder B. Course of carotid artery occlusions with impaired cerebrovascular reactivity. *Stroke* 1992; **23**:171–174.
39. Markus H, Cullinane M. Severely impaired cerebrovascular reactivity predicts stroke and TIA risk in patients with carotid artery stenosis and occlusion. *Brain* 2001; **124**:457–467.
40. Gur AY, Bova I, Bornstein NM. Is impaired cerebral vasomotor reactivity a predictive factor of stroke in asymptomatic patients? *Stroke* 1996; **27**:2188–2190.
41. Roy CS, Sherrington CS. On the regulation of the blood supply of the brain. *J Physiol (Lond)* 1890; **11**:85–108.
42. Gjedde A. Brain energy metabolism and the haemodynamic response. In: Jezzard P, Matthews PM, Smith SM, eds. *Functional MRI*. Oxford: Oxford University Press, 2001:37–65.

43. Siesjo BK. Pathophysiology and treatment of focal cerebral ischaemia. Part 1. Pathophysiology. *J Neurosurg* 1992; **77**: 169–184.
44. Branston NM, Symon L, Crockard HA, *et al.* Relationship between the cortical evoked potential and local cortical blood flow following acute middle cerebral artery occlusion in the baboon. *Exp Neurol* 1974; **45**: 195–208.
45. Heiss WD, Hayakawa T, Waltz AG. Cortical neuronal function during ischaemia. Effects of occlusion of one middle artery on single unit activity in cats. *Arch Neurol* 1976; **33**:813–820.
46. Sharborough FW, Messick JM, Sundt TM. Correlation of continuous electroencephalograms with cerebral blood flow measurements during carotid endarterectomy. *Stroke* 1973; **4**:674–683.
47. Harris RJ, Symon L, Branston NM, *et al.* Changes in extracellular calcium activity in cerebral ischaemia. *J Cereb Blood Flow Metab* 1981; **1**:203–209.
48. Hossmann KA. Viability thresholds and the penumbra of focal ischemia. *Ann Neurol* 1994; **36**:557–565.
49. Astrup J, Siesjo BK, Symon L. Thresholds in cerebral ischemia — the ischemic penumbra. *Stroke* 1981; **12**:723–725.
50. Jones TH, Moraweta RB, Crowell RM, *et al.* Thresholds of focal cerebral ischemia in awake monkeys. *J Neurosurg* 1981; **54**:773–782.
51. Baron JC. Mapping the ischaemic penumbra with PET: implications for acute stroke treatment. *Cerebrovasc Dis* 1999; **9**: 193–201.
52. Baron JC, Bousser MG, Comar D, *et al.* Noninvasive tomographic study of cerebral blood flow and oxygen metabolism in vivo: Potentials, limitations and clinical applications to cerebral ischemic disorders. *Eur Neurol* 1981; **20**: 273–284.
53. Wise RJS, Bernadi S, Frackowiak RSJ, *et al.* Serial observations on the pathophysiology of acute stroke: the transition from ischaemia to infarction as reflected in regional oxygen extraction. *Brain* 1983; **106**:197–222.
54. Marchal G, Beaudouin V, Rioux P, *et al.* Prolonged persistence of substantial volumes of potentially viable brain tissue after stroke: a correlative PET–CT study with voxel-based data analysis. *Stroke* 1996; **27**:599–606.
55. rt-PA Stroke Study Group. Tissue plasminogen activator for ischemic stroke. *N Engl J Med* 1995; **333**:1581–1587.
56. Young AR, Touzani O, Derlon JM, *et al.* Early reperfusion in the anaesthetized baboon reduces brain damage following middle cerebral artery occlusion. *Stroke* 1997; **28**:632–638.
57. Heiss WD, Graf R, Wienhard K, *et al.* Dynamic penumbra demonstrated by sequential multitracer PET after middle cerebral artery occlusion in cats. *J Cereb Blood Flow Metab* 1994; **14**:892–902.

58. Fisher M. The ischemic penumbra and the therapeutic time window. In: Fisher M, Bogousslavsky J, eds. *Current review of cerebrovascular disease*, 4th ed. Philadelphia: Current Medicine, 2002:35–43.
59. Baird A, Benfield A, Schlaug G, *et al*. Enlargement of human cerebral ischemic lesion volumes measured by diffusion-weighted magnetic imaging. *Ann Neurol* 1997; **41**:581–589.
60. Barber P, Darby D, Desmond PM, *et al*. Prediction of stroke outcome with echoplanar perfusion- and diffusion-weighted magnetic resonance imaging. *Neurology* 1998; **51**:418–426.
61. Schellinger PD, Jansen O, Fiebach JB, *et al*. Monitoring intravenous recombinant tissue plasminogen activator thrombolysis for acute ischemic stroke with diffusion and perfusion MRI. *Stroke* 2000; **31**:1318–1328.
62. Kidwell CS, Saver JL, Mattiello J, *et al*. Thrombolytic reversal of acute human cerebral ischemic injury shown by diffusion/perfusion magnetic resonance imaging. *Ann Neurol* 2000; **47**:462–469.
63. Kidwell CS, Saver JL, Starkman S, *et al*. Late secondary ischemic injury in patients receiving intra-arterial thrombolysis. *Ann Neurol* 2002; **52**:698–703.
64. Wintermark M, Maeder P, Thiran J-P, *et al*. Simultaneous measurements of regional cerebral blood flows by perfusion-CT and stable xenon-CT: a validation study. *Am J Neuroradiol* 2001; **22**:905–914.
65. Wintermark M, Reichhart M, Cuisenaire O, *et al*. Comparison of admission perfusion computed tomography and qualitative diffusion- and perfusion-weighted magnetic resonance imaging in acute stroke patients. *Stroke* 2002; **33**:2025–2031.
66. Na DG, Ryoo JW, Lee KH, *et al*. Multiphasic perfusion computed tomography in hyperacute ischemic stroke: comparison with diffusion and perfusion magnetic resonance imaging. *J Comput Assist Tomogr* 2003; **27**:194–206.
67. Warach S. Stroke neuroimaging. *Stroke* 2003; **34**:345–347.
68. Wong KS, Gao S, Chan YL, *et al*. Mechanisms of acute cerebral infarctions in patients with middle cerebral artery stenosis: a diffusion-weighted imaging and microemboli monitoring study. *Ann Neurol* 2002; **52**:74–81.
69. del Zoppo GJ, Hallenback JM. Advances in the vascular pathophysiology of ischemic stroke. *Thromb Res* 2000; **98**:V73–V81.
70. Kogure T, Kogure K. Molecular and biochemical events within the brain subjected to cerebral ischemia. *Clin Neurosci* 1997; **4**: 179–183.
71. Choi DW, Lobner D, Dugan LL. Glutumate receptor-mediated neuronal death in the ischemic brain. In: Hsu C, ed. *Ischemic Stroke: From Basic Mechanisms to New Drug Development*. Basel: Karger, 1998:2–13.

72. Mody I, MacDonald JF. NMDA receptor-dependent excitotoxicity. The role of intercellular Ca^{++} release. *Trends Pharmacol Sci* 1995; **16**:356–359.
73. Samdani AF, Dawson TM, Dawson VI. Nitric oxide synthase in models of focal ischaemia. *Stroke* 1997; **28**:1283–1288.
74. Castillo J, Davalos A, Noya M. Progression of ischaemic stroke and excitotoxic aminoacids. *Lancet* 1997; **349**:79–83.
75. Hossman KA. Peri-infarct depolarizations. *Cerebrovasc Brain Metab Rev* 1996; **8**:195–208.
76. Takano K, Latour LL, Formato JE, *et al*. The role of spreading depression in focal ischemia evaluated by diffusion mapping. *Ann Neurol* 1996; **39**:308–318.
77. Vila N, Castillo, Davalos A, *et al*. Proinflammatory cytokines and early neurological worsening in ischemic stroke. *Stroke* 2000; **31**:2325–2329.

13

A Review of Structural Magnetic Resonance Neuroimaging

*M Symms, H R Jäger, K Schmierer and T A Yousry**

Magnetic resonance imaging (MRI) is often divided into structural MRI and functional MRI (fMRI) with the former being the most widely used imaging technique in research as well as in clinical practice. This review describes the more important developments in structural MRI, including high resolution imaging, T2 relaxation measurement, T2*-weighted imaging, T1 relaxation measurement, magnetisation transfer imaging, and diffusion imaging. The principles underlying these techniques, as well as their use in research and in clinical practice, will be discussed.

The phenomenon of nuclear magnetic resonance (NMR) was first observed in 1945.[1,2] It was the injection of radiofrequency engineering expertise and the availability of stable new frequency sources, both byproducts of the wartime development of radar, which probably made the demonstration of NMR possible. NMR imaging was first reported in 1973,[3,4] and the first human *in vivo* MR images were produced by the end of that decade. Compared with images from other

*Lysholm Department of Neuroradiology, National Hospital of Neurology and Neurosurgery, Institute of Neurology, Queen Square, London WC1N3BG, UK; t.yousry@ion.ucl.ac.uk

modalities, MR images of the head provided excellent anatomical detail and strong grey/white matter contrast. Flow-sensitive techniques, developed in the mid-1980s, measured blood flow velocity[5]: these were the first MR images considered not to be purely structural in nature.' The spectacular advent of functional MR imaging (fMRI)[6] a decade ago revolutionised MRI to such an extent that in current MRI practice, the definition of structural MRI seems to have shifted to mean "not functional" MRI.

DEFINITION(S) OF STRUCTURAL MRI

The division between structural and functional imaging is difficult to make because structure and function are often inextricably intertwined in the brain. Definitions of functional imaging are varied and often broad[6] and will always be arbitrary to a certain extent. On the basis of biological considerations, functional imaging can be regarded as the method providing dynamic physiological information, whereas structural imaging provides static anatomical information. fMRI therefore includes BOLD (blood oxygen level dependent) technique, perfusion (whether by endogenous or exogenous contrast), blood flow, and cerebrospinal fluid (CSF) pulsation measurements. Phase contrast flow measurements can be considered as fMRI. Magnetic resonance angiography is harder to classify, but given that its purpose is to determine how well the vessels carry blood, it can be considered a functional technique. Spectroscopy and chemical shift imaging aim to measure chemical concentrations, and these should be considered separately from other MR techniques. Note that spectroscopy is usually structural/static in nature, though some functional spectroscopy studies have been done.

It is obviously beyond the scope of any article to assess the whole range of techniques and applications of structural MRI. This review will cover some of the new developments in structural MRI and discuss their principles with examples of their role in research and clinical applications.

APPLICATIONS OF STRUCTURAL MRI

T2- and T1-weighted sequences form the core of almost every clinical MRI protocol. Pathological processes are therefore most often described in terms of T1 and T2 signal behaviour, in addition to contrast enhancement, anatomical location, and morphological characteristics.

More recently, fluid attenuated inversion recovery (FLAIR) has been introduced as a complement of, or even a replacement for, the conventional T2-weighted sequence. Over the past years new techniques have been developed to improve spatial resolution (high resolution imaging) and sensitivity (T2* imaging), establish quantitative methods (T1 maps, T2 maps, magnetisation transfer imaging (MTI)), and develop new qualitative and quantitative contrast techniques (diffusion imaging).[7,8] We will describe the principles underlying these methods and address their role in research and clinical practice.

High Resolution Imaging

Principle

The two main characteristics which govern image quality are spatial resolution and signal to noise ratio (SNR). In-plane resolution is chiefly determined by the number of picture elements (pixels) in the frequency and phase encoding directions, and through-plane resolution by the slice thickness. SNR is determined by pixel size, slice thickness, scan time (including the number of phase encoding steps) and the sequence used. Image quality is therefore ultimately determined by scan time, which is constrained by patient compliance.

Any motion occurring during the MRI scan can cause motion artefacts. Cooperative subjects can be persuaded to keep their head still, and head restraints can minimise head motion, but there are other sources of movement. The regular brain pulsation caused by the cardiac cycle can be mostly averaged out over a long scan or explicitly removed by cardiac triggering. The effects of the respiratory cycle are small in most neuroimaging experiments, but random motions such as eye movement cannot be removed.

Applications

Research

Coronal T1-weighted, three dimensional, high resolution images are used to measure the volume of the hippocampus, usually by means of manually tracing its outline. They are also the basis for many cross-sectional and longitudinal studies determining the volume of, and assessing changes in, the hippocampus over time in hippocampal sclerosis and Alzheimer's disease (AD).[9]

Clinical environment

- Epilepsy

High resolution T1-weighted images have become a standard tool in epilepsy. The ability to identify the hippocampus and its structure is paramount for assessing atrophy in hippocampal sclerosis. An example is shown in Fig. 1. Furthermore, accurate visualisation of the cortex is important for the diagnosis of cortical dysplasia.

- Cranial nerves

Identification of subtle structures within the CSF is facilitated by using sequences with a cisternographic effect. These sequences (3D CISS, DRIVE, FIESTA) enable the reliable identification of fine structures such as the anterior choroidal artery, the abducens nerve and even the trochlear nerve.[10,11] Visualisation of the trochlear nerve has helped to elucidate the pathogenesis of "superior oblique myokymia" by revealing the presence of an arterial neurovascular contact at the root exit zone of the nerve.[12]

High-resolution imaging is the "purest" form of structural imaging, and the need to see the brain's structures in evermore detail provides the ongoing impetus to increase MRI resolution.

T2 Relaxation Measurement

Principle

T2-weighted images are commonly used in neuroimaging because they are very sensitive to a wide range of pathology. However, the signal in T2-weighted images contains not only T2 weighting, but also information about the amount of water (proton density) in each pixel. While this extra information can sometimes aid diagnosis, it can also confound it. Using two or more images with different echo times, or by curve-fitting to a long spin echo train, it is possible to calculate T2 maps of the brain that are independent of proton density.

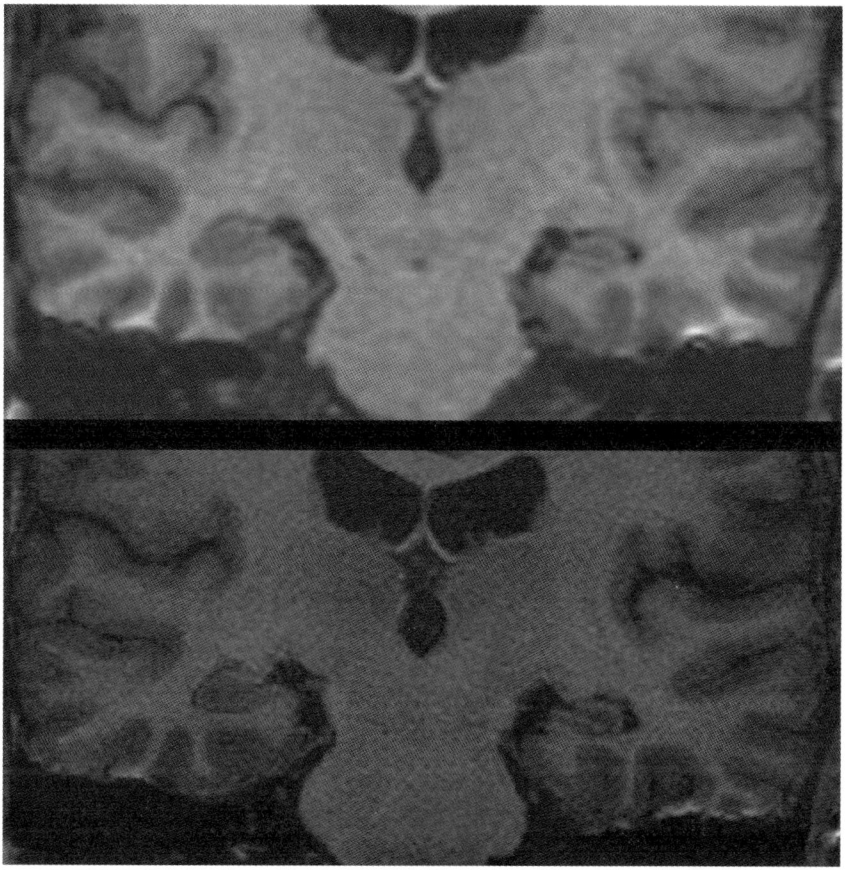

Fig. 1 Coronal section of a T1-weighted three dimensional volume of a subject with epilepsy at (a) conventional (1 mm × 1 mm × 1.5 mm) and (b) high (0.5 mm × 0.5 mm × 1 mm) resolution. Hippocampal sclerosis can be seen as the loss of volume in the left hippocampus (right side of image).

Applications in research

Hippocampal sclerosis can be detected in patients with epilepsy by comparing the hippocampal T2 values with a range of values of normal controls.[13] T2-weighted images of a subject with left hippocampal sclerosis are shown in Fig. 2.

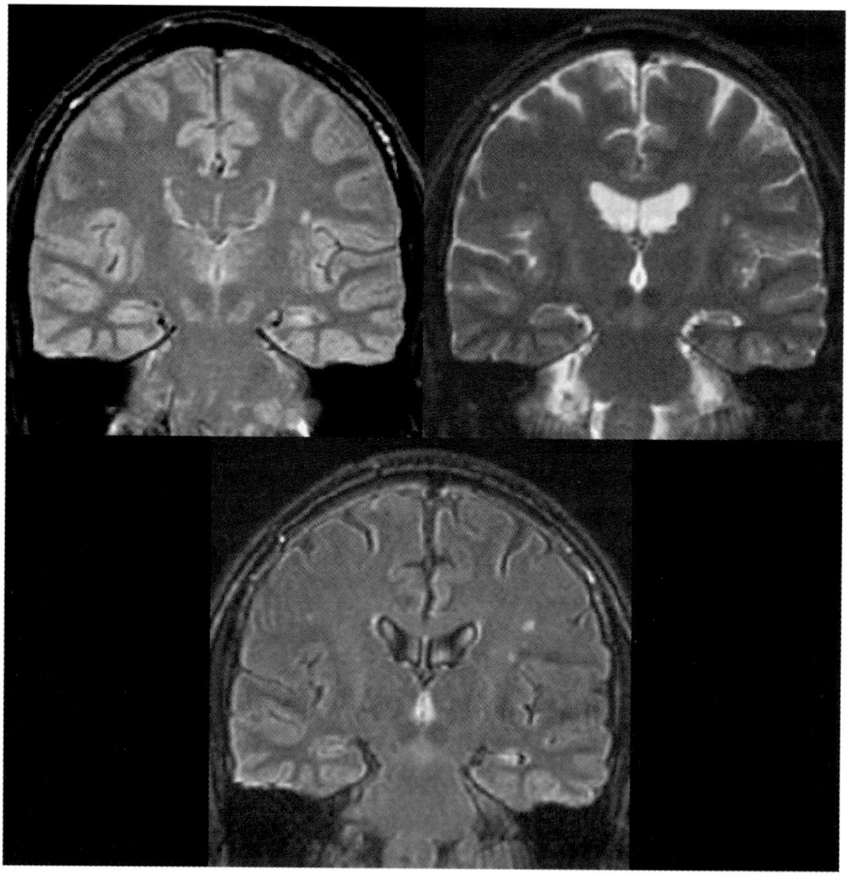

Fig. 2 Proton density, T2-weighted and fast FLAIR images of a subject with left hippocampal sclerosis. The left hippocampus (right side of image) has a brighter T2-weighted signal, which is particularly evident on the fast FLAIR image where the surrounding cerebrospinal fluid has been suppressed.

T2*-Weighted Imaging

Principle

T2-weighted imaging usually uses images derived from a spin echo, because this refocuses any field inhomogeneities that are present — these are usually caused by changes in magnetic susceptibility between neighbouring tissues. If gradient echo (GRE) images are acquired instead, they will reflect the effects of field inhomogeneity and provide additional contrast: T2* weighting.

Applications

Research

Using modified T2*-weighted sequences, it is possible to examine relative levels of iron concentration in the brain: iron content has been observed to be elevated in the substantia nigra in patients with Parkinson's disease.[14]

Clinical environment

• Acute cerebral haemorrhage

Reliable identification of acute haemorrhage is important for differentiation between a haemorrhagic and an ischaemic stroke, especially if thrombolytic treatment is being considered. Routine spin echo MR images are less sensitive than computer tomography (CT) for detection of acute brain haemorrhage, which is one of the reasons why CT scanning is currently the first choice of imaging modality for acute stroke in clinical practice. However, the detection of acute parenchymal haemorrhage on MRI can be considerably improved by T2*-weighted GRE imaging, which exploits the magnetic susceptibility artefact from deoxyhaemoglobin. Sensitivity of MRI protocols that include T2*-weighted GRE images to acute brain haemorrhage has been shown to be similar to that of CT.[15] T2-weighted images associated with a diffusion-weighted imaging (DWI) sequence are prone to susceptibility artefacts[16] and it was therefore suggested to replace the T2*-weighted GRE images thereby saving examination time. However, T2*-weighted GRE images have been found to be more sensitive to haemorrhage than the DWI associated T2-weighted images and should therefore be an integral part of an MRI protocol for acute stroke.

• Subarachnoid haemorrhage

CT scanning and lumbar puncture are the current reference standards for detection of subarachnoid haemorrhage. T2*-weighted GRE has been shown to be the most useful MRI sequence with 94% sensitivity for acute subarachnoid haemorrhage and 100% sensitivity for subacute subarachnoid haemorrhage (after four days).[17] This imaging modality can therefore be of importance in the delayed diagnosis of subarachnoid haemorrhage, which is difficult with CT.

• Cerebral microhaemorrhages

Small hypointense areas on T2 and T2*-weighted images correspond histopathologically to extravasation of blood and have been dubbed "microhaemorrhages".[18] These lesions are much more conspicuous on T2*-weighted GRE images than on T2-weighted spin echo images due to the increased effects of magnetic susceptibility from haemosiderin deposits. Microhaemorrhages are a feature of hypertensive small vessel disease but have also been described in cerebral amyloid angiopathy,[19] cerebral autosomal dominant arteriopathy with subcortical infarcts, and leukoencephalopathy (CADASIL)[20] and trauma.

On T2*-weighted GRE images, microhaemorrhages occurred in 6.4% of a randomly selected elderly population without neurological disease[21] and 3.7% of patients without previous strokes but in 18.1% of patients with a history of ischaemic stroke and in 71% of patients with a history of haemorrhagic stroke.[22] The strong correlation between the presence of microhaemorrhages and a history of haemorrhagic stroke suggests that microhaemorrhages are not only a direct marker of bleeding-prone small vessel disease but could also be a predictor of further haemorrhagic stroke. Examples of the appearance of microhaemorrhages on T2*-weighted images are shown in Fig. 3.

Furthermore, in trauma, microhaemorrhages are a histological hallmark of diffuse axonal injury, and are therefore important to detect. It is interesting to note that the T2* lesion load has been found to correlate positively with the period of loss of consciousness and Glasgow Outcome Scale.[23] The most sensitive way to detect these microbleeds is to use T2*-weighted sequences as they detect three times as many lesions as the standard MRI sequences at 1.5 T and at 3 T.[23]

T1 Relaxation Measurement

Principle

T1 mapping follows a rationale similar to T2 mapping — remove variations in image intensity due to proton density by generating an image depending on just one parameter. T1 maps can be generated from two or more images with different repetition times or flip angles, and so have different T1 weightings, or from two or more images with different inversion times,

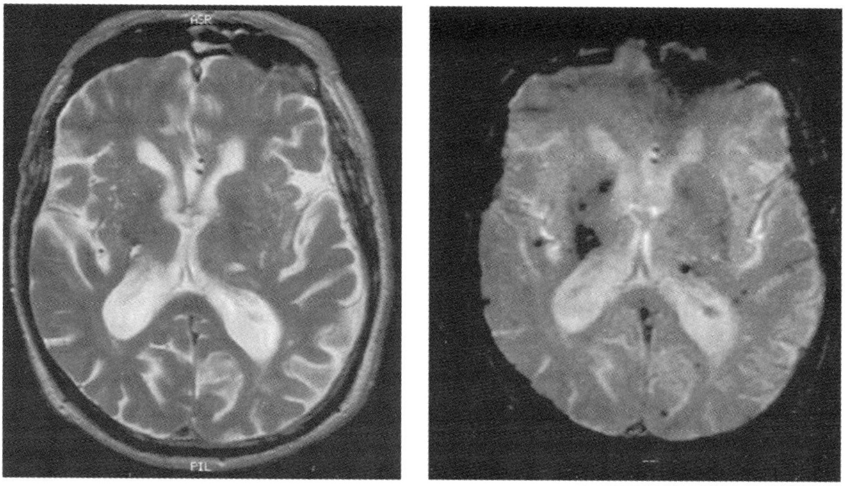

(a) (b)

Fig. 3 (a) T2-weighted fast spin echo (FSE) and (b) T2*-weighted gradient echo (GRE) images of a 66 year old patient with an old right basal ganglia haemorrhage and extensive cerebral microhaemorrhages. On the T2-weighted FSE (a) only the right basal ganglia haemorrhage is visible as a small low signal intensity area. This haemorrhage causes much more susceptibility artefact on the T2*-weighted GRE sequence (b), which reveals many additional central and peripheral foci of low signal intensity in both hemispheres, consistent with cerebral microhaemorrhages.

which again have different T1 weightings. T1 mapping is much more susceptible than T2 mapping to inhomogeneities of the applied radiofrequency (RF) field, so it is advisable to acquire extra images to measure these imperfections and correct for them.[24] This is rarely possible at present in the clinical setting.

Applications in research

In patients with multiple sclerosis (MS) the T1 of normal appearing white matter is significantly different between the infratentorial and supratentorial regions.[25] This has important consequences for the optimisation of T1 weighting of sequences used for detecting lesions in MS; optimal and consistent lesion detection may require different T1 weightings at different levels of the brain.

Magnetisation Transfer Imaging

Principle

The signal in most MR images arises from the protons in the "mobile pool" — that is, from water in the liquid phase in the body. However, many protons in relatively solid phases are bound in proteins and large biomolecules (the "bound pool") and have a very broad NMR resonance that normally decays too quickly for the MR scanner to detect. These bound protons are in a state of chemical exchange and diffusion with protons in the mobile pool, providing a link between the two pools of protons. If a strong RF pulse is applied which is far enough away from the resonance of the liquid pool, but still excites the bound pool, some of the excited magnetisation is transferred from the bound pool to the liquid pool. The resulting magnetisation transfer (MT)-weighted image is attenuated, and the magnetisation transfer ratio (MTR) can be calculated from this image and a similar one without MT weighting. While MTR is difficult to relate to a real physical measure and MTR measures vary according to the imaging sequence and scanner used, it has nevertheless proved to be a sensitive marker of pathological change in many neurological conditions in the sense that it decreases with increasing histopathological changes.

Applications

Research

- Schizophrenia

In a group of subjects with chronic schizophrenia group mapping showed changes in MTR in both white and grey matter areas relative to a group of matched controls.[26] These changes may reflect subtle changes involving neurones or neuronal processes, and are therefore an additional indication for the neuropathological background in schizophrenia.

- Dementia

MTI and regional MTR measurements have been studied in several subtypes of dementia. In vascular dementia, MTR of periventricular white matter lesions has been found to be lower in patients with Binswanger's disease than in patients with vascular disease but no dementia.[27] Furthermore, MTR values correlate with neuropsychological measures, thereby suggesting that MTR reflects the severity of histopathological changes in vascular dementia.

MTR measurements of the hippocampus appear promising for the differentiation between AD and other types of dementia.[27] Normal MTR values of the medial temporal lobe appear to be an early feature of AD[27] and are already seen in patients with mild cognitive impairment.[28] It was found that the MTR of grey matter was abnormally low both in patients with AD and patients with mild cognitive impairment, but MTR changes in the white matter were only present in patients with AD, indicating a more advanced stage of the disease.[28] These findings suggest that MTR could be a novel MR measure for early diagnosis of AD.

- Trauma

Traumatic brain injury may result in cerebral (cortical) contusions and/or diffuse axonal injury. The latter can cause subtle disruption of the structural integrity of brain tissue difficult to detect with conventional MRI. However, even in patients with only mild head injury abnormal MTR values have been found in the corpus callosum.[29] A decreased MTR has also been shown in white matter, which appeared normal on conventional MRI.[30] The presence of abnormalities on MTI is often associated with persistent neurological deficits.

- Tumours

Experience with MT imaging in brain tumours is limited, but the significantly higher MTR of brain abscesses may help to differentiate them from cystic tumours.[31] MTR can also distinguish between brain tumours which are hard and those that are soft on palpation during operation, the latter having lower MTR values.[32]

- Multiple sclerosis

In the past decade a number of studies have used MTI as a probe for tissue damage in MS.[33] Evidence from correlative MTI/pathological studies suggests that the myelin content and the axonal count are the most relevant substrates of MTR changes in patients with MS.[34,35]

Hitherto, the use of MTI in MS has centred around three main fields: (a) its use in conjunction with the contrast agent gadolinium-DTPA (Gd) to improve the detection of new MS lesions, (b) assessment of MTR to distinguish lesions of different severity, and (c) assessment of changes in brain tissue that appears normal on conventional MRI. Due to their significance in MS research, we will focus on the latter two aspects.

Cross-sectional studies revealed a broad range of MTR values in MS lesions. However, it has consistently been shown that MS lesions returning a hypointense signal on T1-weighted MRI have a lower MTR than lesions which are T1-isointense.[36]

Correlation studies of standard T2-weighted MRI, T1-weighted images with and without Gd enhancement, and MTR maps have revealed that the pathological substrate of even an apparently homogeneous group such as Gd enhancing lesions varies considerably.[37,38] Overall the MTR in lesions drops considerably when enhancement occurs,[36,37] but recovery of a lowered MTR may occur over the following months.[38] Ring enhancing lesions, particularly their centre, display MTR values as low as in chronic T1-hypointense lesions,[36,37] which result from severe demyelination and axonal loss,[39] whereas homogeneously enhancing lesions have a significantly higher MTR.[36]

The severity of tissue damage — as reflected by changes in lesion MTR — has been shown to be modestly correlated with the course of MS. Patients with secondary progressive MS (SPMS) display lower lesion MTR than patients with benign MS.[40]

A possible association between the MTR in brain tissue and the course of MS has also triggered a number of studies into tissue compartments appearing normal on conventional MRI. The MTR of the whole brain as well as of the segmented normal appearing white matter has consistently been shown to be lower in patients with MS than in healthy subjects.[41–45]

A number of studies have used MTI to investigate the time course of lesion development. They have shown that at least a subset of MS lesions develop against the background of tissue changes in the normal appearing white matter — the latter being a steady decrease of MTR in the respective region — occurring up to 24 months before Gd enhancement becomes visible.[46,47] Lesions in regions showing a drop in MTR prior to Gd enhancement appeared to be less likely to recover than lesions that did not.[48] Although early recovery of lesions takes place as shown by the increase of an initially decreased lesion MTR during follow up,[38] an overall decrease of brain MTR occurs over time, be it assessed in lesions, normal appearing white matter, or the whole brain.[33,48–50] This has been demonstrated for a period of up to 4.5 years.[50]

Some MTR derived indices are associated with clinical disability.[51,52] A reduction of the whole brain MTR over one year has been an independent predictor of disability more than three years later[40] and a low MTR in normal appearing white matter predicted disability over five years better than the MTR in lesions, suggesting that, in the long term, tissue pathology in normal appearing white matter may be more important than the lesions for the functional decline of patients with MS.

At least two factors exert significant influence on the associations between MTR derived indices and disability: the tissue compartment investigated and the MS subtype. In a study encompassing all subtypes of MS, the patients with SPMS had the lowest whole brain and lesion MTR values.[53] A study including patients with primary progressive MS (PPMS), and patients with relapsing remitting (RR)-onset MS and disability, reported modest correlations of whole brain MTR with disability in both groups.[54] However, in a study of patients with PPMS, MTR in the segmented grey matter only — but not in normal appearing white matter — correlated with disability, suggesting that the grey matter may be particularly relevant for the development of disability in this subgroup.[45]

Recently, a novel quantitative MT technique allowing the estimation of the macromolecular proton fraction (f) and the bound pool T2 relaxation time (T2b) has been applied in patients with MS. Preliminary results showed significant differences for f and T2b between MS lesions, normal appearing white matter, and control white matter. This technique may achieve relative independence from the MT acquisition protocol while offering more pathologically specific information.[55]

After a decade of research and more than 200 studies into MT changes due to MS, guidelines for the implementation of MTR in large multicentre trials have been developed that should allow this technique to be used as a method to monitor treatment effects in patients with MS.[56]

Clinical environment

MTR is not firmly established as a clinical tool at present, which may be due to its restricted commercial availability. However, some of the research areas mentioned above may soon develop into clinical applications.

Diffusion Imaging

Diffusion is often considered a functional technique. We argue that diffusion should be considered as a structural technique: diffusion indices such as mean diffusivity (MD) are measures of the cellular state, while diffusion anisotropy and tractography provide information about the structure of white matter.

Principle

MRI can be used to measure the self-diffusion of water — that is, the random motion of water molecules. If "diffusion encoding" gradients are applied, it can be shown that the NMR signal in the presence of diffusion experiences attenuation of amplitude due to the diffusion of water. Acquiring images with differing amounts of diffusion weighting (known as diffusion-weighted imaging, DWI) allows the diffusion coefficient to be measured. Diffusion in most brain tissues is restricted, so DWI measures the apparent diffusion coefficient (ADC). The most commonly used measure is MD, where the diffusion coefficient is averaged over all directions. Diffusion in brain tissues can be isotropic, such as in CSF, in which the water can diffuse equally easily in each direction, or it can be anisotropic, such as in white matter tracts where diffusion is less restricted along the long axis of the white matter tract than it is perpendicular to the tract. The most commonly used index of diffusion anisotropy is probably fractional anisotropy, though there are various other indices.

Diffusion tensor imaging (DTI) allows the production of maps which show the principal direction of diffusion in any voxel. In structures with diffusion anisotropy, it has been shown that the principal direction of diffusion is parallel to the direction of the tract. Thus, tractography can be performed, where the path of these tracts can be traced through the brain. The tracts can be visualised by means of probability maps. This technique may be useful to investigate white matter changes in MS and other diseases of the central nervous system.[57] At present, the effectiveness of tractography is limited by several factors, notably the low resolution of MRI compared with the size of most white matter tracts.

Applications

Research

- Epilepsy

Both decreases and increases in MD have been observed in patients with epilepsy scanned soon after a seizure.[58] It has been suggested that the same mechanisms for diffusion change following acute stroke can occur in the ictal and postictal states. Changes in fractional anisotropy have been observed in patients with malformations of cortical development.[59] These changes may partly be due to the high contrast between grey and white matter in the anisotropy maps, however, some changes observed on the fractional anisotropy maps have not been detected on the corresponding T1-weighted images.

- Dementia

A number of studies have explored the use of DWI in dementia, yielding at times contradictory results. Whereas some studies reported no significant regional differences between patients with MCI, probable AD, and healthy subjects,[60] others found a significant increase of the ADC in the hippocampal region in both MCI and AD compared with control subjects.[61] Increased ADC values were also found in other brain regions including the temporal stem, posterior cingulate gyrus, corpus callosum, and parietal white matter.[61]

A selective involvement of white matter tracts has been found in patients with AD. DTI revealed a reduction in the integrity of association fibres such as the splenium of the corpus callosum, the superior longitudinal fasciculus, and the cingulum, while the integrity of the pyramidal tracts remained unchanged.[62] At 3 T, fractional anisotropy has been found to be significantly reduced in the temporal white matter, posterior part of the corpus callosum, and anterior and posterior cingulated bundles in patients with AD.[63] Involvement of these white matter tracts correlated well with the degeneration of the cortical structures they are known to connect with, namely the posterior cerebral cortices and the hippocampus.

• Trauma

DWI is being used to assess the integrity of the brain tissue in the acute as well as in the subacute phase following head trauma. Decreased ADC values, reflecting restricted diffusion, have been found 1–18 days after head trauma, most frequently in the corpus callosum.[64] The restricted diffusion could reflect cellular swelling (cytotoxic oedema), which persists into the subacute phase. Alternatively, the movement of water molecules could be impeded by the presence of fragmented membranes from ruptured axons. Conversely, repeated or longstanding head trauma can lead to an increase in ADC values, reflecting an increase in free water movement. Thus, histogram analysis of the whole brain in professional boxers revealed significantly higher average ADC values than in age matched normal controls.[65] Furthermore, the average ADC correlated with the number of hospitalisations for boxing injuries, suggesting a cumulative effect of repeated blows to the head. Similarly, an increase in MD has also been shown several months after blunt head injury in patients who had unremarkable conventional MR images.[66] The increase in MD extended to regions distant from the site of impact, suggesting an expansion of the extracellular space followed by neuronal or glial cell loss as a possible mechanism.

• Multiple sclerosis

DW-MRI has been used over the past 10 years to probe MS.[33,67] A consistent finding has been an increased diffusivity compared with normal appearing white matter in MS lesions visible on T2-weighted MRI, either expressed by an elevated average ADC[68] or MD.[69] Similarly fractional anisotropy has been shown to be lower in MS lesions than in normal appearing white matter.[69,70]

The degree to which changes in DWI occur appears to depend on the clinical course. Hence, MD has been found to be higher in lesions of patients with a secondary progressive course compared with those with RRMS[71] or PPMS.[72] The increased diffusivity in lesions of patients with SPMS corroborates the finding in SPMS of a higher proportion of hypointense lesions on T1-weighted MRI ("black holes") which consistently display more abnormalities of diffusion than T1-isointense lesions[73,74] as long as they do not enhance after application of Gd; Gd enhancing T1-hypointense lesions have been reported as having either different[73,74] or similar[75] diffusion properties compared with their non-enhancing counterparts. This may reflect a variable tissue composition of Gd enhancing lesions, depending on their age and severity.

Changes in the normal appearing brain tissue of patients with MS have also been detected by DWI. Such changes in water diffusion appear to occur very early in the course of MS and may precede the formation of new lesions.[69,76] The abnormalities of diffusion in normal appearing white matter appear to be more pronounced in periplaque normal appearing white matter than in remote regions[77] and have been reported to be associated with the diffusivity in lesions[78] as well as with the lesion load on conventional T2-weighted MRI.[79,80]

So far results regarding diffusion changes in patients with PPMS are conflicting. Evidence is there to suggest that in this subgroup of patients diffusion changes in the normal appearing white matter develop rather independently of lesions[81] whereas other authors have detected an association between the extent of lesions and changes in the diffusivity of normal appearing white matter.[80] The association of changes in diffusion of normal appearing white matter with disability in patients is still a matter of debate. Several recent studies, some of which used histogram analysis, reported such associations in patients with RRMS, SPMS,[71,82] and PPMS.[80]

Some studies have focused on the grey matter of patients with MS by means of DWI. No changes were detected in the basal ganglia of MS patients[83] or in the grey matter of RRMS patients[84] whereas one study reported abnormalities of grey matter MD histograms in patients with PPMS or SPMS.[85]

- Surgical applications

It may be possible to use fractional anisotropy to better observe wallerian degeneration following surgical resection and thereby improve the assessment of postoperative sequelae,[86] especially since a correlation of reduced fractional anisotropy with clinical deficit has been observed.[87] Furthermore, DTI and tractography have the potential to yield important information for preoperative management such as patient counselling and surgical planning. DTI can assess the effect of the tumour on the adjacent tracts by determining whether they are infiltrated or displaced; and tractography has the potential to trace the path of these tracts (Fig. 4).[88] This information will enable both safer as well as more radical surgery.

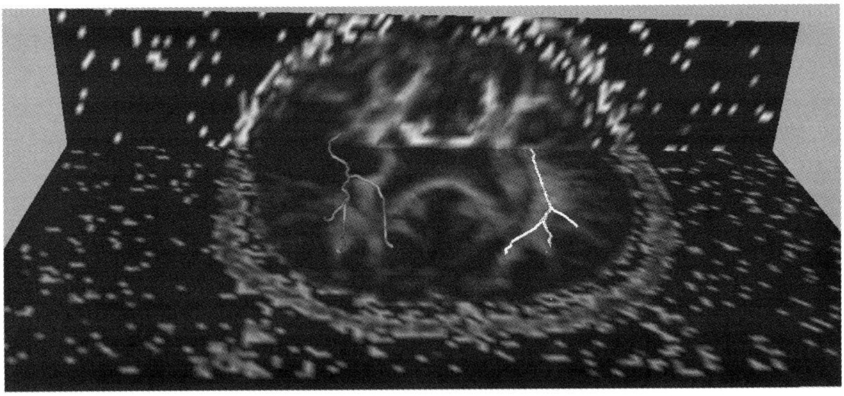

Fig. 4 Composite three dimensional multisection representation of a diffusion fractional anisotropy image. Overlaid are two traces derived from tractography of the corresponding diffusion tensor data, showing the pathways taken by the superior longitudinal fasciculus (SLF) on each side. The SLF takes a normal path on the left (right side of image) but the right SLF (left side of image) takes a more circuitous route, because it has been displaced by a large tumour. Tractography here shows the preservation of the right SLF in spite of the tumour's presence, and also provides important additional information for any possible surgical intervention.

- Amyotrophic lateral sclerosis

In subjects with amyotrophic lateral sclerosis, ADC and fractional anisotropy of the corticospinal tract correlate with disease duration and with disease severity, respectively.[89] This opens the possibility of using these quantitative measures to monitor disease progression or possible drug effect in treatment trials.

- Schizophrenia

Conflicting results have been reported so far, possibly due to the varying methodologies used. A region of interest approach found increases in MD and decreases in fractional anisotropy in the splenium but not the genu of the corpus callosum.[90] Voxel-by-voxel approaches based on spatial normalisation and group mapping have either been negative[91] or have reported widespread differences in fractional anisotropy and MD, particularly in the prefrontal regions.[92]

- Normal brain maturation and ageing

From early life until adolescence ADC decreases and fractional anisotropy increases,[93] whereas in the ageing brain ADC increases and fractional anisotropy decreases.[94] These findings, however, have been reported in cross-sectional studies and are yet to be confirmed in longitudinal studies.

Clinical environment

- Infarction

DWI is far more sensitive for the detection of acute ischaemic changes than conventional MRI or CT.[95] DWI has been shown to detect ischaemic lesions within minutes in experimental animal studies and within 30–90 minutes of an acute stroke in humans.[96] Areas of acute ischaemia show restricted water diffusion and appear bright on DWI and dark on ADC maps. The precise mechanisms leading to a reduction in diffusion are still a matter of debate, but redistribution of extracellular water into the intracellular compartment (cytotoxic oedema) resulting in shrinkage of the extracellular space appears to be the most likely explanation.

Initially, changes on DWI were regarded as a marker of irreversible tissue damage and were used to define the ischaemic penumbra in combination with MR perfusion imaging (perfusion/diffusion mismatch). This assumption was subsequently proved to be erroneous, first by anecdotal reports of reversibility of DWI changes and later by quantitative ADC analysis,[97] which demonstrated that tissue with ADC ratios of 0.9 (ADC reduced to 90% of the ADC of normal brain) and above was likely to recover whereas tissue with ADC ratios below that showed transition to infarction.

Appearances on DWI following stroke are time dependent: ADC values are low in the first week, become "pseudonormal" in the second week, and increase above the ADC of normal brain parenchyma thereafter. Decreased ADC values indicate with good sensitivity and specificity that an infarct is less than 10 days old.[98] MRI diffusion imaging can therefore be useful in identifying subacute infarcts. In the presence of several abnormalities suggestive of ischaemic stroke on T2-weighted images, DWI can help localise the acute lesion and determine its vascular territory, which may influence the clinical management. Acute ischaemia also causes a reduction of diffusion anisotropy. This effect is more marked in white matter than grey matter and fractional anisotropy may be a more sensitive measure of white matter ischaemia than DWI.[99] DTI in the subacute phase can distinguish whether white matter tracts are distorted around the infarct or disrupted by it, with implications for functional recovery.[100]

DWI may also be useful to differentiate arterial infarcts from posterior reversible encephalopathy syndrome (PRES) and from venous infarcts. PRES is due to a breakthrough of the cerebral autoregulation and endothelial dysfunction, affecting predominately the posterior white matter with some involvement of the overlying cortex. The lesions are bright on T2-weighted images and may occasionally be difficult to distinguish from posterior circulation infarcts. However, on DWI these lesions are not hyperintense as acute arterial infarcts but appear isotense or hypointense, indicating the presence of vasogenic oedema, which is reversible.[101] Venous infarcts initially cause vasogenic oedema with increased ADC values. At a later stage, the appearances are more complex and variable. There may be co-existence of vasogenic and cytotoxic oedema and the appearances may be further complicated by the presence of haematomas causing signal loss on DWI.[102]

- Tumours

While DWI has become an established tool in the clinical management of stroke, evidence is mounting of its usefulness in the clinical investigation of cerebral mass lesions. DWI of cerebral mass lesions is currently in transition from a research to a clinical application. We therefore decided to include this section among the clinical applications of diffusion imaging.

It must be emphasised that ADC measurements are essential for assessment of tumours, as the T2 shine-through effects may be predominant on DWI. Earlier studies[103] established the inverse relationship between ADC values and histological cell count (tumour cellularity). Higher grade tumours, which have a higher density of cell nuclei, therefore have lower ADC values than low grade tumours. Statistically significant differences between the ADC values of high grade and low grade gliomas have been shown in a number of studies.[103–105] Similarly, atypical or malignant (WHO grade 2 or 3) meningiomas have lower ADC values than benign meningiomas.[106]

It may be difficult to determine the degree of peritumoral neoplastic infiltration by gliomas on DWI alone[104] but a recent study at 3 T demonstrated that DTI is capable of detecting subtle disruption of white matter tracts in high grade gliomas, consistent with tumour invasion.[107] Primary central nervous system lymphomas tend to have lower ADC measurements than high grade gliomas, because they are highly cellular tumours.[108]

Of far greater importance is the potential to differentiate inflammatory lesions from tumours. The finding that lesions due to toxoplasmosis display significantly higher ADC values than lymphomas can be important for their differentiation in patients with AIDS.[109] A similar problem can occur in the distinction between a purulent brain process and cystic brain tumours; a diagnosis that can be challenging on conventional MRI. Using DWI cerebral abscesses demonstrate a much greater restriction of than in tumour cysts, resulting in significantly lower ADC values.[110]

Finally, arachnoid cysts and epidermoid tumours both have signal characteristics close to CSF on conventional MRI and are difficult to differentiate with this modality. This problem has now been definitively addressed by DWI, in which arachnoid cysts return a low signal, similar to CSF whereas epidermoid tumours return a high signal intensity, indicating restricted diffusion and thereby establishing the diagnosis.

- Creutzfeldt–Jakob disease (CJD)

DWI has become an essential tool in the diagnosis of CJD. The sporadic form shows areas of high signal intensity in the basal ganglia and cerebral cortex, which may precede changes on standard MRI, whereas the variant form is associated with high signal in the basal ganglia and mainly the pulvinar. Early changes become progressively hyperintense on DWI, unlike acute ischaemic infarcts, which tend to "pseudonormalise" within two weeks. The signal alteration corresponds histologically with areas of spongiform change and neuronal loss.[111] DWI not only helps make an early diagnosis of CJD possible but has also been reported to have a high specificity and can differentiate between different forms of CJD.[112]

THE NEXT TEN YEARS

The continual development in MRI technology over the past two decades has produced both steady progress and dramatic breakthroughs. Given such a fertile field of invention, predictions of the future are difficult, but it can be hoped that there will be improvements in the hardware to further decrease imaging time and increase SNR, as well as enhancements in software for post-processing and data management.

Acquisition Development

Conventional MRI uses one RF receiving coil to pick up the NMR signal from the body, and highly efficient configurations of transmitter coil (if there is a separate one), receiver coil, and RF receiver electronics have been used for many years in commercial MR scanners. A recent development, parallel imaging,[113] combines an array of receiver coils in such a way that each coil oversamples the data required to reconstruct the MR image. The coils receive the signal simultaneously, resulting in effective time saving. Some SNR is lost in the process, but nevertheless, parallel imaging represents a new way of optimising MRI acquisition. Many commercial MR scanners are now available with parallel imaging ability, and as acquisition and reconstruction techniques improve, this may well become the modality of choice for many clinical MR examinations.

High Field Imaging

As the main magnetic field is increased, the amount of net magnetisation induced increases. From this consideration alone, it would seem advantageous to operate at the highest possible field and resonant frequency. Several obstacles to high field imaging are related to patients' safety and image quality.

Patients' safety

- RF penetration: As frequency increases, rather than passing into the centre of the body, the applied RF has a tendency to dissipate near the surface, causing adverse heating effects.
- Peripheral magnetic stimulation: If gradients are increased to overcome internal magnetic gradient fields, eventually the applied gradients start generating electrical currents in the patient causing tingling and even physical pain.
- Acoustic noise: As applied magnetic field and gradients increase, the motor effect means more acoustic noise is generated, which may become uncomfortable or even harmful for the patient.

Image degradation

- RF homogeneity: Little RF reaches the centre of the body, so little MR signal is recovered from there.
- Main field inhomogeneity: As the applied magnetic field increases, internal magnetic gradient fields generated by differences between tissue and air cause larger image artefacts. These can be best overcome by larger field gradients.

As technology and MR engineering move on, so imaging at higher fields is becoming more feasible. Recent images obtained on prototype 4.7 T (Fig. 5)[114] and 8 T scanners bear this out. It is now possible to identify *in*

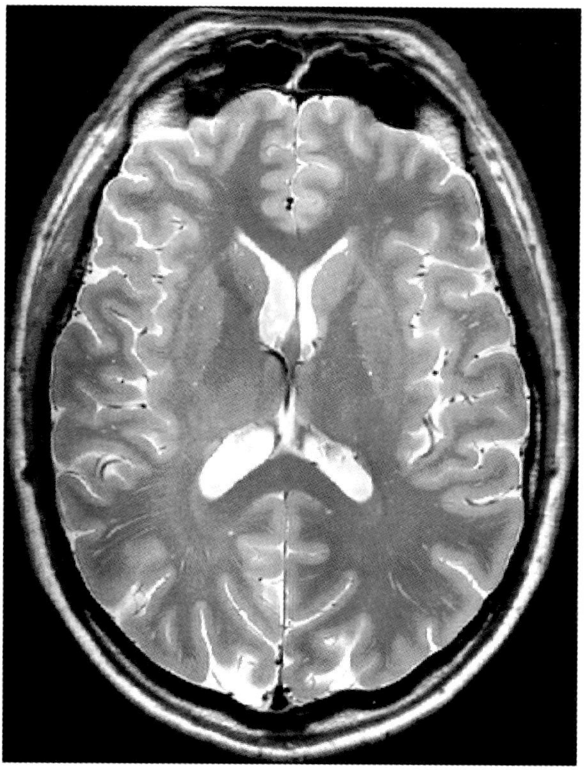

Fig. 5 High resolution fast spin echo image, with a slice thickness of 2 mm, and an in-plane resolution of 470 × 470 μm obtained at 4.7 T. The signal to noise is very good, and allows visualisation of normally occurring Virchow–Robin spaces beyond the extent usually seen at 1.5 T imaging.

vitro different types of cortex and even identify the corresponding cortical layers.[115] It is hoped that this kind of resolution will be approached *in vivo* in the next decade. The current installation of 7 T and 11 T scanners may be opening new horizons in MRI and a first step in that direction.

Post-processing and Data Management

A 10 year old MR scanner can produce excellent structural MR images, but new MR scanners will soon be able to do much more. The modern MR scanner can be integrated with a picture archiving and communicating system (PACS), and further developments will lead to more integrated patient databases, image registration based acquisition prescription, comparison of newly acquired images with previous images of the same patient (serial comparisons) or against a group of similar subjects (group mapping techniques), and the automatic removal of extraneous image features (image segmentation to remove scalp, meninges etc.). Given the power of computing available on new scanners, more advanced image processing techniques such as feature recognition may become possible. Interventional MRI, in which an MR scanner becomes truly interfaced with the surgical theatre, will also be developed further.

CONCLUSION

Structural MRI, using conventional T1, T2, and T2* contrasts, has become the accepted standard for routine examination of the brain, offering exquisite anatomical detail and high sensitivity to pathological changes. New quantitative measures of magnetisation transfer and diffusion are being widely used in research and some will soon become standard clinical tools. Future developments promise an increase in both sensitivity and specificity.

ACKNOWLEDGEMENTS

We thank Philippa Bartlett, Sofia Eriksson, Roger Ordidge, and Geoff Parker for providing some of the figures.

REFERENCES

1. Bloch F. Nuclear Induction. *Phys Rev* 1946; **70**:460–474.
2. Purcell EM, Torrey HC, Pound RV. Resonance absorption by nuclear magnetic moments in a solid. *Phys Rev* 1946; **69**:37–38.
3. Mansfield P, Grannell PK. NMR "diffraction" in solids? *J Phys C: Solid State Phys* 1973; **6**:L422–L426.
4. Lauterbur P. Image formation by local induced interactions: examples employing nuclear magnetic resonance. *Nature* 1973; **242**:190–191.
5. Moran PR. A flow velocity zeugmatographic interlace for NMR imaging in humans. *Magn Reson Imaging* 1982; **1**:197–203.
6. Jezzard P, Matthews PM, Smith SM. *Functional MRI: An Introduction to Methods*. Oxford University Press, 2001.
7. McRobbie DW, Moore EA, Graves MJ, *et al*. *MRI: From Picture to Proton*. Cambridge University Press, 2003.
8. Tofts PS. Quantitative MRI of the Brain: Measuring Changes Caused by Disease. Wiley, 2003.
9. Fox NC, Freeborough PA. Brain atrophy progression measured from registered serial MRI: validation and application to Alzheimer's disease. *J Magn Reson Imaging* 1997; **7**:1069–1075.
10. Yousry I, Camelio S, Wiesmann M, *et al*. Detailed magnetic resonance imaging anatomy of the cisternal segment of the abducent nerve: Dorello's canal and neurovascular relationships and landmarks. *J Neurosurg* 1999; **91**:276–283.
11. Yousry I, Moriggl B, Dieterich M, *et al*. MR anatomy of the proximal cisternal segment of the trochlear nerve: neurovascular relationships and landmarks. *Radiology* 2002; **223**:31–38.
12. Yousry I, Dieterich M, Naidich TP, *et al*. Superior oblique myokymia magnetic resonance imaging support for the neurovascular compression hypothesis. *Ann Neurol* 2002; **51**:361–368.
13. Jackson GD, Connelly A, Duncan JS, *et al*. Detection of hippocampal pathology in intractable partial epilepsy: increased sensitivity with quantitative magnetic resonance T2 relaxometry. *Neurology* 1993; **43**:1793–1799.
14. Graham JM, Paley MNJ, Grunewald RA, *et al*. Brain iron deposition in Parkinson's disease imaged using the PRIME magnetic resonance sequence. *Brain* 2000; **123**:2423–2431.
15. Perl J 2nd, Tkach JA, Porras-Jimenez M, *et al*. Hemorrhage detected using MR imaging in the setting of acute stroke: an in vivo model. *AJNR Am J Neuroradiol* 1999; **20**:1863–1870.
16. Lin DD, Filippi CG, Steever AB, *et al*. Detection of intracranial hemorrhage: comparison between gradient-echo images and b(0) images obtained

from diffusion-weighted echo-planar sequences. *AJNR Am J Neuroradiol* 2001; **22**:1275–1281.

17. Mitchell P, Wilkinson ID, Hoggard N, *et al.* Detection of subarachnoid haemorrhage with magnetic resonance imaging. *J Neurol Neurosurg Psychiatry* 2001; **70**:205–211.
18. Fazekas F, Kleinert R, Roob G, *et al.* Histopathologic analysis of foci of signal loss on gradient-echo T2*-weighted MR images in patients with spontaneous intracerebral hemorrhage: evidence of microangiography-related microbleeds. *AJNR Am J Neuroradiol* 1999; **20**:637–642.
19. Greenberg SM, O'Donnell HC, Schaefer PW, *et al.* MRI detection of new hemorrhages: potential marker of progression in cerebral amyloid angiopathy. *Neurology* 1999; **53**:1135–1138.
20. Dichgans M, Holtmannspotter M, Herzog J, *et al.* Cerebral microbleeds in CADASIL: a gradient-echo magnetic resonance imaging and autopsy study. *Stroke* 2002; **33**:67–71.
21. Roob G, Schmidt R, Kapeller P, *et al.* MRI evidence of past cerebral microbleeds in a healthy elderly population. *Neurology* 1999; **52**:991–994.
22. Tsushima Y, Aoki J, Endo K. Brain microhemorrhages detected on T2*-weighted gradient-echo MR images. *AJNR Am J Neuroradiol* 2003; **24**:88–96.
23. Scheid R, Preul C, Gruber O, *et al.* Diffuse axonal injury associated with chronic traumatic brain injury: evidence from T2*-weighted gradient-echo imaging at 3 T. *AJNR Am J Neuroradiol* 2003; **24**:1049–1056.
24. Parker GJM, Barker GJ, Tofts PS. Accurate multislice gradient echo T1 measurement in the presence of non-ideal RF pulse shape and RF field nonuniformity. *Magn Reson Med* 2001; **45**:838–845.
25. Stevenson VL, Parker GJM, Barker GJ, *et al.* Variations in T1 and T2 relaxation times of normal appearing white matter and lesions in multiple sclerosis. *J Neurol Sci* 2000; **178**:81–87.
26. Foong J, Symms MR, Barker GJ, *et al.* Investigating regional white matter in schizophrenia using diffusion tensor imaging. *NeuroReport* 2002; **13**:333–336.
27. Hanyu H, Asano T, Iwamoto T, *et al.* Magnetization transfer measurements of the hippocampus in patients with Alzheimer's disease, vascular dementia, and other types of dementia. *AJNR Am J Neuroradiol* 2000; **21**:1235–1242.
28. Kabani NJ, Sled JG, Chertkow H. Magnetization transfer ratio in mild cognitive impairment and dementia of Alzheimer's type. *Neuroimage* 2002; **15**: 604–610.
29. McGowan JC, Yang JH, Plotkin RC, *et al.* Magnetization transfer imaging in the detection of injury associated with mild head trauma. *AJNR Am J Neuroradiol* 2000; **21**:875–880.

30. Sinson G, Bagley LJ, Cecil KM, *et al.* Magnetization transfer imaging and proton MR spectroscopy in the evaluation of axonal injury: correlation with clinical outcome after traumatic brain injury. *Am J Neuroradiol* 2001; **22**: 143–151.
31. Pui MH. Magnetization transfer analysis of brain tumor, infection, and infarction. *J Magn Reson Imaging* 2000; **12**:395–399.
32. Okumura A, Takenaka K, Nishimura Y, *et al.* The characterization of human brain tumor using magnetization transfer technique in magnetic resonance imaging. *Neurol Res* 1999; **21**:250–254.
33. Filippi M, Rocca MA, Comi G. The use of quantitative magnetic-resonance-based techniques to monitor the evolution of multiple sclerosis. *Lancet Neurol* 2003; **2**:337–346.
34. van Waesberghe JH, Kamphorst W, De Groot CJ, *et al.* Axonal loss in multiple sclerosis lesions: magnetic resonance imaging insights into substrates of disability. *Ann Neurol* 1999; **46**:747–754.
35. Mottershead JP, Schmierer K, Clemence M, *et al.* High field MRI correlates of myelin content and axonal density in multiple sclerosis: a post-mortem study of the spinal cord. *J Neurol* 2003; **250**:1293–1301.
36. Petrella JR, Grossman RI, McGowan JC, *et al.* Multiple sclerosis lesions: relationship between MR enhancement pattern and magnetization transfer effect. *AJNR Am J Neuroradiol* 1996; **17**:1041–1049.
37. Rovira A, Alonso J, Cucurella G, *et al.* Evolution of multiple sclerosis lesions on serial contrast-enhanced T1-weighted and magnetization-transfer MR images. *AJNR Am J Neuroradiol* 1999; **20**:1939–1945.
38. Lai HM, Davie CA, Gass A, *et al.* Serial magnetisation transfer ratios in gadolinium-enhancing lesions in multiple sclerosis. *J Neurol* 1997; **244**: 308–311.
39. van Walderveen MA, Kamphorst W, Scheltens P, *et al.* Histopathologic correlate of hypointense lesions on T1-weighted spin-echo MRI in multiple sclerosis. *Neurology* 1998; **50**:1282–1288.
40. Davie CA, Silver NC, Barker GJ, *et al.* Does the extent of axonal loss and demyelination from chronic lesions in multiple sclerosis correlate with the clinical subgroup? *J Neurol Neurosurg Psychiatry* 1999; **67**:710–715.
41. van Buchem MA, McGowan JC, Kolson DL, *et al.* Quantitative volumetric magnetization transfer analysis in multiple sclerosis: estimation of macroscopic and microscopic disease burden. *Magn Reson Med* 1996; **36**:632–636.
42. Cercignani M, Bozzali M, Iannucci G, *et al.* Magnetisation transfer ratio and mean diffusivity of normal appearing white and grey matter from patients with multiple sclerosis. *J Neurol Neurosurg Psychiatry* 2001; **70**:311–317.

43. Griffin CM, Chard DT, Parker GJ, *et al*. The relationship between lesion and normal appearing brain tissue abnormalities in early relapsing remitting multiple sclerosis. *J Neurol* 2002; **249**:193–199.
44. De Stefano N, Narayanan S, Francis SJ, *et al*. Diffuse axonal and tissue injury in patients with multiple sclerosis with low cerebral lesion load and no disability. *Arch Neurol* 2002; **59**:1565–1571.
45. Dehmeshki J, Chard DT, Leary SM, *et al*. The normal appearing grey matter in primary progressive multiple sclerosis: a magnetisation transfer imaging study. *J Neurol* 2003; **250**:67–74.
46. Goodkin DE, Rooney WD, Sloan R, *et al*. A serial study of new MS lesions and the white matter from which they arise. *Neurology* 1998; **51**: 1689–1697.
47. Filippi M, Rocca MA, Martino G, *et al*. Magnetization transfer changes in the normal appearing white matter precede the appearance of enhancing lesions in patients with multiple sclerosis. *Ann Neurol* 1998; **43**:809–814.
48. Laule C, Vavasour IM, Whittall KP, *et al*. Evolution of focal and diffuse magnetisation transfer abnormalities in multiple sclerosis. *J Neurol* 2003; **250**: 924–931.
49. van Waesberghe JH, van Walderveen MA, Castelijns JA, *et al*. Patterns of lesion development in multiple sclerosis: longitudinal observations with T1-weighted spin-echo and magnetization transfer MR. *AJNR Am J Neuroradiol* 1998; **19**: 675–683.
50. Rovaris M, Agosta F, Sormani MP, *et al*. Conventional and magnetization transfer MRI predictors of clinical multiple sclerosis evolution: a medium-term follow-up study. *Brain* 2003; **126**:2323–2332.
51. Gass A, Barker GJ, Kidd D, *et al*. Correlation of magnetization transfer ratio with clinical disability in multiple sclerosis. *Ann Neurol* 1994; **36**:62–67.
52. Udupa JK, Nyul LG, Ge Y, *et al*. Multiprotocol MR image segmentation in multiple sclerosis: experience with over 1000 studies. *Acad Radiol* 2001; **8**: 1116–1126.
53. Filippi M, Iannucci G, Tortorella C, *et al*. Comparison of MS clinical phenotypes using conventional and magnetization transfer MRI. *Neurology* 1999; **52**: 588–594.
54. Dehmeshki J, Silver NC, Leary SM, *et al*. Magnetisation transfer ratio histogram analysis of primary progressive and other multiple sclerosis subgroups. *J Neurol Sci* 2001; **185**:11–17.
55. Davies GR, Ramani A, Dalton CM, *et al*. Preliminary magnetic resonance study of the macromolecular proton fraction in white matter: a potential marker of myelin? *Mult Scler* 2003; **9**:246–249.

56. Horsfield MA, Barker GJ, Barkhof F, *et al.* Guidelines for using quantitative magnetization transfer magnetic resonance imaging for monitoring treatment of multiple sclerosis. *J Magn Reson Imaging* 2003; **17**:389–397.
57. Ciccarelli O, Toosy AT, Parker GJ, *et al.* Diffusion tractography based group mapping of major white-matter pathways in the human brain. *Neuroimage* 2003; **19**:1545–1555.
58. Wieshmann UC, Symms MR, Shorvon SD. Diffusion changes in status epilepticus. *Lancet* 1997; **350**:493–494.
59. Eriksson SH, RuggGunn EJ, Symms MR, *et al.* Diffusion tensor imaging in patients with epilepsy and malformations of cortical development. *Brain* 2001; **124**:617–626.
60. Bozzao A, Floris R, Baviera ME, *et al.* Diffusion and perfusion MR imaging in cases of Alzheimer's disease: correlations with cortical atrophy and lesion load. *AJNR Am J Neuroradiol* 2001; **22**:1030–1036.
61. Kantarci K, Jack CR Jr, Xu YC, *et al.* Mild cognitive impairment and Alzheimer disease: regional diffusivity of water. *Radiology* 2001; **219**:101–107.
62. Rose SE, Chen F, Chalk JB, *et al.* Loss of connectivity in Alzheimer's disease: an evaluation of white matter tract integrity with colour coded MR diffusion tensor imaging. *J Neurol Neurosurg Psychiatry* 2000; **69**:528–530.
63. Takahashi S, Yonezawa H, Takahashi J, *et al.* Selective reduction of diffusion anisotropy in white matter of Alzheimer disease brains measured by 3.0 Tesla magnetic resonance imaging. *Neurosci Lett* 2002; **332**:45–48.
64. Liu AY, Maldjian JA, Bagley LJ, *et al.* Traumatic brain injury; diffusion-weighted MR imaging finding. *AJNR Am J Neuroradiol* 1999; **20**:1636–1641.
65. Zhang L, Ravdin LD, Relkin N, *et al.* Increased diffusion in the brain of professional boxers: a preclinical sign of traumatic brain injury? *AJNR Am J Neuroradiol* 2003; **24**:52–57.
66. Rugg-Gunn FJ, Symms MR, Barker GJ, *et al.* Diffusion imaging shows abnormalities after blunt head trauma when conventional magnetic resonance imaging is normal. *J Neurol Neurosurg Psychiatry* 2001; **70**:530–533.
67. Horsfield MA, Jones DK. Applications of diffusion-weighted and diffusion tensor MRI to white matter diseases — a review. *NMR Biomed* 2002; **15**: 570–577.
68. Christiansen P, Gideon P, Thomsen C, *et al.* Increased water self-diffusion in chronic plaques and in apparently normal white matter in patients with multiple sclerosis. *Acta Neurol Scand* 1993; **87**:195–199.
69. Werring DJ, Brassat D, Droogan AG, *et al.* The pathogenesis of lesions and normal-appearing white matter changes in multiple sclerosis: a serial diffusion MRI study. *Brain* 2000; **123**:1667–1676.

70. Cercignani M, Bozzali M, Iannucci G, *et al*. Magnetisation transfer ratio and mean diffusivity of normal appearing white and grey matter from patients with multiple sclerosis. *J Neurol Neurosurg Psychiatry* 2001; **70**:311–317.
71. Castriota SA, Tomaiuolo F, Sabatini U, *et al*. Demyelinating plaques in relapsing-remitting and secondary-progressive multiple sclerosis: assessment with diffusion MR imaging. *AJNR Am J Neuroradiol* 2000; **21**:862–868.
72. Rovaris M, Bozzali M, Iannucci G, *et al*. Assessment of normal-appearing white and gray matter in patients with primary progressive multiple sclerosis: a diffusion-tensor magnetic resonance imaging study. *Arch Neurol* 2002; **59**: 1406–1412.
73. Nusbaum AO, Lu D, Tang CY, *et al*. Quantitative diffusion measurements in focal multiple sclerosis lesions: correlations with appearance on TI-weighted MR images. *AJR Am J Roentgenol* 2000; **175**:821–825.
74. Roychowdhury S, Maldjian JA, Grossman RI. Multiple sclerosis: comparison of trace apparent diffusion coefficients with MR enhancement pattern of lesions. *AJNR Am J Neuroradiol* 2000; **21**:869–874.
75. Filippi M, Iannucci G, Cercignani M, *et al*. A quantitative study of water diffusion in multiple sclerosis lesions and normal-appearing white matter using echo-planar imaging. *Arch Neurol* 2000; **57**:1017–1021.
76. Rocca MA, Cercignani M, Iannucci G, *et al*. Weekly diffusion-weighted imaging of normal-appearing white matter in MS. *Neurology* 2000; **55**:882–884.
77. Guo AC, Jewells VL, Provenzale JM. Analysis of normal-appearing white matter in multiple sclerosis: comparison of diffusion tensor MR imaging and magnetization transfer imaging. *AJNR Am J Neuroradiol* 2001; **22**:1893–1900.
78. Werring DJ, Clark CA, Droogan AG, *et al*. Water diffusion is elevated in widespread regions of normal-appearing white matter in multiple sclerosis and correlates with diffusion in focal lesions. *Mult Scler* 2001; **7**:83–89.
79. Iannucci G, Rovaris M, Giacomotti L, *et al*. Correlation of multiple sclerosis measures derived from T2-weighted, T1-weighted, magnetization transfer, and diffusion tensor MR imaging. *AJNR Am J Neuroradiol* 2001; **22**:1462–1467.
80. Schmierer K, Altmann DR, Kassim N, *et al*. Progressive change in primary progressive MS normal-appearing white matter: a serial diffusion MRI study. *Mult Scler* 2004 (in press).
81. Rocca MA, Iannucci G, Rovaris M, *et al*. Occult tissue damage in patients with primary progressive multiple sclerosis is independent of T2-visible lesions — a diffusion tensor MR study. *J Neurol* 2003; **250**:456–460.
82. Cercignani M, Inglese M, Pagani E, *et al*. Mean diffusivity and fractional anisotropy histograms of patients with multiple sclerosis. *AJNR Am J Neuroradiol* 2001; **22**:952–958.

83. Holtmannspotter M, Inglese M, Rovaris M, *et al.* A diffusion tensor MRI study of basal ganglia from patients with ADEM. *J Neurol Sci* 2003; **206**:27–30.
84. Griffin CM, Chard DT, Ciccarelli O, *et al.* Diffusion tensor imaging in early relapsing-remitting multiple sclerosis. *Mult Scler* 2001; **7**:290–297.
85. Bozzali M, Cercignani M, Sormani MP, *et al.* Quantification of brain gray matter damage in different MS phenotypes by use of diffusion tensor MR imaging. *AJNR Am J Neuroradiol* 2002; **23**:985–988.
86. Pierpaoli C, Barnett A, Pajevic S, *et al.* Water diffusion changes in Wallerian degeneration and their dependence on white matter architecture. *Neuroimage* 2001; **13**:1174–1185.
87. Wieshmann UC, Symms MR, Clark CA, *et al.* Wallerian degeneration in the optic radiation after temporal lobectomy demonstrated in vivo with diffusion tensor imaging. *Epilepsia* 1999; **40**:1155–1158.
88. Clark CA, Barrick TR, Murphy MM, Bell BA. White matter fiber tracking in patients with space-occupying lesions of the brain: a new technique for neurosurgical planning? *Neuroimage* 2003; **20**:1601–1608.
89. Ellis CM, Simmons A, Jones DK, *et al.* Diffusion tensor MRI assesses corticospinal tract damage in ALS. *Neurology* 1999; **53**:1051–1058.
90. Foong J, Maier M, Clark C, *et al.* Neuropathological abnormalities of the corpus callosum in schizophrenia: a diffusion tensor imaging study. *J Neurol Neurosurg Psychiatry* 2000; **68**:242–244.
91. Foong J, Symms MR, Barker GJ, *et al.* Investigating regional white matter in schizophrenia using diffusion tensor imaging. *NeuroReport* 2002; **13**:333–336.
92. Buchsbaum MS, Tang CY, Peled S, *et al.* MRI white matter diffusion anisotropy and PET metabolic rate in schizophrenia. *NeuroReport* 1998; **9**:425–430.
93. Mukherjee P, Miller JH, Shimony JS, *et al.* Normal brain maturation during childhood: developmental trends characterized with diffusion-tensor MR imaging. *Radiology* 2001; **221**:349–358.
94. Pfefferbaum A, Sullivan EV, Hedehus M, *et al.* Age-related decline in brain white matter anisotropy measured with spatially corrected echo-planar diffusion tensor imaging. *Magn Reson Med* 2000; **44**:259–268.
95. Mullins ME, Schaefer PW, Sorensen AG, *et al.* CT and conventional and diffusion-weighted MR imaging in acute stroke: study in 691 patients at presentation to the emergency department. *Radiology* 2002; **224**:353–360.
96. Mullins ME, Schaefer PW, Sorensen AG, *et al.* Magnetic resonance imaging of cerebral infarction. *Top Magn Reson Imaging* 1998; **9**:199–207.
97. Desmond PM, Lovell C, Rawlinson A, *et al.* The value of apparent diffusion coefficient maps in early cerebral ischemia. *AJNR Am J Neuroradiol* 2001; **22**: 1260–1267.

98. Lansberg MG, Thijs VN, O'Brien MW, *et al.* Evolution of apparent diffusion coefficient, diffusion-weighted, and T2-weighted signal intensity of acute stroke. *AJNR Am J Neuroradiol* 2001; **22**:637.

99. Mukherjee P, Bahn MM, McKinstry, *et al.* Differences between gray matter and white matter water diffusion in stroke; diffusion-tensor MR imaging in 12 patients. *Neuroradiol* 2000; **215**:211–220.

100. Gillard JH, Papadakis NG, Martin K, *et al.* MR diffusion tensor imaging of white matter tract disruption in stroke at 3T. *Br J Radiol* 2001; **74**:642–647.

101. Lamy C, Oppenheim C, Meder JF, *et al.* Neuroimaging in posterior reversible encephalopathy syndrome. *J Neuroimaging* 2004; **14**:89–96.

102. Lovblad KO, Bassetti C, Schneider J, *et al.* Diffusion-weighted MR in cerebral venous thrombosis. *Cerebrovasc Dis* 2001; **11**:169–176.

103. Sugahara T, Korogi Y, Kochi M, *et al.* Usefulness of diffusion-weighted MRI with echo-planar technique in the evaluation of cellularity in gliomas. *J Magn Reson Imaging* 1999; **9**:53–60.

104. Kono K, Inoue Y, Nakayama K, *et al.* The role of diffusion-weighted imaging in patients with brain tumors. *AJNR Am J Neuroradiol* 2001; **22**: 1081–1088.

105. Yang D, Korogi T, Sugahara T, *et al.* Cerebral gliomas: prospective comparison of multivoxel 2D chemical-shift imaging proton MR spectroscopy, echoplanar perfusion and diffusion-weighted MRI. *Neuroradiol* 2002; **44**:656–.666.

106. Filippi CG, Edgar MA, Ulug AM, *et al.* Appearance of meningiomas on diffusion-weighted images: correlating diffusion constants with histopathologic findings. *AJNR Am J Neuroradiol* 2001; **22**:65–72.

107. Price SJ, Burnet NG, Donovan T, *et al.* Diffusion tensor imaging of brain tumours at 3T: a potential tool for assessing white matter tract invasion? *Clin Radiol* 2003; **58**:455–462.

108. Guo AC, Cummings TJ, Dash RC, *et al.* Lymphomas and high-grade astrocytomas: comparison of water diffusibility and histologic characteristics. *Radiology* 2002; **224**:177–183.

109. Camacho DL, Smith JK, Castillo M. Differentiation of toxoplasmosis and lymphoma in AIDS patients by using apparent diffusion coefficients. *AJNR Am J Neuroradiol* 2003; **24**:633–637.

110. Guzman R, Barth A, Lovblad KO, *et al.* Use of diffusion-weighted magnetic resonance imaging in differentiating purulent brain processes from cystic brain tumors. *J Neurosurg* 2002; **97**:1101–1107.

111. Mittal S, Farmer P, Kalina P, *et al.* Correlation of diffusion-weighted magnetic resonance imaging with neuropathology in Creutzfeldt-Jakob disease. *Arch Neurol* 2002; **59**:128–134.

112. Demaerel P, Sciot R, Robberecht W, *et al*. Accuracy of diffusion-weighted MR imaging in the diagnosis of sporadic Creutzfeldt–Jakob disease. *J Neurol* 2003; **250**:222–225.
113. Pruessmann KP, Weiger M, Scheidegger MB, *et al*. SENSE: Sensitivity Encoding for Fast MRI. *Magn Reson Med* 1999; **42**:952–962.
114. De Vita E, Thomas DL, Roberts S, *et al*. High resolution MRI of the brain at 4.7 Tesla using fast spin echo imaging. *Br J Radiol* 2003; **76**:631–637.
115. Fatterpekar GM, Naidich TP, Delman BN, *et al*. Cytoarchitecture of the human cerebral cortex: MR microscopy of excised specimens at 9.4 Tesla. *AJNR Am J Neuroradiol* 2002; **23**:1313–1321.

14

Applications of Positron Emission Tomography (PET) in Neurology

*Y F Tai and P Piccini**

Positron emission tomography (PET) is a powerful imaging technique which enables *in vivo* examination of brain functions. It allows non-invasive quantification of cerebral blood flow, metabolism, and receptor binding. In the past PET has been employed mainly in the research setting due to the relatively high costs and complexity of the support infrastructure, such as cyclotrons, PET scanners, and radiochemistry laboratories. In recent years, because of advancements in technology and proliferation of PET scanners, PET is being increasingly used in clinical neurology to improve our understanding of disease pathogenesis, to aid with diagnosis, and to monitor disease progression and response to treatment. This article aims to provide an overview of the principles of PET and its applications to clinical neurology.

*Dr P Piccini, MRC Cyclotron Building, Imperial College School of Medicine, Hammersmith Hospital, Du Cane Road W12 0NN, London, UK;
paola. piccini@csc.mrc.ac.uk

PRINCIPLES OF PET

Positron emission tomography (PET) involves introduction, usually via an intravenous injection, of a radioactive tracer into the human body. A tracer is essentially a biological compound of interest labelled with a positron emitting isotope, such as ^{11}C, ^{18}F, and ^{15}O. These isotopes are used because they have relatively short half-lives (minutes to less than two hours), allowing the tracers to reach equilibrium in the body, but without exposing the subjects to prolonged periods of radiation.

The cyclotron accelerates a beam of protons using two high voltage electrodes and directs it towards the target nuclei, thereby incorporating an extra proton into them. This generates new isotopes with a neutron-to-proton ratio that is energetically unstable. The isotopes are then attached to the compound of interest — that is, the tracer. Most of these unstable isotopes undergo a process of decay whereby the excess proton is converted into a neutron, a positron, and a neutrino. A positron is similar to an electron, except that it carries a positive charge. The emitted positron travels up to a range of a few millimetres in tissue before being annihilated along with an electron from the surroundings. This mutual annihilation process produces two photons of equal energy (511 keV) travelling in opposite directions. PET scanners contain several rings of scintillation detectors, usually made of bismuth germanate (BGO). The pair of photons produced from a single annihilation will register almost simultaneously on opposing pairs of BGO detectors as a "coincidence event". The paths of these two photons, called lines of response (LORs), can thus be traced (Fig. 1). The rings of BGO detectors register thousands of coincidence events emitted from the subject per second. The data gathered from the coincidence events and LORs are used to determine the source of positron annihilation at a given time. These are then converted into a tomographic image using standard reconstruction software.[1]

Acquisition of data in the three dimensional mode using a state-of-the-art scanner greatly increases the spatial resolution of PET images.[2] The improved signal-to-noise ratio and the accompanying enhanced sensitivity of three dimensional scanners also allow a lower effective dose of radiation to be administered to subjects (typically 1–5 mSv per scan at our centre compared with around 4 mSv administered in a head computed tomography (CT) scan). PET images taken with most modern scanners

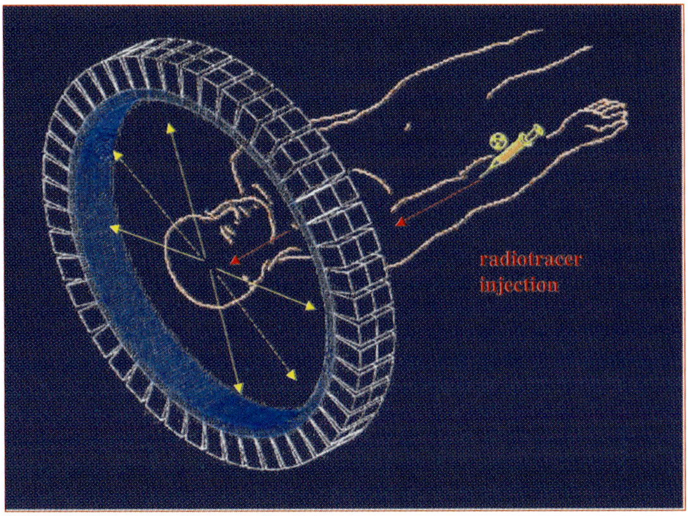

Fig. 1 Schematic representation of detectors in a tomograph, and the annihilation of a positron with an electron and the resulting pair of 511 keV γ rays released at 180° (yellow arrows).

have a reconstructed spatial resolution of about 4 mm. Coregistration of PET images to high resolution magnetic resonance images (MRI) enables the accurate anatomical localisation of functional changes displayed on PET.

For full quantification of the acquired PET data, it is necessary to make certain corrections. The photons or γ rays generated by the positron annihilation process, especially those originating from the centre of the brain, may be attenuated by the tissue before reaching the detectors. On an average, about 80% of photon pairs emitted from the centre of the brain will be lost to tissue attenuation.[1] This can be overcome by performing a transmission scan using an external positron emitting source before injecting the tracer. The tissue attenuation factor can then be calculated and applied.

The tracer is distributed to various body compartments following administration to the subject. In order to model the kinetics of the tracer accurately, we need to know its plasma concentration (often by continuous arterial blood sampling to obtain arterial input function) and cerebral concentration (from dynamic sequences of PET images) over time, correcting for metabolism and radioactive decay of the compound. However, if a study is undertaken to compare two different groups of subjects or two different conditions, a more qualitative approach can be adopted. For example, there is a negligible number of dopamine receptors in the cerebellum, and hence a dopamine D_2 receptor ligand such as ^{11}C-raclopride will not have significant specific binding in the cerebellum. The kinetics of ^{11}C-raclopride in this region will closely resemble those in the plasma. Therefore, the cerebellum can act as a "reference region", providing tissue input function and obviating the need for invasive arterial cannulation.[3] The data can be analysed by defining specific regions of interest on the images if there is an a priori hypothesis regarding expected changes to be observed, or by comparing values throughout the whole brain volume on a voxel-by-voxel basis using statistical parametric mapping.[4]

Specific tracers are chosen to illustrate the particular brain functions the investigators are interested in. For example, ^{18}F-2-deoxyglucose (^{18}FDG) is used to investigate cerebral glucose metabolism, whereas $H_2^{15}O$ is used to examine cerebral blood flow. Table 1 lists some of the commonly used tracers and their specific applications. The tracer used usually has very high specific activity (ratio of radioactive to cold compound) such that only a very small amount (in the range of nanomoles to picomoles per gram) is administered. Therefore the tracer will have minimal effect on the subject's biological system.[1]

PET scans may be carried out at rest, while performing certain tasks or after administration of challenge compounds. Several recent studies have suggested that using the principle of competition between endogenous

Table 1 Common PET tracers used to study neurological disorders.

Table Width = A

Application	Tracer
Cerebral blood flow	$H_2^{15}O$
Oxygen metabolism	$^{15}O_2$
Glucose metabolism	^{18}F-2-Fluoro-2-deoxyglucose (^{18}FDG)
Dopamine storage	^{18}F-6-Fluorodopa (^{18}F-dopa)
Dopamine D_1 receptors	^{11}C-SCH23390
Dopamine D_2 receptors	^{11}C-Raclopride
Central benzodiazepine binding	^{11}C-Flumazenil
Opioid binding	^{11}C-Diprenorphine
Cellular amino acid uptake	^{11}C-Methionine
Activated microglia	^{11}C(R)-PK11195

neurotransmitters and administered tracers for the same neuroreceptors, it is possible to image acute fluctuations in the concentration of synaptic neurotransmitters.[5] One of the most commonly performed studies is the ^{11}C-raclopride displacement study. ^{11}C-Raclopride, a competitive reversible ligand, competes with endogenous dopamine for postsynaptic dopamine D_2 receptors. According to the classic occupancy model, when there is an increased synaptic release of dopamine after performing a task mediated by dopamine or after administration of a dopamine depleting agent such as methamphetamine, most of the D_2 receptors will be occupied by endogenous dopamine. Fewer receptors will be available to ^{11}C-raclopride, and this is evidenced by a reduction in ^{11}C-raclopride binding. Other mechanisms such as receptor internalisation after agonist stimulation may also contribute to the signal change observed.[5] Data from a microdialysis study in rhesus monkeys suggest that after administering 0.2 mg/kg of amphetamine, a 1% reduction in striatal ^{11}C-raclopride binding corresponds to at least an 8% increase in striatal extracellular dopamine levels.[6] This technique has a significant advantage over $H_2^{15}O$ activation studies and functional MRI, which are modelled on haemodynamic changes following a task, since a ligand displacement study enables researchers to study the role of a particular neurotransmitter in mediating certain tasks or functions, such as the release of dopamine while performing a goal-directed motor task with a reward.[7] Pharmacological challenge using methamphetamine is used to show the ability of the surviving fetal striatal grafts in Parkinson's disease to release normal levels of dopamine (see section on Movement disorders).

APPLICATION OF PET TO CLINICAL NEUROLOGY

Movement Disorders

Dopamine is the key neurotransmitter in the nigro-striatal-pallidal-thalamo-cortical circuit. ^{18}F-6-Fluorodopa (^{18}F-dopa) is one of the most commonly used ligands for studying the dopaminergic system in movement disorders. Following intravenous injection ^{18}F-dopa is taken up by the terminals of dopaminergic neurones and converted to ^{18}F-dopamine by dopa decarboxylase, and subsequently to other dopamine metabolites. The influx constant (Ki) of ^{18}F-dopa reflects dopa transport into the terminals, dopa decarboxylase activity and dopamine storage capacity.[8] ^{18}F-dopa PET can therefore provide an *in vivo* indicator of the function and integrity of presynaptic dopaminergic terminals. Tracers that bind to presynaptic dopamine transporters, such as ^{11}C-methylphenidate, and dopamine terminal vesicle monoamine transporters, such as ^{11}C-dihydrotetrabenazine, have also been developed as markers of presynaptic dopaminergic function.

Differentiating various types of parkinsonian syndromes clinically, especially in the early stages of the disease, can be difficult. Conventional imaging methods such as MRI often do not reveal any abnormality. PET may be employed as an adjunct to clinical diagnosis in equivocal cases. Parkinson's disease (PD) is characterised by loss of dopaminergic neurones in the pars compacta of the substantia nigra. The greatest loss of neurones is seen in the ventrolateral tier of the pars compacta, with lesser involvement in the dorsomedial tier. Dopaminergic neurones in these regions project to the putamen and head of the caudate nucleus, respectively.[9] These changes are detected by ^{18}F-dopa PET, as evidenced by progressive decline in ^{18}F-dopa Ki in the putamen in a caudal-rostral pattern. The biggest decrease is seen in the putamen contralateral to the side with the most severe symptoms. The caudate nucleus is also affected later on.[10] Diffuse loss of nigrostriatal dopaminergic projection is seen in multiple system atrophy (MSA) and progressive supranuclear palsy (PSP), as reflected by the symmetrical loss of ^{18}F-dopa signal in the entire striatum (Fig. 2). Corticobasal degeneration (CBD) shows asymmetric and equivalent reduction in ^{18}F-dopa Ki of the caudate and putamen. ^{18}F-Dopa PET is able to discriminate PD from the striatonigral degeneration form of MSA in 70% of cases and from PSP in 90% of cases[11]; it is, however, less effective in discriminating between the atypical parkinsonian syndromes.

The various parkinsonian syndromes also exhibit different patterns of cerebral glucose metabolism. ^{18}FDG PET in PD reveals normal or increased glucose metabolism in the striatum but decreased metabolism in temporoparietal areas.[12] PSP shows bilateral striatal and frontal hypometabolism, whereas decreases in striatal, brainstem, and cerebellar metabolism are found in MSA. In CBD, there is asymmetric hypometabolism of the striatum, thalamus, frontal and temporoparietal cortex, with the hemisphere contralateral to the most affected limb

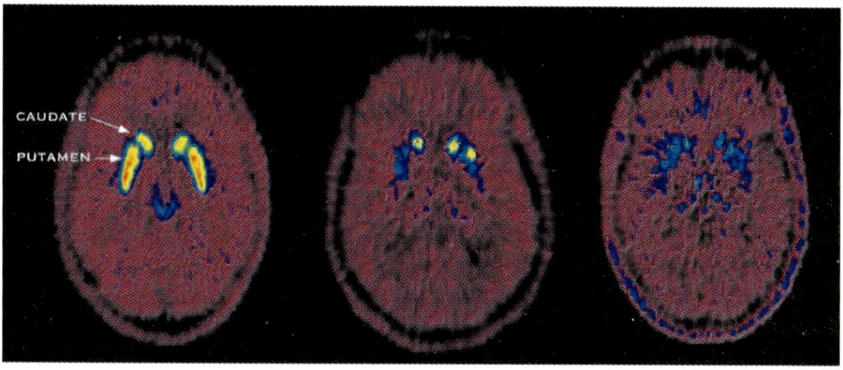

Fig. 2 ^{18}F-Dopa PET in a healthy volunteer (normal); in a subject with Parkinson's disease (PD) showing asymmetrical loss of putaminal ^{18}F-dopa Ki, with relative preservation of the caudate Ki earlier on in the disease; and in a subject with progressive supranuclear palsy (PSP) showing diffuse and symmetrical decrease in striatal Ki.

displaying the greatest reduction. However, ^{18}FDG PET does not provide additional discriminatory information to ^{18}F-dopa PET.[10] Table 2 summarises the major PET findings in parkinsonian syndromes.

According to current models of basal ganglia connectivity, the internal segment of the globus pallidus (GPi) receives nigrostriatal projections via distinct direct and indirect pathways. In PD, the loss of nigral dopaminergic cells results in loss of inhibitory output to GPi in the direct pathway. In contrast, the inhibitory striatal output to the external segment of globus pallidus (GPe) becomes overactive in the indirect pathway. This, in turn, causes a reduction of inhibitory output from GPe to the subthalamic nucleus (STN). There is, therefore, increased excitatory activity from the STN to the GPi, and the consequence of all this is that the GPi, which provides major inhibitory output to the ventral thalamus, becomes disinhibited or overactive. Consequently, there is a reduction of excitatory projections from the ventral thalamus to the supplementary motor area (SMA) and prefrontal cortex,[13] both of which are involved in motor planning and preparation. It has been proposed that their underactivation in PD results in hypokinetic symptoms such as akinesia. This model has been supported by $H_2^{15}O$ activation studies of PD patients performing paced

Table 2 Summary of positron emission tomography (PET) findings in parkinsonian syndromes.

Table Width = E

PET tracer	Parkinson's disease	Progressive supranuclear palsy	Multiple system atrophy	Corticobasal degeneration
^{18}F-Dopa	Asymmetric reduction (putamen > caudate)	Symmetrical reduction (caudate = putamen)	Symmetrical reduction (caudate = putamen)	Asymmetric reduction (caudate = putamen)
^{18}FDG	Normal/raised in striatum. Reduced in temporoparietal cortex	Reduced in bilateral striatum and frontal cortex	Reduced in striatum, brainstem, and cerebellum	Asymmetric reduction in striatum, thalamus, frontal and temporo-parietal cortex

joystick movements in freely chosen directions. Decreased cerebral blood flow in the SMA and prefrontal cortex compared with controls improved after subcutaneous administration of apomorphine, a dopamine D_1 and D_2 receptor agonist, with concomitant reduction in the subjects' akinesia.[14]

Surgical treatments for advanced PD with motor complications have been developed on the basis of the above understanding of basal ganglia circuitry. The aim of these treatments is to reduce inhibitory output from the GPi to the ventral thalamus, either directly by deactivating GPi or indirectly by deactivating STN, thereby reducing its excitatory output to GPi. This can be accomplished structurally by stereotactic thermocoagulation (for example medial pallidotomy) or functionally using high frequency electrical deep-brain stimulation (DBS), which causes a depolarising conduction block. The advantages of the latter are that it is reversible, and the frequency of electrical stimulation can be adjusted according to patient's response. However, it does require insertion of electrodes into appropriate targets with the accompanying surgical and infectious risks. Several randomised controlled trials have shown the efficacy of these techniques in improving "off" medication motor function and "on" medication dyskinesia in advanced PD.[15,16] As predicted, $H_2^{15}O$ activation studies of PD patients following STN DBS or medial pallidotomy showed improvement in SMA and prefrontal cortex activation when performing paced joystick movements in freely chosen directions.[17,18]

PET has been developed as a biological marker of disease severity and progression in PD. Striatal ^{18}F-dopa Ki is shown to correlate with postmortem dopaminergic cell density in the substantia nigra.[19] The reduction in putaminal Ki in PD also correlates with cross-sectional motor disability.[20] A longitudinal progression study of PD found a 9–12% annual decline in striatal dopa Ki.[21]

Several neuroprotective/restorative trials have used ^{18}F-dopa PET as a biological marker of response to treatment. The clinical improvement seen in PD patients receiving human fetal neural transplantation and intraputaminal infusion of glial cell line derived neurotrophic factor is accompanied by increases in striatal dopa Ki.[22–24] At the postmortem examination of two PD patients with transplants, who died of unrelated causes, increased striatal ^{18}F-dopa uptake was associated with the survival of grafts and dopaminergic reinnervation of the striatum.[25] "Off" phases dyskinetic involuntary movements observed in some PD patients following transplantation procedures are not associated with abnormal increases in ^{18}F-dopa uptake indicating that this side effect does not result from excessive growth of grafted dopaminergic neurones.[26,27]

$H_2^{15}O$ activation scans of four PD patients 18 months after receiving bilateral fetal transplantation showed restoration of premotor and prefrontal activation whilst performing a paced motor task, suggesting functional integration of the grafts into host neuronal circuitry.[28] In one PD patient who had sustained clinical improvement 10 years after unilateral fetal striatal implantation, the implanted striatum exhibited normal levels of basal and methamphetamine induced dopamine release as evidenced by his ^{11}C-raclopride PET displacement study.[29]

PET has also been widely used to study hyperkinetic movement disorders. Huntington's disease (HD) is an autosomal dominant disorder arising from expanded CAG repeats in the *IT15* gene on chromosome 4. Medium spiny neurones in the striatum, which express dopamine D_1 and D_2 receptors, bear the brunt of HD pathology and are progressively lost. Using ^{11}C-SCH23390 and ^{11}C-raclopride PET, parallel reduction in striatal D_1 and D_2 receptor binding was found in HD patients.[30] Striatal D_2 binding decreases by approximately 5% per year in HD, and the reduction correlates with the duration and clinical severity of the disease.[31] ^{18}FDG PET showed striatal glucose hypometabolism in HD, with the cortex becoming progressively involved with increasing severity of disease, reflecting the widespread nature of HD pathology.[32] ^{11}C-Raclopride and ^{18}FDG PET have both been used as markers of graft survival in HD fetal striatal transplantation trials. One recent study reported increased striatal glucose metabolism in HD patients who experienced clinical improvement following bilateral striatal implantation but not in those without benefits, suggesting such improvement can be attributed to the surviving grafts.[33]

Although HD can be diagnosed accurately using genetic tests, there is, as yet, no reliable way to predict disease onset in presymptomatic carriers. Several PET studies have found reduced striatal D_2 binding and glucose metabolism in some HD carriers.[34,35] Larger trials are ongoing to ascertain the accuracy of PET in identifying carriers nearing the onset of disease, since intervention at this early stage with putative neuroprotective agents such as minocycline and riluzole may yield most benefits.

Striatal D_2 receptor binding and glucose metabolism are also reduced in chorea due to other degenerative conditions (for example neuroacanthocytosis), but are preserved in non-degenerative chorea (for example systemic lupus erythematosus, Sydenham's chorea).[36]

Epilepsy

Complex partial seizures in a significant proportion of patients remain uncontrolled despite optimal medical therapy. Surgical removal of epileptogenic foci in partial seizures such as intractable temporal lobe epilepsy results in significant improvement in control of the seizures and the quality of life.[37] Modern MRI is able to identify the source of the seizure in the majority of patients with partial seizures. However, 20–30% of potential surgical candidates with focal epilepsy have normal MRI.[38] These patients are also less likely to become seizure free if they do undergo epilepsy surgery.[39] Microscopic structural abnormalities, which may only be evident on histological examination, may not be detected on MRI.[40] The main clinical uses of PET in epilepsy are localisation of epileptogenic foci in potential surgical candidates with partial seizures and corroborating findings from other investigational modalities such as electroencephalography (EEG).

In partial seizures, there is an increase in glucose metabolism and cerebral blood flow in the region of the epileptogenic focus during the ictal period.[41] Post-ictally, the hyperperfusion gradually returns to baseline, but the glucose metabolism remains elevated for another 24–48 hours after the seizure.[42] seizures changed to seizure here PET shows decreased glucose metabolism and blood flow in the epileptogenic focus. It is important to perform PET with concomitant scalp EEG recordings to correlate the PET findings with the clinical status of the patient. Interictal studies in patients with temporal lobe epilepsy using ^{18}FDG PET have found a 60–90% incidence of temporal lobe hypometabolism.[38] However, the area with abnormal cerebral blood flow and metabolism seen on PET is considerably larger than the actual structural abnormality, possibly due to reduced synaptic inhibition or deafferentation of neighbouring neurones in areas of epileptic propagation.[38] Therefore, false localisations may occur, the probability of which can be lessened by using quantitative rather than qualitative assessment of regional cerebral metabolism. Overall, for epileptogenic foci ^{18}FDG PET may be more suitable as a lateralising rather than localising tool.

γ-Aminobutyric acid (GABA) is the principal inhibitory neurotransmitter in the brain, acting at up to 40% of synapses. A decreased number of GABAergic inhibitory interneurones was found in epileptogenic cortex.[43] Flumazenil (FMZ) is a specific reversible antagonist that binds to the benzodiazepine binding site of the $GABA_A$–central benzodiazepine receptor complex. ^{11}C-FMZ PET therefore provides an *in vivo* marker of $GABA_A$ receptor binding.[44] ^{11}C-FMZ binding is reduced by 30% in epileptogenic foci.[45] An autoradiographic and histopathological study of sclerotic hippocampi revealed that the decreased ^{11}C-FMZ binding is due to reduced number of neurones and decreased density of central benzodiazepine receptors per neurone.[46] There is also a good correlation between quantitative *in vivo* hippocampal ^{11}C-FMZ PET and *ex vivo* ^{3}H-FMZ autoradiographic studies in individual patients with hippocampal sclerosis.[47] One study examined 100 patients with partial seizures who had undergone presurgical evaluation including ^{18}FDG and ^{11}C-FMZ PET. The latter demonstrated abnormalities in 94% of the patients with temporal lobe epilepsy. ^{11}C-FMZ abnormality coincided with MRI abnormality in 81% of the cases. The area with abnormal ^{11}C-FMZ binding is usually smaller than that seen on ^{18}FDG PET but larger than the abnormality detected on MRI.[48] This suggests that the area of neuronal loss is more circumscribed than the region with reduced metabolism. Another study which examined the focus localising abilities of ^{11}C-FMZ and ^{18}FDG PET, using extra- and intracranial EEG recordings as reference, found the former to be more sensitive and accurate.[49]

Hammers *et al.* investigated 18 patients with refractory temporal lobe epilepsy and normal MRI using ^{11}C-FMZ PET. Sixteen patients showed abnormalities in temporal lobe ^{11}C-FMZ binding, in seven of whom the findings were concordant with clinical and EEG data[50] (Fig. 3). Three patients subsequently underwent anterior temporal lobe resection with significant clinical improvement. Neuropathological findings from the surgical specimens from these patients suggest that some of the abnormalities seen on ^{11}C-FMZ PET are likely to be due to microdysgenesis, which is not often detected even with optimal MRI.

PET may reduce the need for invasive EEG as part of the preoperative localisation of surgical targets in the future. Some centres have proposed using PET as a routine preoperative evaluation tool for patients undergoing epilepsy surgery. However, PET does not provide additional information if

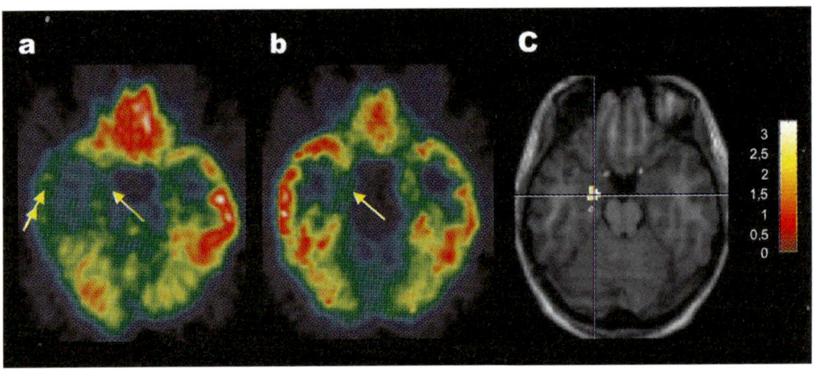

Fig. 3 Comparison of ^{18}FDG PET (a) and ^{11}C-FMZ PET (b) in a 28 year-old patient with complex partial seizures, interictal left temporal epileptiform discharges and left temporal seizure onset on video telemetry. MRI including T1 weighted and T2 weighted images, FLAIR and quantitative hippocampal volumetry and T2 mapping was normal. (a) ^{18}FDG PET shows extensive hypometabolism in the left temporal lobe, affecting both the medial temporal cortex (arrow) as well as the inferior and lateral (double arrow) temporal neocortex. (b) ^{11}C-FMZ-PET shows a circumscribed decrease of ^{11}C-FMZ volume-of-distribution (Vd) in the left medial temporal lobe (arrow). (c) Statistical analysis with statistical parametric mapping (SPM99) localises a significant decrease of FMZ-Vd in the left anterior hippocampus, compared with 21 controls. The statistical map is overlaid on the patient's own MRI; the colour scale shows the *t*-value. (Images courtesy of A Hammers.)

the MRI has identified the obvious cause of the epilepsy, such as hippocampal sclerosis.[51] Therefore, PET is likely to be most useful in situations where MRI is equivocal or normal.

PET is less useful clinically for primary generalised seizures. During seizures there is a global increase in cerebral glucose metabolism, but the interictal pattern is usually normal.[38]

Brain Tumours

Tumour cells, especially of higher histological grades, typically have increased metabolic and mitotic rates compared with normal brain tissue. ^{18}FDG PET can provide important prognostic information as increased glucose metabolism of gliomas correlates with higher histological grades (III and IV) and shorter survival period.[52,53] It can also detect transformation of low grade glioma into a high grade variety.[54] The information obtained may influence the choice of the therapeutic approach. Response to chemotherapy and radiotherapy is associated with a significant reduction in tumour glucose metabolism.[55,56] Therefore, ^{18}FDG PET may provide an objective subclinical evidence of response to treatment. The European Organisation for Research and Treatment of Cancer PET Study Group recommends that a reduction of 15–25% in glucose uptake after one cycle of chemotherapy would be classified as partial metabolic response; a complete metabolic response occurs when there is complete resolution of glucose uptake within the tumour volume so that it is indistinguishable from surrounding normal tissue.[57] However, there is often a transient increase in tumour glucose uptake within 24 hours of treatment, which is not present in studies done 7–14 days after treatment.[56,58] Therefore, a one to two week post-treatment interval is recommended prior to performing ^{18}FDG PET to accurately assess response to treatment.[57] Increased glucose metabolism in recurrent cerebral glioma can help differentiate it from post radiotherapy changes which may be indistinguishable with conventional imaging techniques such as CT or MRI.[59,60]

Similarly, increased uptake of ^{11}C-methionine, which reflects cellular amino acid uptake, is associated with high grade glioma and poorer survival.[61] ^{11}C-Methionine is useful for discriminating between recurrences of local or metastatic tumours and radiation induced changes. In one recent study, ^{11}C-methionine PET had a sensitivity of 77.8% and specificity of 100% for differentiating recurrence of metastatic brain tumours from post radiotherapy changes.[62] However, ^{11}C-methionine uptake may also be elevated in other conditions where there is a disruption of the blood–brain barrier, such as cerebral haematoma or even necrotic areas caused by radiotherapy,[63] whereas glucose metabolism may be normal or low in lower grade tumours compared with surrounding cortex. Combined use of ^{11}C-methionine and ^{18}FDG PET enhances the accuracy of discrimination between recurrent tumour and post radiotherapy changes.[64] The higher glucose metabolism in cerebral lymphoma also helps to distinguish it from cerebral infections (toxoplasmosis and tuberculoma) in patients with AIDS.[65,66]

Dementia

^{18}FDG PET has been used extensively to study dementia and it may be an effective tool for early diagnosis and differentiation of various types of dementia. Alzheimer's disease (AD) patients exhibit characteristic temporoparietal glucose hypometabolism. With progression of disease, there may also be frontal involvement.[67] The degree of hypometabolism correlates with the severity of dementia,[68] The glucose hypometabolism in AD is likely to be due to a combination of neuronal cell loss and decreased synaptic activity.[69] In one recent study which included 138 patients with symptoms of dementia and postmortem histopathological examination, ^{18}FDG PET was able to identify AD with a sensitivity of 94% and specificity of 73%. It could also correctly predict a progressive course of dementia with 91% sensitivity, and a non-progressive course with a specificity of 75%.[70] ^{18}FDG PET has been used to detect subjects at risk for AD even before onset of symptoms. Asymptomatic carriers of the apolipoprotein E type 4 allele, who are at increased risk for familial AD, showed similar pattern of glucose hypometabolism as AD patients.[71,72] After a mean follow up of two years, the cortical metabolic abnormality continued to decline despite preservation of cognitive performance.[73,74] Entorhinal cortex hypometabolism on ^{18}FDG PET in elderly people with normal cognition can predict the progression to mild cognitive impairment or even AD.[75] The identification of asymptomatic individuals at risk could be important should an effective neuroprotective agent that can delay or prevent progression to AD become available in the future.

^{18}FDG PET in dementia with Lewy bodies (DLB) reveals changes similar to those seen in AD, plus additional hypometabolism in the primary and associative visual cortices.[76] In a PET study with postmortem confirmatory diagnosis, the antemortem occipital glucose hypometabolism could help distinguish DLB from AD with 90% sensitivity and 80% specificity.[77] ^{18}FDG PET in multi-infarct dementia shows multiple focal areas of hypometabolism, the extent of which is greater than the actual pathology seen on postmortem examination. This is probably due to the degeneration of axons following the infarct with disconnection of remote structures.[78,79] Frontotemporal dementia is associated with hypometabolism in the frontal and temporal lobes.[79]

^{11}C(R)-PK11195 is a selective ligand for the peripheral benzodiazepine binding sites (PBBS). PBBS are present in the normal brain at very low levels, but they are selectively expressed and upregulated by activated microglia. It has been proposed that activated microglia play a role in the pathogenesis of neurodegenerative diseases such as AD by mediating neuroinflammation.[80] ^{11}C(R)-PK11195 PET in AD patients showed increased binding in the entorhinal, temporoparietal and cingulate cortices, corresponding to postmortem distribution of AD pathology.[81] The ability to detect microglial activation and neuroinflammatory response in vivo may be deployed to monitor disease activity in the many proposed and ongoing neuroprotective studies in AD using anti-inflammatory agents.[80]

Despite the characteristic patterns seen in many of the dementia syndromes, there can be considerable overlap in their PET findings. Therefore, PET findings should be interpreted in the context of each individual's clinical information. Recent efforts to develop a specific ligand for β-amyloid plaques may further enhance the sensitivity of PET for early diagnosis of AD and provide a biological marker of disease progression.[82]

Stroke and Neuronal Plasticity

Knowledge acquired from PET cerebral blood flow and metabolism studies has contributed significantly to the development of thrombolysis as a therapeutic approach in ischaemic stroke. Following stroke, PET can identify a "core" region of irreversibly damaged tissue with profoundly depressed cerebral blood flow and metabolism. This core region is surrounded by the "penumbra", an area of hypoperfused tissue but with relatively normal oxygen consumption, which may yet be salvaged by reperfusion.[83] Survival of the penumbra correlates with the degree of recovery after ischaemic stroke.[84] The incidence and extent of the penumbra decrease with time since onset of stroke. One study showed that in 90% of patients studied within six hours after onset of stroke, there still is a substantial amount of cortical penumbra. Such findings are detected in about a third of patients even at 5–18 hours after onset.[85] The variability in the survival of the penumbra suggests that the therapeutic window for reperfusion strategies may be different for certain subsets of patients, and this should be investigated in future trials of thrombolysis trials.

Results from several PET studies suggest that recruitment of remote areas and functional reorganisation are amongst the mechanisms responsible for the recovery of cerebral functions in adult brains after insults such as stroke. One $H_2^{15}O$ activation study showed that in patients who recovered from hemiplegic stroke there was bilateral activation of motor cortices when moving the fingers of the affected hand, whereas movement of fingers of the normal hand resulted in the activation of only the contralateral motor cortex and the ipsilateral cerebellum.[86] In another study, patients with non-fluent aphasia due to left anterior perisylvian infarction including the left pars opercularis (PO_p) with subsequent recovery were compared with two control groups: normal subjects and anterior aphasic patients with sparing of the left POp.[87] During production of propositional speech, the left PO_p infarct group showed increased activation of the homotopic right PO when compared with the two control groups. Further understanding of the mechanisms underlying neuroplasticity will help design appropriate strategies for rehabilitation and identify patients who are most likely to benefit from such therapy.

Similar plasticity is seen in neurodegenerative conditions. In PD, there is a decrease of contralateral putaminal ^{18}F-dopa Ki by about 50% before development of symptoms.[21] Whone *et al.* showed in early PD a lack of significant clinical progression despite continuing loss of nigrostriatal projections.[88] This is probably due to the upregulation of nigropallidal dopaminergic projection to GPi, as evidenced by increased GPi ^{18}F-dopa Ki compared with healthy volunteers, which reduces the inhibitory output from GPi to the thalamus. Such compensatory changes are not seen in more advanced PD with motor complications. This may partially explain why in early PD there is often a good response to dopaminergic medications with little clinical fluctuations. Loss of nigropallidal upregulation may result in alteration of the firing pattern of GPi from tonic to burst firing, heralding the onset of motor complications.

Neuropharmacology

PET receptor ligand studies have generated a wealth of knowledge about disease pathogenesis and potential therapeutic targets for novel pharmaceutical agents. PET offers the opportunity to use an *in vivo* technique to study the pharmacodynamics and biodistribution of new agents and to ensure they target the organs or compartments of interest — for example, in the case of neuropharmacology, the ability of drug to cross the blood-brain barrier and bind to specific receptors in the brain. The study of drug occupancy can provide information about the occupancy of the binding sites for a particular dose of the drug and its pharmacokinetics. This will help determine optimal drug dosing regimens.[89]

POTENTIAL FUTURE APPLICATIONS

The UK government has recently announced a plan to increase the number of PET centres in the country for clinical use. The mainstay of clinical application of PET in neurology is likely to be in the domains of epilepsy surgery and neuro-oncology.

Potential future applications include:

- early diagnosis of brain metastasis; distinguish local recurrences from radiotherapy induced changes; and detect malignant transformation of low grade tumours
- preoperative localisation of seizure foci in potential candidates for epilepsy surgery, especially in those with equivocal MRI findings
- as an adjunct to clinical diagnosis in atypical cases of parkinsonian syndromes and dementia
- early and presymptomatic diagnosis of individuals at risk for neurodegenerative disorders such as AD and PD if an effective neuroprotective agent becomes available.

REFERENCES

1. Cherry SR, Phelps ME. Imaging brain function with positron emission tomography. In: Toga AW, Mazziotta JC, eds. *Brain Mapping — The Methods*. San Diego: Academic Press, 1996:191–222.
2. Spinks TJ, Jones T, Bloomfield PM, *et al*. Physical characteristics of the ECAT EXACT3D positron tomograph. *Phys Med Biol* 2000; **45**:2601–2618.

3. Gunn RN, Lammertsma AA, Hume SP, *et al.* Parametric imaging of ligand-receptor binding in PET using a simplified reference region model. *Neuroimage* 1997; **6**:279–287.
4. Friston KJ, Frith CD, Liddle PF, *et al.* The relationship between global and local changes in PET scans. *J Cereb Blood Flow Metab* 1990; **10**:458–466.
5. Laruelle M. Imaging synaptic neurotransmission with in vivo binding competition techniques: a critical review. *J Cereb Blood Flow Metab* 2000; **20**:423–451.
6. Breier A, Su TP, Saunders R, *et al.* Schizophrenia is associated with elevated amphetamine-induced synaptic dopamine concentrations: evidence from a novel positron emission tomography method. *Proc Natl Acad Sci USA* 1997; **94**:2569–2574.
7. Koepp MJ, Gunn RN, Lawrence AD, *et al.* Evidence for striatal dopamine release during a video game. *Nature* 1998; **393**:266–268.
8. Brooks DJ. Imaging end points for monitoring neuroprotection in Parkinson's disease. *Ann Neurol* 2003; **53**(Suppl 3):S110–S118; discussion S118–S119.
9. Lang AE, Lozano AM. Parkinson's disease. First of two parts. *N Engl J Med* 1998; **339**:1044–1053.
10. Brooks DJ. PET studies on the early and differential diagnosis of Parkinson's disease. *Neurology* 1993; **43**(12 suppl 6):S6–S16.
11. Burn DJ, Sawle GV, Brooks DJ. Differential diagnosis of Parkinson's disease, multiple system atrophy, and Steele-Richardson-Olszewski syndrome: discriminant analysis of striatal 18F-dopa PET data. *J Neurol Neurosurg Psychiatry* 1994; **57**:278–284.
12. Hu MT, Taylor-Robinson SD, Chaudhuri KR, *et al.* Cortical dysfunction in non-demented Parkinson's disease patients: a combined (31)P-MRS and (18)FDG-PET study. *Brain* 2000; **123**(Pt 2):340–352.
13. Alexander GE, Crutcher MD, DeLong MR. Basal ganglia-thalamocortical circuits: parallel substrates for motor, oculomotor, "prefrontal" and "limbic" functions. *Prog Brain Res* 1990; **85**:119–146.
14. Jenkins IH, Fernandez W, Playford ED, *et al.* Impaired activation of the supplementary motor area in Parkinson's disease is reversed when akinesia is treated with apomorphine. *Ann Neurol* 1992; **32**:749–757.
15. Fine J, Duff J, Chen R, *et al.* Long-term follow-up of unilateral pallidotomy in advanced Parkinson's disease. *N Engl J Med* 2000; **342**:1708–1714.
16. Krack P, Batir A, Van Blercom N, *et al.* Five-year follow-up of bilateral stimulation of the subthalamic nucleus in advanced Parkinson's disease. *N Engl J Med* 2003; **349**:1925–1934.
17. Samuel M, Ceballos-Baumann AO, Turjanski N, *et al.* Pallidotomy in Parkinson's disease increases supplementary motor area and prefrontal activation during performance of volitional movements an H2(15)O PET study. *Brain* 1997; **120**(Pt 8):1301–1313.

18. Ceballos-Baumann AO, Boecker H, Bartenstein P, *et al.* A positron emission tomographic study of subthalamic nucleus stimulation in Parkinson disease: enhanced movement-related activity of motor-association cortex and decreased motor cortex resting activity. *Arch Neurol* 1999; **56**: 997–1003.
19. Snow BJ, Tooyama I, McGeer EG, *et al.* Human positron emission tomographic [18F]fluorodopa studies correlate with dopamine cell counts and levels. *Ann Neurol* 1993; **34**:324–330.
20. Morrish PK, Sawle GV, Brooks DJ. Clinical and [18F] dopa PET findings in early Parkinson's disease. *J Neurol Neurosurg Psychiatry* 1995; **59**: 597–600.
21. Morrish PK, Sawle GV, Brooks DJ. An [18F]dopa-PET and clinical study of the rate of progression in Parkinson's disease. *Brain* 1996; **119**(Pt 2):585–591.
22. Brooks DJ, Piccini P. Non-invasive in vivo imaging of transplant function. In: Anders A, Dunnett S, eds. Functional transplantation. *Prog Brain Res* 2000; **127**:321–332.
23. Hagell P, Schrag A, Piccini P, *et al.* Sequential bilateral transplantation in Parkinson's disease: effects of the second graft. *Brain* 1999; **122**(Pt 6): 1121–1132.
24. Gill SS, Patel NK, Hotton GR, *et al.* Direct brain infusion of glial cell line-derived neurotrophic factor in Parkinson disease. *Nat Med* 2003; **9**:589–595.
25. Kordower JH, Freeman TB, Snow BJ, *et al.* Neuropathological evidence of graft survival and striatal reinnervation after the transplantation of fetal mesencephalic tissue in a patient with Parkinson's disease. *N Engl J Med* 1995; **332**:1118–1124.
26. Piccini P. Dyskinesias after transplantation in Parkinson's disease [review]. *Lancet Neurol* 2002; **1**:472.
27. Hagell P, Piccini P, Bjorklund A, *et al.* Dyskinesias following neural transplantation in Parkinson's disease. *Nat Neurosci* 2002; **5**:627–628.
28. Piccini P, Lindvall O, Bjorklund A, *et al.* Delayed recovery of movement-related cortical function in Parkinson's disease after striatal dopaminergic grafts. *Ann Neurol* 2000; **48**:689–695.
29. Piccini P, Brooks DJ, Bjorklund A, *et al.* Dopamine release from nigral transplants visualized in vivo in a Parkinson's patient. *Nat Neurosci* 1999; **2**: 1137–1140.
30. Turjanski N, Weeks R, Dolan R, *et al.* Striatal D1 and D2 receptor binding in patients with Huntington's disease and other choreas. A PET study. *Brain* 1995; **118**(Pt 3):689–696.
31. Pavese N, Andrews TC, Brooks DJ, *et al.* Progressive striatal and cortical dopamine receptor dysfunction in Huntington's disease: a PET study. *Brain* 2003; **126**(Pt 5):1127–1135.

32. Martin WR, Clark C, Ammann W, *et al.* Cortical glucose metabolism in Huntington's disease. *Neurology* 1992; **42**:223–229.
33. Bachoud-Levi AC, Remy P, Nguyen JP, *et al.* Motor and cognitive improvements in patients with Huntington's disease after neural transplantation. *Lancet* 2000; **356**:1975–1979.
34. Antonini A, Leenders KL, Spiegel R, *et al.* Striatal glucose metabolism and dopamine D2 receptor binding in asymptomatic gene carriers and patients with Huntington's disease. *Brain* 1996; **119**:2085–2095.
35. Weeks RA, Piccini P, Harding AE, *et al.* Striatal D1 and D2 dopamine receptor loss in asymptomatic mutation carriers of Huntington's disease. *Ann Neurol* 1996; **40**:49–54.
36. Brooks DJ. Positron emission tomography studies in movement disorders. *Neurosurg Clin North Am* 1998; **9**:263–282.
37. Wiebe S, Blume WT, Girvin JP, *et al.* A randomized, controlled trial of surgery for temporal-lobe epilepsy. *N Engl J Med* 2001; **345**:311–318.
38. Duncan JS. Imaging and epilepsy. *Brain* 1997; **120**(Pt 2):339–377.
39. Berkovic SF, McIntosh AM, Kalnins RM, *et al.* Preoperative MRI predicts outcome of temporal lobectomy: an actuarial analysis. *Neurology* 1995; **45**: 1358–1363.
40. Kuzniecky R, Garcia JH, Faught E, *et al.* Cortical dysplasia in temporal lobe epilepsy: magnetic resonance imaging correlations. *Ann Neurol* 1991; **29**: 293–298.
41. Engel J, Jr, Kuhl DE, Phelps ME. Regional brain metabolism during seizures in humans. *Adv Neurol* 1983; **34**:141–148.
42. Leiderman DB, Albert P, Balish M, *et al.* The dynamics of metabolic change following seizures as measured by positron emission tomography with fludeoxyglucose F 18. *Arch Neurol* 1994; **51**:932–936.
43. Marco P, Sola RG, Pulido P, *et al.* Inhibitory neurons in the human epileptogenic temporal neocortex. An immunocytochemical study. *Brain* 1996; **119**(Pt 4): 1327–1347.
44. Maziere M, Hantraye P, Prenant C, *et al.* Synthesis of ethyl 8-fluoro-5,6-dihydro-5-[11C]methyl-6-oxo-4H-imidazo [1,5-a] [1,4]benzodiazepine-3-carboxylate (RO 15.1788-11C): a specific radioligand for the in vivo study of central benzodiazepine receptors by positron emission tomography. *Int J Appl Radiat Isot* 1984; **35**:973–976.
45. Savic I, Persson A, Roland P, *et al.* In-vivo demonstration of reduced benzodiazepine receptor binding in human epileptic foci. *Lancet* 1988; **2**:863–866.
46. Johnson EW, de Lanerolle NC, Kim JH, *et al.* "Central" and "peripheral" benzodiazepine receptors: opposite changes in human epileptogenic tissue. *Neurology* 1992; **42**:811–815.

47. Koepp MJ, Hand KS, Labbe C, *et al.* In vivo [11C]flumazenil-PET correlates with ex vivo [3H]flumazenil autoradiography in hippocampal sclerosis. *Ann Neurol* 1998; **43**:618–626.
48. Ryvlin P, Bouvard S, Le Bars D, *et al.* Clinical utility of flumazenil-PET versus [18F]fluorodeoxyglucose-PET and MRI in refractory partial epilepsy. A prospective study in 100 patients. *Brain* 1998; **121**(Pt 11):2067–2081.
49. Savic I, Ingvar M, Stone-Elander S. Comparison of [11C]flumazenil and [18F]FDG as PET markers of epileptic foci. *J Neurol Neurosurg Psychiatry* 1993; **56**:615–621.
50. Hammers A, Koepp MJ, Hurlemann R, *et al.* Abnormalities of grey and white matter [11C]flumazenil binding in temporal lobe epilepsy with normal MRI. *Brain* 2002; **125**(Pt 10):2257–2271.
51. Gaillard WD, Bhatia S, Bookheimer SY, *et al.* FDG-PET and volumetric MRI in the evaluation of patients with partial epilepsy. *Neurology* 1995; **45**: 123–126.
52. Di Chiro G, DeLaPaz RL, Brooks RA, *et al.* Glucose utilization of cerebral gliomas measured by [18F] fluorodeoxyglucose and positron emission tomography. *Neurology* 1982; **32**:1323–1329.
53. Patronas NJ, Di Chiro G, Kufta C, *et al.* Prediction of survival in glioma patients by means of positron emission tomography. *J Neurosurg* 1985; **62**:816–822.
54. Francavilla TL, Miletich RS, Di Chiro G, *et al.* Positron emission tomography in the detection of malignant degeneration of low-grade gliomas. *Neurosurgery* 1989; **24**:1–5.
55. Brock CS, Young H, O'Reilly SM, *et al.* Early evaluation of tumour metabolic response using [18F]fluorodeoxyglucose and positron emission tomography: a pilot study following the phase II chemotherapy schedule for temozolomide in recurrent high-grade gliomas. *Br J Cancer* 2000; **82**:608–615.
56. Rozental JM, Levine RL, Mehta MP, *et al.* Early changes in tumor metabolism after treatment: the effects of stereotactic radiotherapy. *Int J Radiat Oncol Biol Phys* 1991; **20**:1053–1060.
57. Young H, Baum R, Cremerius U, *et al.* Measurement of clinical and subclinical tumour response using [18F]-fluorodeoxyglucose and positron emission tomography: review and 1999 EORTC recommendations. European Organization for Research and Treatment of Cancer (EORTC) PET Study Group. *Eur J Cancer* 1999; **35**:1773–1782.
58. Rozental JM, Levine RL, Nickles RJ, *et al.* Glucose uptake by gliomas after treatment. A positron emission tomographic study. *Arch Neurol* 1989; **46**: 1302–1307.
59. Valk PE, Budinger TF, Levin VA, *et al.* PET of malignant cerebral tumors after interstitial brachytherapy. Demonstration of metabolic activity and correlation with clinical outcome. *J Neurosurg* 1988; **69**:830–838.

60. Barrington SF. Clinical uses of PET in neurology. *Nucl Med Commun* 2000; **21**:237–240.
61. De Witte O, Goldberg I, Wikler D, *et al.* Positron emission tomography with injection of methionine as a prognostic factor in glioma. *J Neurosurg* 2001; **95**:746–750.
62. Tsuyuguchi N, Sunada I, Iwai Y, *et al.* Methionine positron emission tomography of recurrent metastatic brain tumor and radiation necrosis after stereotactic radiosurgery: is a differential diagnosis possible? *J Neurosurg* 2003; **98**: 1056–1064.
63. Dethy S, Goldman S, Blecic S, *et al.* Carbon-11-methionine and fluorine-18-FDG PET study in brain hematoma. *J Nucl Med* 1994; **35**:1162–1166.
64. Ogawa T, Kanno I, Shishido F, *et al.* Clinical value of PET with 18F-fluorodeoxyglucose and L-methyl-11C-methionine for diagnosis of recurrent brain tumor and radiation injury. *Acta Radiol* 1991; **32**:197–202.
65. O'Doherty MJ, Barrington SF, Campbell M, *et al.* PET scanning and the human immunodeficiency virus-positive patient. *J Nucl Med* 1997; **38**:1575–1583.
66. Villringer K, Jager H, Dichgans M, *et al.* Differential diagnosis of CNS lesions in AIDS patients by FDG-PET. *J Comput Assist Tomogr* 1995; **19**:532–536.
67. Salmon E. Functional brain imaging applications to differential diagnosis in the dementias. *Curr Opin Neurol* 2002; **15**:439–444.
68. Mazziotta JC, Frackowiak RS, Phelps ME. The use of positron emission tomography in the clinical assessment of dementia. *Semin Nucl Med* 1992; **22**: 233–246.
69. Salmon E, Gregoire MC, Delfiore G, *et al.* Combined study of cerebral glucose metabolism and [11C]methionine accumulation in probable Alzheimer's disease using positron emission tomography. *J Cereb Blood Flow Metab* 1996; **16**: 399–408.
70. Silverman DH, Small GW, Chang CY, *et al.* Positron emission tomography in evaluation of dementia: regional brain metabolism and long-term outcome. *JAMA* 2001; **286**:2120–2127.
71. Small GW, Mazziotta JC, Collins MT, *et al.* Apolipoprotein E type 4 allele and cerebral glucose metabolism in relatives at risk for familial Alzheimer disease. *JAMA* 1995; **273**:942–947.
72. Reiman EM, Caselli RJ, Yun LS, *et al.* Preclinical evidence of Alzheimer's disease in persons homozygous for the epsilon 4 allele for apolipoprotein E. *N Engl J Med* 1996; **334**:752–758.
73. Small GW, Ercoli LM, Silverman DH, *et al.* Cerebral metabolic and cognitive decline in persons at genetic risk for Alzheimer's disease. *Proc Natl Acad Sci USA* 2000; **97**:6037–6042.

74. Reiman EM, Caselli RJ, Chen K, *et al.* Declining brain activity in cognitively normal apolipoprotein E epsilon 4 heterozygotes: A foundation for using positron emission tomography to efficiently test treatments to prevent Alzheimer's disease. *Proc Natl Acad Sci USA* 2001; **98**:3334–3339.
75. de Leon MJ, Convit A, Wolf OT, *et al.* Prediction of cognitive decline in normal elderly subjects with 2-[(18)F]fluoro-2-deoxy-D-glucose/positron-emission tomography (FDG/PET). *Proc Natl Acad Sci USA* 2001; **98**: 10966–10971.
76. Albin RL, Minoshima S, D'Amato CJ, *et al.* Fluoro-deoxyglucose positron emission tomography in diffuse Lewy body disease. *Neurology* 1996; **47**:462–466.
77. Minoshima S, Foster NL, Sima AA, *et al.* Alzheimer's disease versus dementia with Lewy bodies: cerebral metabolic distinction with autopsy confirmation. *Ann Neurol* 2001; **50**:358–365.
78. Feeney DM, Baron JC. Diaschisis. *Stroke* 1986; **17**:817–830.
79. Herholz K. FDG PET and differential diagnosis of dementia. *Alzheimer Dis Assoc Disord* 1995; **9**:6–16.
80. Aisen PS. The potential of anti-inflammatory drugs for the treatment of Alzheimer's disease. *Lancet Neurol* 2002; **1**:279–284.
81. Cagnin A, Brooks DJ, Kennedy AM, *et al.* In-vivo measurement of activated microglia in dementia. *Lancet* 2001; **358**:461–467.
82. Bacskai BJ, Klunk WE, Mathis CA, *et al.* Imaging amyloid-beta deposits in vivo. *J Cereb Blood Flow Metab* 2002; **22**:1035–1041.
83. Guadagno JV, Calautti C, Baron JC. Progress in imaging stroke: emerging clinical applications. *Br Med Bull* 2003; **65**:145–157.
84. Furlan M, Marchal G, Viader F, *et al.* Spontaneous neurological recovery after stroke and the fate of the ischemic penumbra. *Ann Neurol* 1996; **40**:216–226.
85. Heiss WD, Kracht LW, Thiel A, *et al.* Penumbral probability thresholds of cortical flumazenil binding and blood flow predicting tissue outcome in patients with cerebral ischaemia. *Brain* 2001; **124**(Pt 1):20–29.
86. Chollet F, DiPiero V, Wise RJ, *et al.* The functional anatomy of motor recovery after stroke in humans: a study with positron emission tomography. *Ann Neurol* 1991; **29**:63–71.
87. Blank SC, Bird H, Turkheimer F, *et al.* Speech production after stroke: the role of the right pars opercularis. *Ann Neurol* 2003; **54**:310–320.
88. Whone AL, Moore RY, Piccini PP, *et al.* Plasticity of the nigropallidal pathway in Parkinson's disease. *Ann Neurol* 2003; **53**:206–213.
89. Gee AD. Neuropharmacology and drug development. *Br Med Bull* 2003; **65**: 169–177.

15

Functional Magnetic Resonance Imaging

P M Matthews and P Jezzard*

Blood oxygenation level dependent (BOLD) functional magnetic resonance imaging (fMRI) is a powerful approach to defining activity in the healthy and diseased human brain. BOLD fMRI detects local increases in relative blood oxygenation that are most probably a direct consequence of neurotransmitter action and thus reflect local neuronal *signalling*. The method allows localisation to volumes of the order of a few to several cubic millimetres and can be used in serial studies of individual subjects. Basic approaches to experimental design and analysis are reviewed briefly, as well as potential clinical applications. The latter include three broad areas: anatomical characterisation of normal or pathological patterns of brain functioning; distinguishing pathological traits; and monitoring treatment responses. New research is emphasising the integration of fMRI with other techniques, particularly electrophysiological. In conjunction with MRI methods for characterising pathological load, fMRI promises

*Centre for Functional Magnetic Resonance Imaging of the Brain, Department of Clinical Neurology, University of Oxford, Oxford, UK.
Centre for Functional Magnetic Resonance Imaging of the Brain, John Radcliffe Hospital, Headington, Oxford OX3 9DU, UK; paul@fmrib.ox.ac.uk

a refined understanding of when disease processes begin and how they can be modified by new treatments.

A variety of methods have been developed over the past few decades to allow mapping of the functioning human brain. Two basic classes of mapping technique have evolved: those that map (or localise) the underlying electrical activity of the brain; and those that map local physiological or metabolic consequences of altered brain electrical activity. Among the former are the non-invasive neural electromagnetic techniques of electroencephalography (EEG) and magnetoencephalography (MEG). These methods allow exquisite temporal resolution of neural processes (typically over a 10–100 ms time scale), but suffer from poor spatial resolution (between 1 and several centimetres). Functional MRI (fMRI) methods are in the second category. They can be made sensitive to the changes in regional blood perfusion, blood volume (for example, using injected magnetic resonance contrast agents), or blood oxygenation that accompany neuronal activity. Blood oxygenation level dependent (BOLD) fMRI, which is sensitive primarily to the last of these variables, allows an image spatial resolution that is of the order of a few millimetres, with a temporal resolution of a few seconds (limited by the haemodynamic response itself). An accessible and more detailed introduction to the technique than is possible in this brief review is found in a recent book.[1]

PRINCIPLES OF FUNCTIONAL MRI

Contrast in a Magnetic Resonance Image

The contrast in a magnetic resonance image (which determines the apparent structure in what we see) depends on how it is acquired. By adding radio frequency or gradient pulses, and by careful choice of their timings, it is possible to highlight different characteristics of the tissue being imaged. While it is generally true that MRI maps the distribution of water in the brain, the useful contrast in MR images comes not just from spatial variations in the density of water but also from differences in fundamental nuclear magnetic processes known as relaxation, which are characterised by distinct rates or "relaxation times". There are three relaxation times that are of primary interest in MRI — T1, T2, and T2*. These describe the time constant for the return of the magnetisation to its equilibrium position aligned along the static magnetic field of the scanner whenever it is disturbed (T1 relaxation) and the time constants associated with loss of signal once the magnetisation has been sampled (T2 and T2* relaxation). T2* is the most relevant relaxation time for understanding contrast in fMRI images.

The Physiological Basis of BOLD fMRI

Most of the energy used for neuronal activity is expended as a result of the postsynaptic neuronal depolarisation and, to a lesser extent, the action potentials generated.[2] The energy cost therefore arises from information transfer and its integration postsynaptically. Substrate delivery for energy metabolism is increased with increased local blood flow. However, it is not the increased energy use itself that directly drives the increase in blood flow.[3] Instead, increased blood flow appears to be a direct consequence of neurotransmitter action and thus reflects local *signalling*. Electrophysiologically, increases in the BOLD signal are correlated most clearly with the local field potential rather than the neuronal firing rate.[4] Blood flow in fact increases over a wider volume and to a greater extent than is necessary simply to provide oxygen and glucose for the increased energy production, so oxygen extraction decreases with greater neuronal activity.

The volume over which blood flow increases associated with neuronal activity is found is determined by the level of local control of perfusion, which is thought to be the feeding arterioles.[5] There may be multiple mediators of the arteriolar response, but nitric oxide (NO) and eicosanoids clearly are important under normal circumstances.[6,7] Binding of glutamate to receptors on astrocytes triggers NO release, and glial cells around the synapse may contribute to controlling the vascular response.[8]

Biophysics of BOLD fMRI

Reduced oxygen extraction leads to an increase in the ratio of oxy- to deoxyhaemoglobin in a region of neuronal activation. The origin of the associated BOLD fMRI signal change lies in the different magnetic properties of haemoglobin-carrying oxygen (oxyHb) and deoxygenated haemoglobin (deoxyHb). DeoxyHb is slightly paramagnetic relative to brain tissue, whereas oxyHb is isomagnetic.[9] Vessels containing oxygenated arterial blood thus cause little or no distortion to the magnetic field in the surrounding tissue, while capillaries and veins containing blood that is partially deoxygenated distort the magnetic field in their vicinity[10,11] (Fig. 1). The microscopic field inhomogeneities associated with the presence of deoxyHb lead to destructive interference from signal within the tissue voxel, a process that tends to shorten the T2* relaxation time. Thus, as oxygen extraction falls with enhanced local blood flow in a region of greater neuronal activity, the T2* becomes longer and the MRI signal intensity increases relative to the baseline state.

The precise amount by which the MRI signal intensity increases depends on several factors. There is a contribution from water molecules in blood (the intravascular compartment) and from water molecules in the tissue space around the vessels (the extravascular compartment). The observed signal is a volume weighted average of signal changes both from intravascular water in local capillaries and veins and water in the immediate extravascular compartment. BOLD signal change increases linearly with the static field strength of the MRI scanner for blood vessels that are

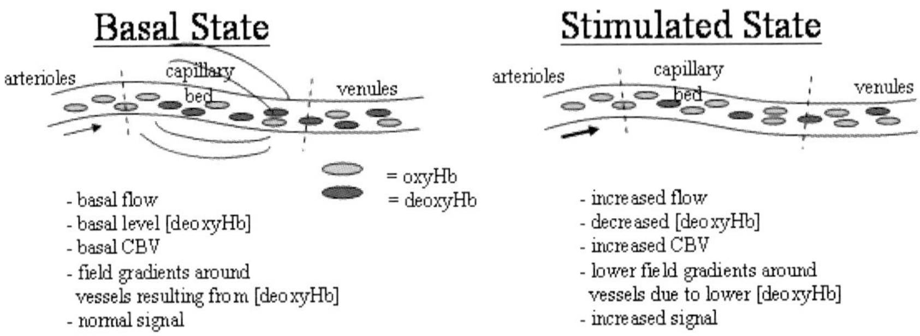

Fig. 1 Schematic diagram showing the haemodynamic variables that change during neuronal activity. In the basal state deoxyhaemoglobin in the capillaries and venules causes microscopic field gradients to be established around the blood vessels. This in turn leads to a decreased signal in a gradient echo magnetic resonance imaging sequence. In the activated state there is a significant increase in flow, but only a modest increase in oxygen consumption. This results in a lower concentration of deoxyhaemoglobin in the capillaries and venules and hence in a reduction in the microscopic field gradients and an increase in the signal intensity. CBV, cerebral blood volume; deoxyHb, deoxyhaemoglobin.

of greater radius than approximately 8 μm and quadratically when considering blood vessels that are smaller than this value.[12,13]

Although only 3–5% of the water molecules in grey matter are in the vascular space (in white matter the value is closer to 2%), the contribution of the intravascular contribution to the BOLD signal change can be substantial.[13,14] Because the T2 and T2* relaxation times of blood at 1.5 Tesla are long compared with tissue, and the extravascular water effects are relatively localised around the vessels, signal from the intravascular water pool has a dominant effect (estimated at 60%) on activity related signal intensity changes at 1.5 Tesla.[15] Signal changes with brain activity thus can sometimes be detected in large draining veins that may be some distance from the site of neuronal activity.

In at least some areas of the brain (for example, visual and primary motor cortex), a small transient decrease in the BOLD signal may be observed after onset of activity before the characteristic signal increase.[16] This has been interpreted as reflecting local deoxygenation of blood in the capillary bed preceding the onset of activation associated hyperaemia. This "initial dip" may provide a more accurate measure of the localisation of activation.[17] However, the magnitude of the change is several-fold smaller than that of the later BOLD signal increase, so it is unlikely to provide a practical approach for improved functional mapping for clinical applications in the immediate future.

Practical Implementation

Many MRI scanner manufacturers now supply add-on features that allow standard fMRI procedures to be performed easily. These include suitable pulse sequences, peripheral devices for presentation of stimuli to the subjects in the scanner, devices for recording responses from the subject, and even statistical analysis and display packages that allow assessment of the data while the subject remains in the magnet. The most common imaging sequence used is the fast method of echo planar imaging (EPI),[18] which allows collection of whole brain data in a few seconds or less. The spatial resolution is considerably lower (typically $4 \times 4 \times 4\,\mathrm{mm}^3$) than for a conventional MRI scan (Fig. 2). Image intensity is also reduced in frontal and temporal regions and there is some distortion of the shape of the

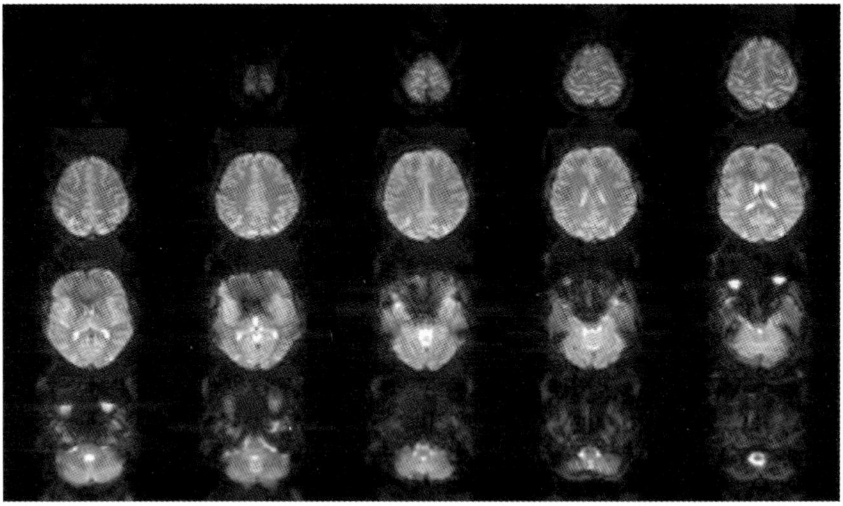

Fig. 2 Example of a whole brain echo planar imaging (EPI) dataset collected in three seconds. Note the signal loss in the frontal and temporal lobes of the brain. Note also the lower spatial resolution.

brain. These problems arise from the sensitivity of the EPI scanning to field gradients caused by magnetic susceptibility differences — for example, at air sinus/tissue interfaces. This problem worsens with increasing field strength.

In an fMRI experiment a large series of images is acquired rapidly while the subject performs a task that shifts brain activity between two or more well defined states. Several hundred such volumes may be collected in a single session while the subject does different tasks. By correlating the signal time course in each volume element (voxel) of the slice stack with the known time course of the task it is possible to identify those voxels in the brain that show changes associated with the brain function under consideration.

Design of fMRI Studies

Methods such as positron emission tomography (PET) provide an absolute measure of tissue metabolism. In contrast, BOLD fMRI can at present be used only for determining *relative* signal intensity changes associated with different cognitive states during a single imaging session. The most time

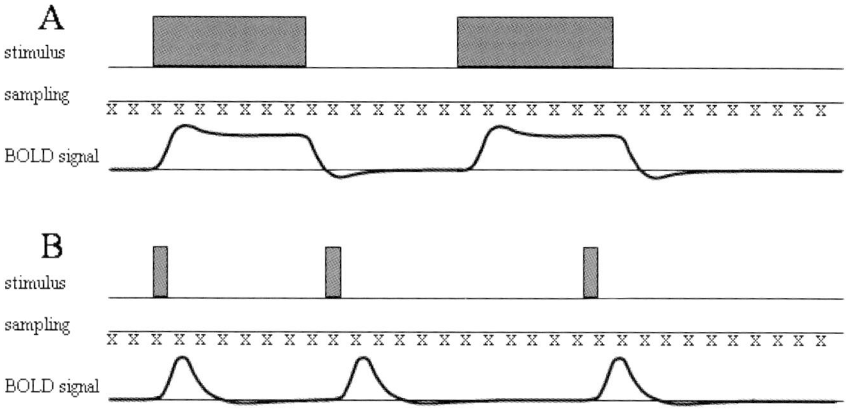

Fig. 3 Schematic representation of a block design functional magnetic resonance imaging (fMRI) paradigm (A) and an event related fMRI paradigm (B). For the block design a relatively long (30 second) stimulation period is alternated with a control period. For the event related design a brief stimulus period is used, which can either be periodic or randomised. In both cases volumes of data (indicated by the crosses) are collected continuously, typically with a repeat time of three to five seconds.

efficient approach for comparing brain responses in different states during the imaging experiment is the "block" design[19] (Fig. 3). This design uses relatively long alternating periods (for example, 30 seconds), during each of which a discrete cognitive state is maintained. In the simplest form, there may only be two such states, which are alternated throughout the experiment in order to ensure that variations arising from fluctuations in scanner sensitivity, patient movement, or changes in attention have a similar impact on the signal responses associated with both states.

However, it can become difficult to control a cognitive state precisely for the relatively long periods of each block, or some tasks may simply be inappropriate for this design (for example, as in an "oddball" paradigm). In such instances an event related design can be used in which data are acquired while discrete stimuli or responses are repeated[20] (Fig. 3). Results from many trials are then averaged to give a measurable response. Event related fMRI demands longer acquisition times than the block design in order to achieve a sufficient signal to noise ratio. A related approach is to present stimuli in a periodic fashion and then to map responses in terms of their temporal phase relative to that of the stimulus presentation.[21]

Analysis of fMRI Studies

The raw BOLD fMRI data can be acquired over periods as short as a few minutes. For simple analyses, near "real-time" viewing of final statistical maps of activation is possible (although — at least in a research environment — full analysis may demand extensive computation and much more substantial analysis times). The basic objective in the analysis of functional imaging experiments is to identify voxels that show signal changes that vary with the changing brain states of interest across the serially acquired images. This is a challenging problem for fMRI data because the signal changes are small (0.5% to 5%) (leading to potential false negative results or type II error) and the number of voxels simultaneously interrogated across the imaged volume is very large (potentially giving potential false positive results or type I error). One approach to enhancing the sensitivity is to undertake studies of groups of individuals: even if the changes are small, consistently activated regions may then be identified.

Different types of statistical analyses can be done. A "fixed effects" analysis gives an expression of changes in the group mean signal relative to the group pooled within-subject variance. This provides a sensitive measure of whether the group is activating on average, but does not look at subject to subject variability and therefore cannot be used to make generalisations about the larger population from which the group was drawn. To do this one would use a "random" or "mixed" effects model. Such models take into account not only the variance in a measurement for an individual subject, but also the variance in measurements between individuals.

The signal changes observed are small, and interpretation of results for single individuals (for example, in a clinical study) demands an appreciation of the reproducibility of a study. The exact volume of significant activation may show considerable variation between sessions, as the low signal to noise voxels on the edge of the activation volume will be included variably, depending on noise fluctuations.[22] Nonetheless, test–retest correlations of activation extent for typical cognitive tasks are good (for example, $r = 0.69$).[23,24] Meta-analyses have confirmed consistent localisations.[25] Specific BOLD signal characteristics (such as the maximum signal change or timing relative to a stimulus) also can be highly reproducible.[26]

The accuracy of localisations must be quantified only with respect to other techniques. This is complicated by uncertainties as to how data from different modalities (or brains of different sizes and shapes) are best aligned (or registered).[27] Comparisons with invasive electrophysiology in non-human primates suggest that the correspondence between direct recordings of local field potentials and fMRI changes may be high.[4] Good agreement has been found between functional localisations based on EEG and fMRI in humans.[28]

One of the most significant confounding factors in fMRI is the extreme sensitivity to *motion*, either of the whole head or even the brain alone (for example, pulsations associated with the respiratory or cardiac cycles). A first step in analysis therefore is post hoc realignment of the brain volumes using automated algorithms that minimise the difference between subsequent images. Following motion correction of the data, spatial smoothing and temporal filtering of the data are often applied, primarily to reduce noise in the data. A variety of statistical tests can then be done to identify voxels in which the signal changes correlate over time with switching between the applied "control" and "stimulus" states. The simplest approach is to generate a map of the t statistic for signal changes on a voxel by voxel basis. A related approach is to correlate the time course of signal change in each voxel with a model time course based on the expected neural response (suitably convolved with a model of the haemodynamic response), which can also be used to generate a t statistic map. The significance threshold in all cases must be made more stringent in proportion to the number of independent comparisons (although, because of spatial correlations in the data, this is smaller than the total number of voxels).

There are now several software packages produced by academic centres that include a full set of tools for analysis of fMRI data and that are distributed at no cost. Examples of these are FSL (www.fmrib.ox.ac.uk/fsl), SPM (www.fil.ion.ucl.ac.uk/spm), and AFNI (afni.nimh.nih.gov/afni).

CLINICAL APPLICATIONS OF fMRI

Localisation of Brain Functions

Lateralisation of language functions in the surgical treatment of epilepsy

Surgery offers the possibility of improved seizure control or cure, especially for patients with temporal lobe epilepsy, but demands an understanding

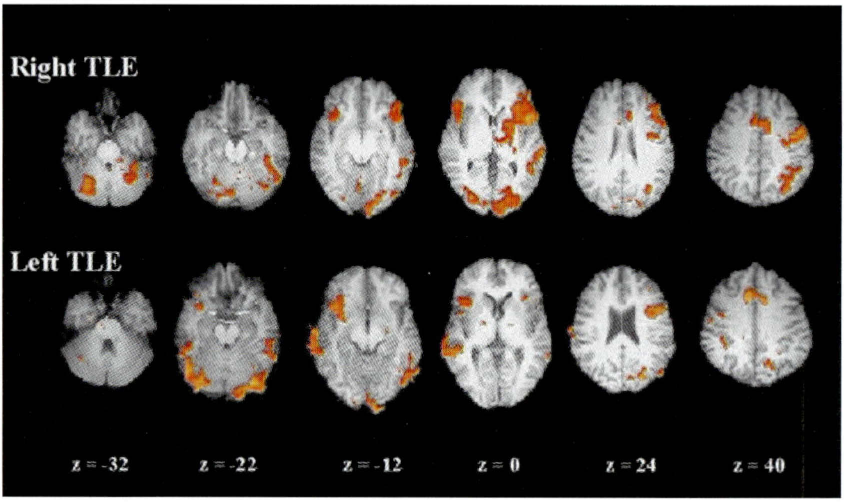

Fig. 4 Differences in relative language lateralisation for a verbal fluency functional magnetic resonance imaging (fMRI) task can be found between patients with right or left temporal lobe epilepsy (TLE). Illustrative activation maps are shown here, coregistered with individual high resolution structural MRI. Cluster detection was done on all voxels above $z = 2.3$ to determine clusters significantly activated (corrected $p < 0.01$) in the experimental task condition. The right TLE patient has predominantly left hemisphere activation. In contrast, the activation map for a patient with left TLE shows bihemispheric activations.

of language lateralisation for surgical planning. However, current clinical methods for language lateralisation (for example, the Wada test) are highly invasive. fMRI offers a promising alternative approach[24] (Fig. 4). While there is good agreement between conventional invasive Wada testing and fMRI results, fMRI is more sensitive to involvement of the non-dominant hemisphere. fMRI also provides more specific anatomical information. The reproducibility of distinct patterns of activation in individual subjects is good, potentially allowing clinical decisions to be made on the basis of results.[24]

Localisation of eloquent cortex before surgery

A general issue in presurgical planning for excisions near regions of eloquent cortex is precise localisation of essential functions. fMRI is an attractive strategy for this functional mapping because of its potentially wide availability in clinical centres and relatively low cost.[29] However, because the fMRI BOLD response is sensitive to signal changes in draining veins, there is potential for mislocalisation of major brain activity. The true correspondence between the BOLD fMRI and electrocortical localisations is difficult to define exactly because of difficulties in accurately registering (that is, aligning) the fMRI data with brain structural images. As described earlier, aberrations in brain geometry are induced by distortions of the magnetic field in the functional images which are not found in the conventional structural images. The brain also may shift position in complex ways when exposed for surgery.

These issues were recently reinvestigated.[30] Distances between the centres of the MEG and fMRI activation regions were measured, and consistent differences (of the order of 10 mm) were identified. For example, localisation of the primary motor cortex from the fMRI data was consistently more posterior than for the MEG localisation. For somatosensory responses, the localisation of the fMRI activation was inferior and lateral to that of the MEG. Thus, while the MEG dipole and the BOLD fMRI response maximum are in *similar* regions, the different physiological responses are slightly displaced.

Considering the possible causes for these localisation discrepancies is informative. Fundamentally different information is provided by the two techniques, not just with respect to the basis of signal change (for example, the potential sensitivity of fMRI to the "draining vein"), but also to the time period over which the response is averaged. The MEG "window" is short (tens of milliseconds). In contrast, the fMRI response is averaged over a much broader time period (seconds). The fMRI response may therefore reflect contributions from more than a single electrophysiologically defined dipole in the region of interest. For example, in the somatosensory cortex the early MEG response is localised in Brodman's area 3b, while later responses may be found in areas 1 and 2. The shift in the fMRI response relative to the MEG response may reflect the fact that fMRI is measuring an *average* localisation across these multiple regions. It is intriguing to consider whether shifts in the relative timing of activation from adjacent regions might contribute to apparent changes in functional localisation with brain injury.[31,32]

Localising spontaneous brain activity: the ictal focus

It is not only induced brain activity (that is, associated with task performance) that is associated with increased local blood flow. The spontaneous electrical discharges of an epileptic focus are also accompanied by increased tissue metabolism, oxygen utilisation, and blood flow. In the absence of information on the phase of the haemodynamic signal change with respect to the scanning acquisition, it is not possible readily to identify such foci using conventional methods (although see Matthews *et al.*[33]), but a model for the time course of the expected signal change can be developed *after* the imaging data are acquired.

Electroencephalography provides a measure of locally coherent cortical field potentials and can give precise information on the timing of ictal spiking activity to inform such a model if EEG and fMRI data are acquired simultaneously. Indeed, with simultaneous EEG and fMRI it has been possible to identify ictal foci in patients with subclinical seizures.[34,35] Direct comparisons between dipole localisation using a simple EEG dipole based model and fMRI showed fair correspondence between the centres of activation for the two subjects, or revealed major foci within a few centimetres of each other but clearly in the same regions of the brain. However, potential applications still may be limited. This approach is time consuming even with an active subclinical ictal focus, and motion from a clinical seizure would probably introduce artefacts that would make the fMRI signals irrecoverable.

Brain Plasticity with Injury or Disease

A direct extension of work for localisation of brain functions is to try to define explicitly ways in which these localisations may change with brain injury or disease. Such changes may be a consequence of so called "adaptive plasticity" (that is, induced changes in functional organisation in the brain), or of recruitment of intact brain regions in compensation for functional deficits arising from disease.

It has long been thought that the younger or developing brain has greater inherent plasticity, but there have been few direct tests of this notion. From an fMRI study of children with hemiplegia acquired either in utero or after birth, Gadian and his colleagues came to the unexpected conclusion that factors *other* than age must dominate the potential for adaptive changes.[36] Perhaps not surprisingly then, even the adult brain shows considerable potential for adaptive plasticity or compensatory recruitment of new brain regions. Several studies of patients after strokes[32,37–40] have confirmed earlier PET observations[41] suggesting that new regions of intact brain are recruited in the motor cortex ipsilateral to the hand moved after injury to the corticospinal tract projecting contralaterally. A key question has been whether such changes are adaptive or whether they represent either epiphenomena or even maladaptive responses. Increased activation in primary sensorimotor cortex relative to premotor cortical areas has been identified in dystonic musicians, for example, emphasising that "greater" activation is not necessarily "better".[42]

To test for a functional role of ipsilateral motor cortex recruitment, Johansen-Berg and colleagues studied a group of healthy controls and patients after stroke using both fMRI and transcranial magnetic stimulation (TMS).[43] TMS transiently interferes with any ongoing brain activity below the stimulation coil. There was a significant slowing of movement reaction time with TMS over ipsilateral premotor cortex with movements of the paretic hand for patients, but no significant effect of TMS applied in the same way in the healthy controls. Thus the greater ipsilateral motor cortex activity shown by fMRI in patients must contribute to function in a unique way after corticospinal tract injury.

fMRI studies are showing that functional reorganisation is a general response to brain injury.[41,44] A wide range of regions within the spatially distributed cortical motor network may contribute to this.[45] Adaptive changes may in fact contribute to maintaining subclinical the expression of pathology in early stages of the disease.[44,46]

fMRI studies of learning in healthy subjects show that changes in functional brain organisation also may be induced.[47] An exciting clinical extension of this concept is to defining functional changes in the brain with neurorehabilitation.[48] Specific regions of the brain change activity with clinical improvements after treatment. With definition of the functional anatomy and mechanisms responsible for recovery, it may be possible to provide improved prognostic markers for better identification of patients who will benefit from a treatment or better tailoring of treatment to individual needs.

fMRI AS A MARKER OF PATHOLOGICAL STATE

Identifying Preclinical Expression of Disease

fMRI can be sensitive to early (and even preclinical) stages of brain pathology. A pioneering illustration of this approach was an fMRI based memory study of a group of apparently healthy subjects at risk of developing earlier onset Alzheimer's disease.[49] One year after fMRI scanning, those who were beginning to develop memory problems in early clinical expression of presumed Alzheimer's disease were identified. A significant difference in the pattern and volume of activated cortex with the memory task was found in these subjects relative to those who did not develop memory impairment.

A related application is the use of fMRI to identify patients with a disease trait. Subjects who have recovered from depression have a substantial risk for recurrence of depression, suggesting that there are persistent abnormalities in brain function associated with vulnerability to depression. Because of the potential interaction between depression and stimuli associated with aversive emotional conditioning, Smith *et al.* applied a pain conditioning paradigm to study a group of patients who had recovered from depression and who were not on drug treatment, but were at risk of recurrence of depression.[50] While the direct response to pain itself was similar between healthy control subjects and the previously depressed patients, and there were no differences in ratings of mood or affective response, the responses to anticipation of painful stimuli were different between the two groups. The recovered depressed patients showed an altered fMRI response in the cerebellum relative to healthy controls. This provides a link between theories of depression and a growing body of work showing that the cerebellum plays a role in conditioning, cognition, and emotional responses. Antidepressant treatment has also been associated with increased cerebellar metabolism in other work. In principle, this type of study might be used to distinguish between different types of depression or to identify healthy subjects at risk of depression.

fMRI in the Development of New Treatments

Using fMRI to guide therapeutic development is clearly one of the most exciting prospects for the technique. Initial work has not just been for drug development and response monitoring. The greatest impact may be on areas in which sensitive and objective end points were previously difficult to define — for example, neurorehabilitation.[48]

A similar example is the assessment of outcomes using behavioural therapy, such as for the control of pain. Distinct mechanisms contribute to the perception of pain, including attention.[49] Bantick and her colleagues reported how the patterns of the fMRI brain response may change with self induced distraction from a thermal pain stimulus. When attention is distracted, the overall brain activation is reduced substantially. There is a particular reduction in regions of limbic cortex associated with emotional response to pain. "Gating" of pain signals to the brain as a result of higher cortical processes may occur through several areas. fMRI studies of brain stem changes associated with pain suggest that the periaquaductal grey matter is an important locus for such action.[51] Similar mechanisms may contribute to the placebo effect.[52]

Identification of the biological basis for cognitive and behavioural changes offers insights into mechanisms of vulnerability and variability of responses to treatments for neurological and psychiatric diseases. The recent evidence that functional neuroimaging methods such as fMRI may have relative sensitivity to systems level brain changes suggests that the method also may provide a powerful strategy for identifying specific genetic factors that may play dominant roles in cognitive processes. A "candidate gene" approach, which investigates the relation between a brain functional phenotype defined by imaging and a specific allele, is one example. Egan and his colleagues have recently shown that the met allele of brain derived neurotrophic factor (BDNF) is associated with poorer episodic memory and abnormal hippocampal activation by fMRI.[53]

LINKING fMRI TO OTHER MRI TECHNIQUES FOR CHARACTERISING PATHOLOGY AND OTHER DIRECTIONS FOR THE FUTURE

fMRI can localise brain functions well, allowing eloquent brain areas to be defined, characterising the reorganisation of patterns of brain activation as a consequence of disease or injury, and potentially identifying differences in brain function between subjects associated with disease susceptibility or other factors causing variation. There are limitations of the technique that remain important to overcome. The specificity of the information provided could be improved, as the BOLD signal includes contributions from multiple factors. It is also a *relative* measure of activity. The genesis of differences between groups of subjects may thus be complex to interpret.

There are several ways in which the technique is currently being advanced. More quantitative approaches to functional mapping are being applied. One promising area is the development of non-invasive methods for direct perfusion measurement, such as arterial spin labelling (ASL).[54] While the sensitivity of these techniques remains low relative to BOLD imaging, in some applications ASL may provide critical adjunctive information — for example, by providing an *absolute* measure baseline blow flow changes against which the task associated changes of BOLD imaging can be calibrated. An entirely different strategy that relies on detecting perturbations of the MRI signal by the very tiny magnetic fields generated by oriented coherently depolarising neuronal aggregates is being explored.[55]

A broader range of methods for characterisation of brain pathology can be combined with fMRI in order to interpret the heterogeneity in responses across patient populations. The importance of this has already been emphasised by various studies[31,44–46] in which differences in brain activation have been demonstrated with variations in the extent of pathological changes, rather than simply in association with a disease (Fig. 5).

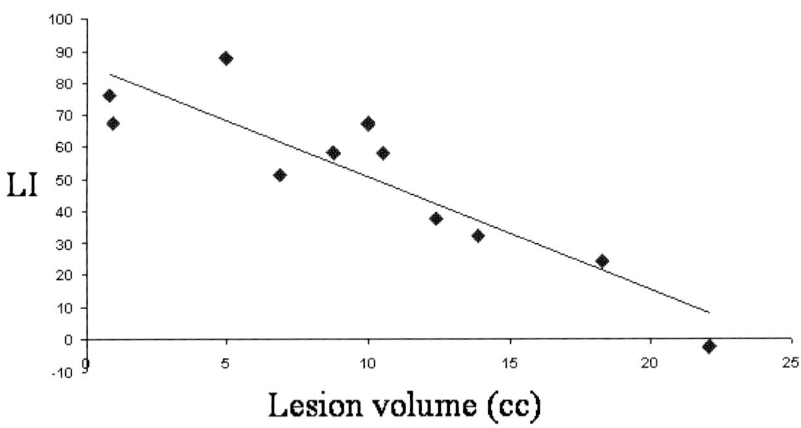

Fig. 5 Patients with multiple sclerosis show greater activation in the motor cortex ipsilateral to the hand moved than do healthy controls. Here activation in the contralateral relative to the ipsilateral hemisphere is expressed as a lateralisation index (LI). LI decreases (that is, activation becomes more bihemispheric) as the lesion load increases.

New techniques may allow much more precise characterisation of neuroanatomical relations with functional changes. The potentially important distinction between adaptive reorganisation and compensatory recruitment could be addressed by defining whether brain regions previously uninvolved in a process are recruited. This ideally would demand cytoarchitectonic characterisation of associated grey matter regions. Recent advances in diffusion tractography have allowed regional connectivity patterns to be used to map local cortical or local grey matter structures.[56] These correspond closely to conventional cytoarchitonic maps, but are obtained entirely non-invasively. Diffusion tractography also provides direct information on axon tracts, constraining solutions to problems of functional connectivities between brain areas (Fig. 6). The approaches available are thus continuing to evolve.

For the clinician, the accessibility of MRI promises the freedom to exploit these new methods rapidly. Already many centres are using fMRI as an adjunct to neurosurgical planning. Functional measures may ultimately be incorporated as surrogate markers of responses in drug trials,[57] in triage

A **B**

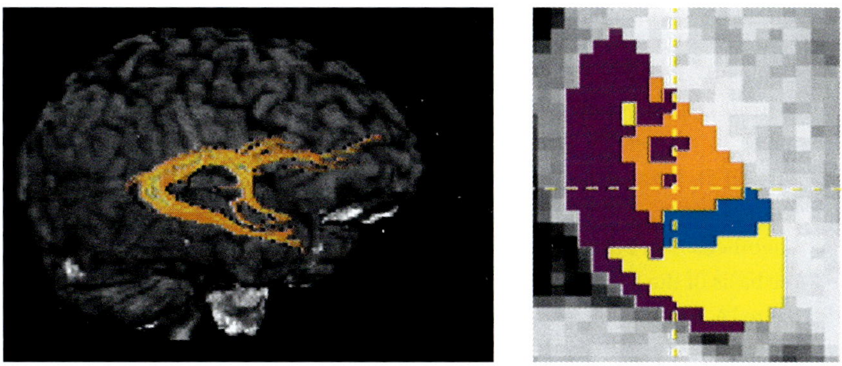

Fig. 6 (A) A probabilistic diffusion tractography path is defined between a voxel in the medial dorsal thalamus and prefrontal and temporal cortex. (B) Local patterns of connectivity in the thalamus can be used to define structures corresponding well to cytoarchitectonically defined nuclei. The axial view of the thalamus shows clusters of common projections to prefrontal (purple), motor, and premotor (orange), somatosensory (blue), and occipital/parietal cortex (yellow) that correspond well with medial dorsal and anterior, ventral lateral, ventral posterolateral nuclei, and the pulvinar, respectively.

for selection or assessment of neurorehabilitation procedures,[48] or used to complement neuropsychological measures in cases of suspected early cognitive impairment.[49] First steps towards understanding how fMRI might be used to define the prognosis for recovery after stroke or other brain injury at an early stage have already been taken.[58] Brain functional characterisation of non-organic impairments may make more confident diagnosis possible.[59] Recent work has shown how vulnerability traits for mental illness or atypical responses to psychoactive drugs can be identified.[50,60] The scope of possibilities is as broad as the range of questions that can be asked. It is a field that will deserve watching for some time to come!

ACKNOWLEDGEMENTS

We acknowledge personal and research support from the Medical Research Council, as well as our many excellent colleagues (particularly Dr S Smith) who have so generously shared their expertise and data. PMM also thanks the MS Society of Great Britain and Northern Ireland.

REFERENCES

1. Jezzard P, Matthews PM, Smith SM. *Functional MRI: an introduction to methods.* Oxford: Oxford University Press, 2001.
2. Attwell D, Laughlin SB. An energy budget for signaling in the grey matter of the brain. *J Cereb Blood Flow Metab* 2001; **21**:1133–1145.
3. Attwell D, Iadecola C. The neural basis of functional brain imaging signals. *Trends Neurosci* 2002; **25**:621–625.
4. Logothetis NK, Pauls J, Augath M, *et al.* Neurophysiological investigation of the basis of the fMRI signal. *Nature* 2001; **412**:150–157.
5. Detre JA, Wang J. Technical aspects and utility of fMRI using BOLD and ASL. *Clin Neurophysiol* 2002; **113**:621–634.
6. Buerk DG, Ances BM, Greenberg JH, *et al.* Temporal dynamics of brain tissue nitric oxide during functional forepaw stimulation in rats. *Neuroimage* 2003; **18**:1–9.
7. St Lawrence KS, Ye FQ, Lewis BK, *et al.* Measuring the effects of indomethacin on changes in cerebral oxidative metabolism and cerebral blood flow during sensorimotor activation. *Magn Reson Med* 2003; **50**:99–106.
8. Faraci FM, Brian JE. Nitric oxide and the cerebral circulation. *Stroke* 1994; **25**: 692–703.

9. Pauling L, Coryell C. The magnetic properties and structure of hemoglobin, oxyhemoglobin and carbon monoxyhemoglobin. *Proc Natl Acad Sci USA* 1936; **22**:210–216.
10. Ogawa S, Lee TM, Kay AR, *et al.* Brain magnetic resonance imaging with contrast dependent on blood oxygenation. *Proc Natl Acad Sci USA* 1990; **87**:9868–9872.
11. Turner R, Le Bihan D, Moonen CT, *et al.* Echo-planar time course MRI of cat brain oxygenation changes. *Magn Reson Med* 1991; **22**:159–166.
12. Ogawa S, Menon RS, Tank DW, *et al.* Functional brain mapping by blood oxygenation level-dependent contrast magnetic resonance imaging. A comparison of signal characteristics with a biophysical model. *Biophys J* 1993; **64**: 803–812.
13. Kennan RP, Zhong J, Gore JC. Intravascular susceptibility contrast mechanisms in tissues. *Magn Reson Med* 1994; **31**:9–21.
14. Jezzard P, Ramsey NP. Quantitative functional MRI. In: Tofts P, ed. *Quantitative Magnetic Resonance in the Brain: Monitoring Disease Progression and Treatment Response.* New York: John Wiley, 2003:413–453.
15. Boxerman JL, Bandettini PA, Kwong KK, *et al.* The intravascular contribution to fMRI signal change: Monte Carlo modeling and diffusion-weighted studies in vivo. *Magn Reson Med* 1995; **34**:4–10.
16. Buxton RB, Wong EC, Frank LR. Dynamics of blood flow and oxygenation changes during brain activation: the balloon model. *Magn Reson Med* 1998; **39**:855–864.
17. Duong TQ, Kim DS, Ugurbil K, *et al.* Spatiotemporal dynamics of the BOLD fMRI signals: toward mapping submillimeter cortical columns using the early negative response. *Magn Reson Med* 2000; **44**:231–242.
18. Mansfield P. Multi-planar image formation using NMR spin echoes. *J Phys Chem* 1977; **10**:L55–L58.
19. Friston KJ, Zarahn E, Josephs O, *et al.* Stochastic designs in event-related fMRI. *Neuroimage* 1999; **10**:607–619.
20. Buckner RL. Event-related fMRI and the hemodynamic response. *Hum Brain Mapp* 1998; **6**:373–377.
21. Sereno MI, Dale AM, Reppas JB, *et al.* Borders of multiple visual areas in humans revealed by functional magnetic resonance imaging. *Science* 1995; **268**:889–893.
22. Cohen MS, DuBois RM. Stability, repeatability, and the expression of signal magnitude in functional magnetic resonance imaging. *J Magn Reson Imaging* 1999; **10**:33–40.
23. Nybakken GE, Quigley MA, Moritz CH, *et al.* Test-retest precision of functional magnetic resonance imaging processed with independent component analysis. *Neuroradiology* 2002; **44**:403–406.

24. Adcock JE, Wise RG, Oxbury JM, *et al.* Quantitative fMRI assessment of the differences in lateralization of language-related brain activation in patients with temporal lobe epilepsy. *Neuroimage* 2003; **18**:423–438.
25. Fox PT, Parsons LM, Lancaster JL. Beyond the single study: function/location metanalysis in cognitive neuroimaging. *Curr Opin Neurobiol* 1998; **8**: 178–187.
26. Neumann J, Lohmann G, Zysset S, *et al.* Within-subject variability of BOLD response dynamics. *Neuroimage* 2003; **19**:784–796.
27. Brett M, Johnsrude IS, Owen AM. The problem of functional localization in the human brain. *Nat Rev Neurosci* 2002; **3**:243–249.
28. Bonmassar G, Anami K, Ives J, *et al.* Visual evoked potential (VEP) measured by simultaneous 64-channel EEG and 3T fMRI. *Neuroreport* 1999; **10**:1893–1897.
29. Sunaert S, Yousry TA. Clinical applications of functional magnetic resonance imaging. *Neuroimaging Clin N Am* 2001; **11**:221–236, viii.
30. Stippich C, Freitag P, Kassubek J, *et al.* Motor, somatosensory and auditory cortex localization by fMRI and MEG. *Neuroreport* 1998; **9**:1953–1957.
31. Lee M, Reddy H, Johansen-Berg H, *et al.* The motor cortex shows adaptive functional changes to brain injury from multiple sclerosis. *Ann Neurol* 2000; **47**: 606–613.
32. Pineiro R, Pendlebury S, Johansen-Berg H, *et al.* Functional MRI detects posterior shifts in primary sensorimotor cortex activation after stroke: evidence of local adaptive reorganization? *Stroke* 2001; **32**:1134–1139.
33. Matthews PM, Adcock J, Chen Y, *et al.* Towards understanding language organisation in the brain using fMRI. *Hum Brain Mapp* 2003; **18**:239–247.
34. Lemieux L, Krakow K, Fish DR. Comparison of spike-triggered functional MRI BOLD activation and EEG dipole model localization. *Neuroimage* 2001; **14**: 1097–1104.
35. Lemieux L, Salek-Haddadi A, Josephs O, *et al.* Event-related fMRI with simultaneous and continuous EEG: description of the method and initial case report. *Neuroimage* 2001; **14**:780–787.
36. Holloway V, Gadian DG, Vargha-Khadem F, *et al.* The reorganization of sensorimotor function in children after hemispherectomy. A functional MRI and somatosensory evoked potential study. *Brain* 2000; **123**:2432–2444.
37. Cao Y, D'Olhaberriague L, Vikingstad EM, *et al.* Pilot study of functional MRI to assess cerebral activation of motor function after poststroke hemiparesis. *Stroke* 1998; **29**:112–122.
38. Cramer SC, Nelles G, Benson RR, *et al.* A functional MRI study of subjects recovered from hemiparetic stroke. *Stroke* 1997; **28**:2518–2527.
39. Reddy H, Narayanan S, Matthews PM, *et al.* Relating axonal injury to functional recovery in MS. *Neurology* 2000; **54**:236–239.

40. Reddy H, De Stefano N, Mortilla M, *et al.* Functional reorganization of motor cortex increases with greater axonal injury from CADASIL. *Stroke* 2002; **33**: 502–508.
41. Weiller C, Ramsay SC, Wise RJ, *et al.* Individual patterns of functional reorganization in the human cerebral cortex after capsular infarction. *Ann Neurol* 1993; **33**:181–189.
42. Pujol J, Roset-Llobet J, Rosines-Cubells D, *et al.* Brain cortical activation during guitar-induced hand dystonia studied by functional MRI. *Neuroimage* 2000; **12**:257–267.
43. Johansen-Berg H, Rushworth MF, Bogdanovic MD, *et al.* The role of ipsilateral premotor cortex in hand movement after stroke. *Proc Natl Acad Sci USA* 2002; **99**:14518–14523.
44. Reddy H, Narayanan S, Arnoutelis R, *et al.* Evidence for adaptive functional changes in the cerebral cortex with axonal injury from multiple sclerosis. *Brain* 2000; **123**:2314–2320.
45. Rocca MA, Matthews PM, Caputo D, *et al.* Evidence for widespread movement-associated functional MRI changes in patients with PPMS. *Neurology* 2002; **58**:866–872.
46. Filippi M, Bozzali M, Rovaris M, *et al.* Evidence for widespread axonal damage at the earliest clinical stage of multiple sclerosis. *Brain* 2003; **126**:433–437.
47. Karni A, Meyer G, Jezzard P, *et al.* Functional MRI evidence for adult motor cortex plasticity during motor skill learning. *Nature* 1995; **377**:155–158.
48. Johansen-Berg H, Dawes H, Guy C, *et al.* Correlation between motor improvements and altered fMRI activity after rehabilitative therapy. *Brain* 2002; **125**:2731–2742.
49. Bookheimer SY, Strojwas MH, Cohen MS, *et al.* Patterns of brain activation in people at risk for Alzheimer's disease. *N Engl J Med* 2000; **343**:450–456.
50. Smith KA, Ploghaus A, Cowen PJ, *et al.* Cerebellar responses during anticipation of noxious stimuli in subjects recovered from depression. Functional magnetic resonance imaging study. *Br J Psychiatry* 2002; **181**:411–415.
51. Tracey I, Ploghaus A, Gati JS, *et al.* Imaging attentional modulation of pain in the periaqueductal gray in humans. *J Neurosci* 2002; **22**:2748–2752.
52. Petrovic P, Kalso E, Petersson KM, *et al.* Placebo and opioid analgesia — imaging a shared neuronal network. *Science* 2002; **295**:1737–1740.
53. Egan MF, Kojima M, Callicott JH, *et al.* The BDNF val66met polymorphism affects activity-dependent secretion of BDNF and human memory and hippocampal function. *Cell* 2003; **112**:257–269.
54. Wong EC, Buxton RB, Frank LR. Implementation of quantitative perfusion imaging techniques for functional brain mapping using pulsed arterial spin labeling. *NMR Biomed* 1997; **10**:237–249.

55. Bodurka J, Bandettini PA. Toward direct mapping of neuronal activity: MRI detection of ultraweak, transient magnetic field changes. *Magn Reson Med* 2002; **47**:1052–1058.
56. Behrens TE, Johansen-Berg H, Woolrich MW, *et al.* Non-invasive mapping of connections between human thalamus and cortex using diffusion imaging. *Nat Neurosci* 2003; **6**:750–757.
57. Parry A, Scott RB, Palace J, *et al.* Potentially adaptive functional changes of cognitive processing in patients with multiple sclerosis and their acute modulation by rivastigmine. *Brain* (in press).
58. Ward NS, Brown MM, Thompson AJ, *et al.* Neural correlates of outcome after stroke: a cross-sectional fMRI study. *Brain* 2003; **126**:1430–1448.
59. Marshall JC, Halligan PW, Fink GR, *et al.* The functional anatomy of a hysterical paralysis. *Cognition* 1997; **64**:B1–B8.
60. Egan MF, Goldberg TE, Kolachana BS, *et al.* Effect of COMT Val108/158 Met genotype on frontal lobe function and risk for schizophrenia. *Proc Natl Acad Sci USA* 2001; **98**:6917–6922.

16

How to Spot Bias and Other Potential Problems in Randomised Controlled Trials

S C Lewis and C P Warlow*

These days, all clinical trials should be reported using the CONSORT guidelines[1] (Table 1); indeed *JNNP* recommends this in its instructions for authors. However, not all trials are reported in this way, and many journals do not insist on it. Thus some trials may have been carried out adequately but reported inadequately, while others have been carried out inadequately. Our aim in this article is to guide clinicians in what to look for in a report of a randomised controlled trial (RCT), so they can assess whether the trial was done adequately; we do not intend it to be a guide on how to do an RCT, as there are many such guides available.[2]

The two crucial principles in clinical research are to minimise bias and to increase precision. If a study is not designed with these two principles in mind, no amount of analysis will sort them out. We will discuss some of the major biases to look out for, issues related to precision, and some other aspects of statistical analysis.

*Dr Steff C Lewis, Neurosciences Trials Unit, Division of Clinical Neurosciences, Western General Hospital, Crewe Road, Edinburgh EH4 2XU, UK; steff.lewis@ed.ac.uk

Table 1 Checklist of items to include when reporting a randomised trial (from the CONSORT statement).

Table width = E

Paper section and topic		Description
Title and abstract		How participants were allocated to interventions (for example, "random allocation," "randomised," or "randomly assigned")
Introduction	Background	Scientific background and explanation of rationale
Methods	Participants	Eligibility criteria for participants and the settings and locations where the data were collected
	Interventions	Precise details of the interventions intended for each group and how and when they were actually administered
	Objectives	Specific objectives and hypotheses
	Outcomes	Clearly defined primary and secondary outcome measures and, when applicable, any methods used to enhance the quality of measurements (for example, multiple observations, training of assessors)
	Sample size	How sample size was determined and, when applicable, explanation of any interim analyses and stopping rules
	Randomisation — sequence generation	Method used to generate the random allocation sequence, including details of any restriction (for example, blocking, stratification)
	Randomisation — allocation concealment	Method used to implement the random allocation sequence (for example, numbered containers or central telephone), clarifying whether the sequence was concealed until interventions were assigned
	Randomisation — implementation	Who generated the allocation sequence, who enrolled participants, and who assigned participants to their groups
	Blinding (masking)	Whether or not participants, those given the interventions, and those assessing the outcomes were blinded to group assignment. When relevant, how the success of blinding was evaluated

Table 1 (*Continued*)

Paper section and topic		Description
	Statistical methods	Statistical methods used to compare groups for primary outcome(s); methods for additional analyses, such as subgroup analyses and adjusted analyses
Results	Participant flow	Flow of participants through each stage (a diagram is strongly recommended). Specifically, for each group report the numbers of participants randomly assigned, receiving intended treatment, completing the study protocol, and analysed for the primary outcome. Describe protocol deviations from study as planned, together with reasons
	Recruitment	Dates defining the periods of recruitment and follow up
	Baseline data	Baseline demographic and clinical characteristics of each group
	Numbers analysed	Number of participants (denominator) in each group included in each analysis and whether the analysis was by "intention to treat" . State the results in absolute numbers when feasible (for example, 10/20, not 50%)
	Outcomes and estimation	For each primary and secondary outcome, a summary of results for each group, and the estimated effect size and its precision (for example, 95% confidence interval)
	Ancillary analyses	Address multiplicity by reporting any other analyses undertaken, including subgroup analyses and adjusted analyses, indicating those prespecified and those exploratory
	Adverse events	All important adverse events or side effects in each intervention group
Discussion	Interpretation	Interpretation of the results, taking into account study hypotheses, sources of potential bias or imprecision, and the dangers associated with multiplicity of analyses and outcomes
	Generalisability	Generalisability (external validity) of the trial findings
	Overall evidence	General interpretation of the results in the context of current evidence

BIAS

Bias is any departure of results from the truth. An RCT is less susceptible to bias than other study designs for assessing therapeutic interventions. However, just because a study is randomised does not mean it is unbiased. There are at least seven important potential sources of bias in RCTs, which are discussed below. When assessing bias, it is important to consider its magnitude as well as its direction. Trials that have shown large treatment effects may still be positive after a small bias has been accounted for.

Poor Allocation Concealment

In a good trial, the treatment allocation is concealed during the randomisation procedure. In other words, at the time a clinician randomises a patient they will have no idea what the next treatment allocation is going to be. If allocation is concealed, it is not possible for a clinician to avoid a particular treatment allocation for a particular patient. For example, consider an RCT of surgical support stockings versus no treatment to prevent deep venous thrombosis after stroke. Say an incontinent patient arrives and the nurse is considering randomising. The nurse has access to the randomisation list and knows that the next random allocation is "stockings". Stockings on incontinent patients are a lot of work, as they need regular changing and washing, so the nurse chooses not to randomise the patient. Because incontinence is linked to stroke severity, in the long run this practice would cause the "stockings" arm of the trial to contain less severe strokes than the "no stockings" arm, which could bias the results of the trial, even though treatment allocation was randomised.

Good methods of allocation concealment include sequentially numbered, opaque, sealed envelopes; tamper-proof, sequentially numbered containers; pharmacy controlled lists; and telephone, fax, email, or internet contact with a central randomisation office.[3,4]

Imbalance in Baseline Prognostic Variables

In all trials, methods should be used to make sure that the treatment groups are as similar as possible. For example, there has been controversy over the results of the National Institute of Neurological Disorders and Stroke (NINDS) trial of thrombolysis for acute ischaemic stroke[5] because the patients in the recombinant tissue plasminogen activator (rt-PA) group had less severe strokes than those in the control group,[6] even though they were randomised. As less severe patients would be expected to have better outcomes than more severe patients, the trial results may have been biased in favour of rt-PA. The trialists adjusted for the imbalance in baseline severity in the analysis, but if the treatment groups had been comparable to start with, the arguments would not have arisen.

In a very large trial, randomisation should ensure that the treatment groups are balanced, but in small trials, imbalance can and does occur. Thus in smaller trials stratification is often used to increase the comparability of the treatment groups. Stratification ensures that roughly equal numbers of participants with a particular prognostic characteristic (perhaps age, or disease severity) will be allocated to each treatment group. It involves using separate randomisation lists for each prognostic subgroup (for example, for age <80 and age ≥ 80).[7] It has been recommended[8] that trials seeking to demonstrate the superiority of one treatment over another should consider stratifying randomisation when the overall sample size is small (for example, <200 patients per treatment arm for a dichotomous outcome), or when interim analyses are planned that will involve small sample sizes (stratification is recommended in all equivalence trials). The stratification factors must be strongly related to outcome. Thus in a trial of a treatment for acute stroke, one would stratify for stroke severity (which is strongly related to outcome), but not for sex (which is only weakly associated with outcome). In practice, it is probably more important that the reader can see that the treatment arms are balanced with respect to important prognostic baseline factors than to know the details of how this was achieved, although it is generally recommended that the stratification variables are presented in trial reports.[1]

Unblinding and no Blinding

If anyone involved in a trial is aware of the allocated treatment, this may affect their judgement. In the Canadian cooperative trial of cyclophosphamide and plasma exchange in multiple sclerosis, neither of the active treatment groups was shown to be superior to placebo when the outcomes were blindly assessed by neurologists.[9] However, in unblinded outcome assessment by neurologists there was an apparent treatment effect in one of the treatment groups. Trials can be designed so that the patient, the treatment team, the outcome assessor, and even the trial statistician and any data monitoring committee are all blinded to the allocated treatment. However, it may be impossible to blind the administering clinician to treatment allocation, particularly in trials of interventions such as surgery or physiotherapy. This can even be difficult in some placebo controlled drug trials; for instance, intravenous rt-PA often causes bruising at the injection site. Probably the most important thing is for the person who assesses the primary outcome to be blinded to treatment allocation.

In general, the more blinding that is achieved, the less biased the trial results should be. It is worth noting that there is no single definition of the phrase "double blind", so trial reports should explain exactly who was blinded and how this was achieved.[10]

Missing Data

In general, the more information from randomised patients that is missing, the more wary one should be of the trial results. For example, in a trial of a drug to prevent severe depression which actually works, more patients in the placebo group would become depressed. These would be more likely to stop taking a treatment that did not seem to work, and also to default from attending follow up appointments. If the results of those patients were missing from the final analysis, it would make the placebo group results look better than they actually were; at worst, it might then appear that the treatment was not working at all, or at best that it was working less well than was really the case. In the presence of much missing data, one never really knows what the true treatment effect is.

There are some circumstances when the exclusion of patients does not bias the results.[11] For instance, it is allowable to exclude the data of a few ineligible patients who were mistakenly randomised into a trial because of human error. However, one must be sure that treatment is not potentially harmful for the ineligible patients, so it would not be appropriate to exclude patients with primary intracerebral haemorrhage who were inadvertently randomised into a trial of thrombolytic therapy. One must also be sure that the study results are not applied to the ineligible patients. For instance, if an acute stroke treatment was to be given before computed tomography was done, then a few people with brain tumours would receive the treatment. Thus it would not be appropriate to exclude such patients from the analysis of a trial in which they had been inadvertently randomised.

Although excluding patients from the analysis in some circumstances does not bias the results, if many patients were excluded from a trial, one should question the quality of the trial design and execution. It should certainly be clearly described why any data are missing, and what effect this may have had on the results.

One way that trials can minimise the problem of missing patient information is to use central randomisation and follow up. For example, the FOOD trial[12] is a family of three RCTs of feeding after stroke that uses this system. Baseline data are collected during a randomisation phone call before the randomisation actually happens, and so are 100% complete. The central office follows up all randomised patients, and in February 2001, of 3012 patients randomised, only 10 had permanently missing primary outcome data.

Lack of Intention to Treat Analysis

Intention to treat means that patients are analysed in the treatment group they were randomised to, whatever happens later. Some trials analyse the data using an on-treatment analysis where patients are only analysed if they received the treatment they were randomised to. An intention to treat analysis preserves the randomisation process. It has the advantage of being more like standard clinical practice (where patients will start on other treatments if the first treatment they are given does not agree with them, or they may choose not to take any treatment at all). It also takes care of unexpected adverse events (patients cannot "drop out" of the trial analysis if they have an adverse effect of treatment) and is less open to fraud (the trialists cannot exclude any patients who did not achieve the hoped-for outcome).

Figure 1 shows an example of how an on-treatment analysis can cause bias in practice. Patients with carotid transient ischaemic attacks are randomised to receive either carotid surgery or no surgery. If there is a delay of a few days between randomisation and surgery, patients may have a primary outcome event (in this case a stroke) after randomisation but before the surgery is done. In the surgery group, patients who suffer a severe stroke or die will not receive surgery (and would therefore be removed from

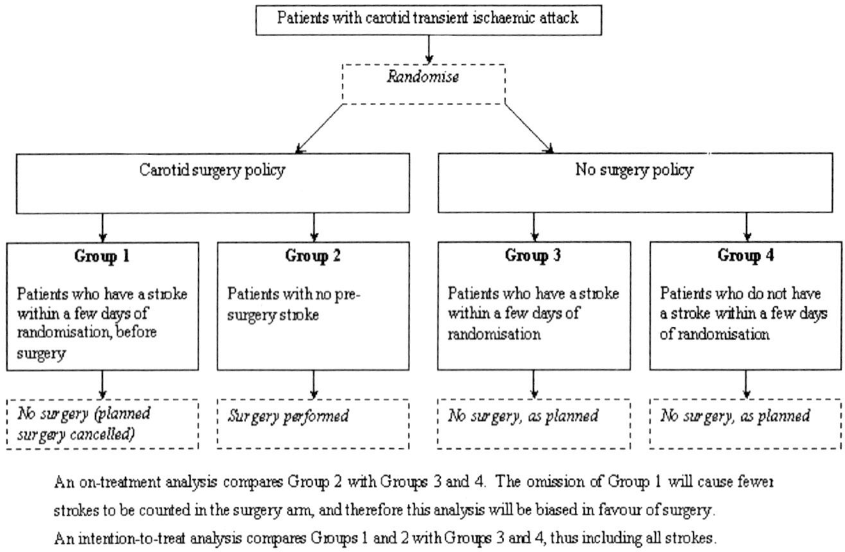

Fig. 1 An example from a carotid surgery trial showing how an intention to treat analysis is less biased than an on-treatment analysis.

the on-treatment analysis). However, in the no surgery group, if patients have a stroke or die within a few days of randomisation, they will have received their allocated treatment (no surgery) and would therefore be counted in the on-treatment analysis. The omission of early strokes from the surgery group would cause fewer strokes to be counted in the surgery arm, and therefore the on-treatment analysis would be biased in favour of surgery. The intention to treat analysis includes all patients and is therefore unbiased.

Counting Death as a Good Outcome

It is important when reading a trial report to consider how death has been analysed and what effect this may have had on the results. For instance, the trial may measure the proportion of patients who were disabled at follow up, using all patients randomised as the denominator. In this case, the trial is really comparing the proportion of patients who were alive and disabled at follow up to the proportion of patients who were not disabled at follow up, and this latter group includes both those who were alive and not disabled, and those who were dead. Thus death has been included as a good outcome. It would be more sensible to measure the proportion of patients who were alive and not disabled at follow up.[13]

Competing Interests

It has been shown that research funded by pharmaceutical companies is more likely to have outcomes favouring the sponsor than research funded from other sources.[14,15] Pharmaceutical company research is certainly not of poorer quality than other research, but the companies may have a tendency not to publish unfavourable results. The number of industry sponsored trials in stroke is increasing,[16] so this problem is not going to go away, but there are now guidelines on the relation between sponsors and investigators that may improve the situation.[17]

PRECISION

A treatment effect estimate is precise when the confidence interval around it is very tight, and we are therefore confident about its magnitude.

Size of Trials

Larger trials provide more precise estimates of treatment effects than small trials, and they may allow a few sensible and predefined subgroup analyses.[18] Small trials, with wide confidence intervals around their estimates of treatment effect, are clinically uninformative (although they may add to an existing meta-analysis or generate enthusiasm to do further bigger trials).

Trials that use continuous outcome measures (for example, blood pressure or time taken to walk 10 metres) generally have greater precision than trials of the same size that use binary outcome measures (for example, dead versus alive, hypertensive versus not). However, with continuous outcome measures there may be more of a problem with missing data, and it may be unclear how to score dead patients.

RCTs that measure the outcomes that really matter to patients (such as death or dependency) often require large numbers of patients to be followed up for long periods. To reduce the size of the trial, some trialists use surrogate outcomes.[19,20] For instance, to investigate whether a neuroprotective drug reduces death and dependency after stroke, one would probably have to randomise thousands of patients, which would take several years. If one used the size of the infarct on magnetic resonance imaging 48 hours post-treatment as a surrogate marker for efficacy, one could substantially reduce the size and duration of the trial. This would mean that a new treatment could be proved efficacious and licensed much earlier, and so benefit many more patients. However, such surrogate outcomes often do not prove to be effective substitutes for the true clinical outcome.[19] One of the reasons for this failure may be that the surrogate marker and the clinical outcome are on different causal pathways. Alternatively, several independent processes may cause the disease, only one of which involves the surrogate marker.[20] For instance, the early "inflammatory" stage of multiple sclerosis may be detected on magnetic resonance imaging, but this does not necessarily relate to progression of the disease or later disability. Thus if a trial used this evidence from MRI as a surrogate marker of clinical outcome in multiple sclerosis, it would only show the effects of treatment on the early inflammatory lesions.

Power Versus Confidence Intervals

The concept of power is very important when designing a study. Assume that a drug reduces the absolute risk of having a stroke by 10%. If a trial to measure this treatment effect was repeated over and over, it would sometimes estimate it to be greater than 10% and sometimes less than 10%. In some of the trials, the estimated treatment effect would be so small that the result would be statistically non-significant (as the confidence interval for the estimate of the treatment effect would overlap "no effect"). If the trial is designed to have 80% power, then, if the treatment effect truly exists, if the trial was repeated 100 times a statistically significant treatment effect would be found in 80 of them. So one in every five trials would falsely show a non-statistically significant result.

However, although this is an essential concept when designing a trial, once a trial has been completed it is more important to concentrate on the width of the confidence interval around the treatment effect than on the power itself.[21] When the study was designed, the power calculation was based on a guess at what the treatment effect might be. After the study is completed, the result is known and it makes no sense to use prestudy guesses to interpret the result. The confidence interval around the treatment effect is based on the actual trial result, and this is what really counts.

Early Stopping

Some trials are planned to be large, but they end up small because they are stopped early owing to an apparently huge beneficial effect or a harmful effect. The results of such trials should be treated with some scepticism, because if they had been allowed to continue, the final estimated treatment effect may well have been much smaller.[22] In the initial stages of a trial, the treatment effect tends to zig and zag through some quite extreme values before settling down, and thus trials that stop early may just have stopped on one of the random highs or lows in the treatment effect estimate. Of course, there may be ethical reasons why a trial has to stop because of early unexpected harmful effects. However, for some treatments (such as thrombolysis for acute stroke) and for many surgical interventions, the trial data monitoring committee should have carefully considered the possibility that any early harm may be outweighed by later benefit.

"Absence of Evidence" and "Evidence of Absence"

Care should be taken in the interpretation of non-statistically significant results. It is quite common for investigators to confuse "absence of evidence of effectiveness" with "evidence of absence of effectiveness." Take, for example, a trial examining the effect of a drug on death or dependency after stroke. Nine of 20 patients treated with the drug are dead or dependent at follow up compared with 10 of 20 untreated patients, giving a p value of 0.8. However, although the point estimate for the absolute treatment effect is 5% (10/20 minus 9/20), the 95% confidence interval for the difference in proportions runs from −24% to +33%. It is therefore quite plausible that the treatment could cause great harm, or great benefit. Thus the conclusion of this trial is that we do not know whether the drug works, and we would need to do a larger trial to find out.

To assess "evidence of absence of effectiveness," a trial needs to be designed as an equivalence trial. RCTs cannot prove that two treatments are of identical efficacy, but they can prove that two treatments are of similar efficacy.[23] In equivalence trials, trialists need to prespecify what they mean by clinical equivalence. Usually a range of equivalence for the treatment difference is defined such that any value within the range is deemed clinically unimportant.[24] When the trial results are published, to show that two treatments are equivalent the confidence interval for the treatment effect must fall wholly within this predefined range of equivalence.[25]

ANALYSIS AND INFERENCES

Baseline Differences

In RCTs, many consider that it is not appropriate to test for differences in the level of baseline factors between treatment groups.[26,27] In an RCT with two treatment groups, such tests are testing the hypothesis that the two groups come from the same population. However, if the randomisation was fair, then the two groups will certainly have come from the same population, and one ends up testing whether the randomisation was fair, not whether the two groups had similar characteristics.[26] One should be very wary if no baseline data are presented at all.

Adjusted Versus Unadjusted Analyses

Results of RCTs can be presented either adjusted or unadjusted for any differences in baseline prognostic variables. Differences between treatment groups may bias the results (as explained earlier), and "adjustment" is any statistical method that alleviates this problem. To understand adjustment, consider the following example. Patients who suffer strokes of moderate severity generally have worse outcomes than those of mild severity. In a trial of a drug in acute stroke, let us assume the proportion of patients with mild stroke was much higher in the treated group than in the untreated group. An unadjusted analysis would overestimate the treatment effect, as the treated patients were more likely to do well before treatment than the untreated patients. However, one can calculate the treatment effect in just the mild patients, and in just the moderate patients, and then average these two (Table 2). The statistical procedures used to adjust results are often more complex than this, but they follow the same basic principle.

When reading a trial report, the unadjusted analyses are easier to understand, as sometimes the adjusted results seem to have come out of a

Table 2 Adjusted and unadjusted analyses — a simple example from a hypothetical trial of a treatment for acute stroke where, for some reason, a greater number of mild patients was randomised to the treatment being tested.

Table width = D

	All patients			**Patients with moderate stroke**			**Patients with mild stroke**		
	Treated	**Untreated**	**Total**	**Treated**	**Untreated**	**Total**	**Treated**	**Untreated**	**Total**
Dead or dependent	250	420	670	100	360	460	150	60	210
Alive and independent	750	580	1330	150	390	540	600	190	790
Total	1000	1000	2000	250	750	1000	750	250	1000

Calculating relative risks of being dead and dependent:
Overall unadjusted treatment effect is 250/1000 ÷ 420/1000 = 0.60
Treatment effect for those with moderate stroke is 100/250 ÷ 360/750 = 0.83
Treatment effect for those with mild stroke is 150/750 ÷ 60/250 = 0.83
So an overall adjusted treatment effect would be 0.83.

statistical "black box" and it is unclear exactly what has been done. However, adjusted analyses have statistical advantages in some cases. The key issue is the correlation between each baseline variable and outcome.[28] If this is high (say >0.5), then adjusting the analyses for the baseline variable is important. This might happen, for instance, if the same variable is measured at baseline and as an outcome after treatment (for example, measuring blood pressure before and after treatment in a trial of a blood pressure lowering drug). However, if the correlation between a baseline variable and outcome is low, then there is probably no point adjusting for it. These arguments apply no matter how large the trial is, and whether or not stratification for the baseline variable has been done in the randomisation process. When reading a trial report, one should be most convinced when both adjusted and unadjusted analyses are presented and agree with one another.

Subgroups and Multiple Testing

A report of an RCT will often contain at least one subgroup analysis, such as the treatment effect in young versus old patients. Although there are many guidelines on appropriate ways to do such analyses, inappropriate analyses are still frequently presented.[29]

In Fig. 2, the effect of aspirin in male and female subjects from the Canadian cooperative study group[30] is shown — a landmark trial in its time. The

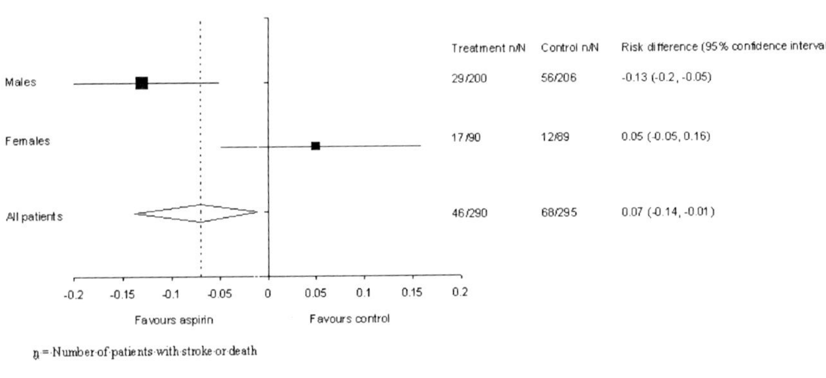

Fig. 2 The dangers of subgroup analysis: the Canadian cooperative study group 1978 — difference in risk of stroke or death in 585 patients with transient ischaemic attacks or stroke treated long term with aspirin versus no aspirin.

results in Fig. 2 are shown in a forest plot,[31] which is a convenient way of presenting subgroup analysis results. In male subjects, the effect of aspirin was statistically significant, and in female subjects it was not. However, this method of inspection of subgroup p values is incorrect. If one group has more subjects (as in this case, where there are twice as many men in the trial as women), then it is far more likely to have a significant p value, and this has nothing to do with the magnitude of the treatment effect in that group. The correct way to do a subgroup analysis is to compare the size of the treatment effect in one subgroup with the size of the treatment effect in the other.[32] In this case, the correct p value is 0.008, which is statistically significant. However, this really shows the dangers of subgroup analysis, because this one was not prespecified, was not biologically plausible, and was later proved wrong by the antiplatelet trialists' collaboration meta-analysis.[33]

If one examines the effects of treatment in 20 subgroups in an RCT, one will obtain, on average, one spurious false positive result at $p = 0.05$ just by chance. This is more likely than the chance of rolling two sixes with a pair of dice, which happens with a probability of 0.028. Sometimes spurious subgroup results can be dramatic.[34] If inappropriate subgroup analyses have been done (such as inspection of subgroup p values), the number of false positive results could increase drastically.[29] If many subgroup analyses are shown, it is quite likely that a couple of "statistically significant" subgroup effects will occur, but this does not mean that they are clinically meaningful. The ISIS 2 trial[35] showed that aspirin was less beneficial to people born under the star sign of Gemini or Libra than other star signs, which puts the results of the other subgroup analyses into perspective. If only a few subgroup analyses are shown (possibly spread over several publications from the same trial) but they are all statistically significant, one should be very cautious about their interpretation, because the investigator may well have only presented the statistically significant results, and may have hidden the hundreds of non-significant results.

The results of subgroup analyses are more believable if there are only a few subgroups, predefined using biologically plausible arguments or from an a priori hypothesis generated from previous studies, from a trial large enough to stand a good chance of finding statistically significant treatment–subgroup interactions,[27] and best of all, confirmed in a completely separate RCT. But in general, one should put most emphasis on the primary result in all patients.

The effect of false positive results from multiple testing occurs for all analyses, not just subgroup analyses. If the investigator has analysed 20 different outcomes, it is likely that one will be statistically significant just by chance.

Sometimes, a trial measures outcome at several time points, and undertakes a separate analysis at each one. Again, it is likely that one will be significant just by chance. Matthews *et al.* describe appropriate methodology for such situations.[36]

Unit of Analysis

Most of the examples discussed so far have related to standard two arm parallel group trials — that is, patients are randomised to one treatment or another, and after a period of time an outcome is measured on each patient. However, there are many other types of trial, and when reading a trial report, it is important to check whether the analysis is appropriate to the design, or whether the design is appropriate at all. For instance, crossover trials are used in some areas of medicine. In these trials each patient receives both treatments, but the order in which they receive them is randomised. Crossover trials have the advantage that each patient acts as their own control. However, they only work in chronic conditions, where the treatment cannot cure the disease, only alleviate the symptoms. Thus they might be suitable in migraine, but not in meningitis.

In most trials, the unit of randomisation and the unit of analysis are the same. Patients are randomised to treatments, and then the outcome is measured and analysed at the level of the patient — that is, the outcome measure is measured once on each patient (for example, death). However, there are trials where the unit of randomisation is not the same as the unit of analysis. In a trial of a drug to reduce the size of secondary brain tumours, the patient would be randomised, but if patients had more than one tumour, the outcome might be recorded at the level of the tumour. The report of such a trial should describe clearly that the analysis has taken this design into account. It is important that this is done, as one would expect the tumours in one patient to behave quite similarly, thus reducing the variability in the estimate of treatment effect. Similar problems occur in cluster randomised trials. These are trials where the unit of randomisation is, for instance, a general practice surgery, or a residential home, but where the outcome is measured at the level of the patient. This design must be taken into account in the analysis.[37]

Recurrent Events

In some areas of medicine, outcomes tend to be measured in terms of rates of events. For instance, in multiple sclerosis one may be interested in the number of relapses that patients have within a fixed period. Similar data are obtained from trials examining numbers of epileptic fits or headaches. These data need to be analysed using appropriate statistical methods which take account of the fact that some patients are much more likely to suffer recurrent events than others.[38] One should think carefully about what such results actually mean. In epilepsy, one of 10 patients having 100 fits is rather different from 10 of 10 patients having 10 fits each. Although the mean number of fits is 10 in both scenarios, they are clearly quite different situations.

Adverse Events

In addition to considering the efficacy of a treatment, it is essential to consider safety.[39] In the elderly, there can be more adverse effects of treatments than in the young (such as postural hypotension and fainting in blood pressure lowering trials). Even minor adverse events can be important when considered across large populations. To take an extreme example, let us assume that it was decided to give the whole population over the age of 50 a statin to lower their cholesterol levels, but that in many people this caused them to have mild numbness in the hands. Such an adverse effect might not be seen as a problem in a person who was seriously ill and very likely to suffer a heart attack in the coming year unless treated. However, in previously healthy people, the adverse effect would not be accepted, and across a whole population it could be disastrous. One should be wary of trials where no adverse effects of treatment are reported, as there are no magic bullets in medicine. One should also be wary of trials where the length of follow up has been very short, as adverse events may only arise after a longer period. Few RCTs will be big enough to detect rare adverse events, but small trials will not even be able to detect common adverse events. In order to detect adverse effects reliably, effective postmarketing surveillance in large numbers of patients, or a case–control study, is required.

The assessment of adverse effects is particularly important in equivalence trials. It is not enough to prove that two treatments are equivalent in terms of efficacy. A new treatment must be as safe as, or safer than, the old one (and if it is not significantly safer than the old treatment then it should be shown to be cheaper or more convenient).

Generalisability and Interpretation

The result of an RCT is only applicable to day to day practice if the patients included in the trial were similar to those who would be treated in practice. Thus if a treatment has only been tested in men aged under 65, it is generally impossible to know whether it will benefit a 95 year old woman. On balance, trials with broad inclusion criteria are more generalisable than those with strict criteria. By examining the inclusion and exclusion criteria, and the baseline data, clinicians should consider whether the trial sample is reasonably representative of the people they wish to treat.[40] However, more emphasis should be placed on the overall outcome of the trial than on the results for one particular subgroup within the trial.[41]

THE FUTURE

Although in the past the quality of trial reporting has been poor, it does appear to be improving.[16,42] The use of the CONSORT guidelines[1] by more journals will increase quality further. In the future, we can hope to be more able to believe what we read.

CONCLUSIONS

Allocation should be concealed during randomisation. Outcomes should be defined carefully. Outcome assessors should be blind to treatment allocation. It should be clear why missing data are missing. Analyses should be on the basis of intention to treat. It is worth bearing in mind that even if a trial is reported as "intention to treat" and that allocation is concealed, it does not mean that these statements are true. It is important to read a trial report carefully to make sure that the authors knew the precise definitions of these terms.

One should be particularly cautious not to overinterpret the results of subgroup analyses. We all want to find treatments that work, and we can get very excited when we think we have found something, but we need to remember that some things are just too good to be true.

Website of interest: CONSORT: http://www.consort-statement.org/

ACKNOWLEDGEMENTS

We would like to thank Jon Stone and Rustam Al Shahi for helpful comments on an earlier draft.

REFERENCES

1. Moher D, Schulz KF, Altman D. The CONSORT statement: revised recommendations for improving the quality of parallel-group randomized trials. *JAMA* 2001; **285**:1987–1991.
2. Pocock SJ. *Clinical Trials — A Practical Approach*. Chichester: John Wiley, 1983.
3. Altman DG, Schulz KF. Concealing treatment allocation in randomised trials. *BMJ* 2001; **323**:446–447.
4. Schulz KF, Grimes DA. Allocation concealment in randomised trials: defending against deciphering. *Lancet* 2002; **359**:614–618.
5. National Institute of Neurological Disorders and Stroke rt-PA stroke study group. Tissue plasminogen activator for acute ischemic stroke. *N Engl J Med* 1995; **333**:1581–1587.
6. Mann J. Truths about the NINDS study: setting the record straight. *West J Med* 2002; **176**:192–194.
7. Roberts C, Torgerson D. Randomisation methods in controlled trials. *BMJ* 1998; **317**:1301.
8. Kernan WN, Viscoli CM, Makuch RW, *et al*. Stratified randomisation for clinical trials. *J Clin Epidemiol* 1999; **52**:19–26.
9. Noseworthy JH, Ebers GC, Vandervoort MK, *et al*. The impact of blinding on the results of a randomised, placebo-controlled multiple sclerosis clinical trial. *Neurology* 1994; **44**:16–20.
10. Schulz KF, Chalmers I, Altman DG. The landscape and lexicon of blinding in randomized trials. *Ann Intern Med* 2002; **136**:254–259.
11. Fergusson D, Aaron SD, Guyatt G, *et al*. Post-randomisation exclusions: the intention to treat principle and excluding patients from analysis. *BMJ* 2002; **325**:652–654.
12. The FOOD trial collaboration. Performance of a statistical model to predict stroke outcome in the context of a large, simple, randomized, controlled trial of feeding. *Stroke* 2003; **34**:127–133.

13. Lubsen J, Kirwan BA. Combined endpoints: can we use them? *Stat Med* 2002; **21**:2959–2970.
14. Bekelman JE, Li Y, Gross CP. Scope and impact of financial conflicts of interest in biomedical research — a systematic review. *JAMA* 2003; **289**:454–465.
15. Lexchin J, Bero LA, Djulbegovic B, *et al*. Pharmaceutical industry sponsorship and research outcome and quality: systematic review. *BMJ* 2003; **326**:1167–1170.
16. Kidwell CS, Liebeskind DS, Starkman S, *et al*. Trends in acute ischaemic stroke trials through the 20th century. *Stroke* 2001; **32**:1349–1359.
17. Donnan GA, Davis SM, Kaste M, for the international trial subcommittee of the international stroke liaison committee, American Stroke Association. Recommendations for the relationship between sponsors and investigators in the design and conduct of clinical stroke trials. *Stroke* 2003; **34**:1041–1045.
18. Warlow C. Advanced issues in the design and conduct of randomized clinical trials: the bigger the better? *Stat Med* 2002; **21**:2797–2805.
19. Fleming TR, DeMets DL. Surrogate endpoints in clinical trials: are we being misled? *Ann Intern Med* 1996; **125**:605–613.
20. Bucher HC, Guyatt GH, Cook DJ, *et al*. For the evidence-based medicine working group. Users' guides to the medical literature, XIX. Applying clinical trial results. A. How to use an article measuring the effect of an intervention on surrogate end points. *JAMA* 1999; **282**:771–778.
21. Goodman SN, Berlin JA. The use of predicted confidence intervals when planning experiments and the misuse of power when interpreting results. *Ann Intern Med* 1994; **121**:200–206.
22. Wheatley K, Clayton D. Be skeptical about unexpected large apparent treatment effects: the case of an MRC AML12 randomization. *Control Clin Trials* 2003; **24**:66–70.
23. Siegel JP. Equivalence and noninferiority trials. *Am Heart J* 2000; **139**:S160–S170.
24. Ebbutt AF, Frith L. Practical issues in equivalence trials. *Stat Med* 1998; **17**: 1691–1701.
25. Jones B, Jarvis P, Lewis JA, *et al*. Trials to assess equivalence: the importance of rigorous methods. *BMJ* 1996; **313**:36–39.
26. Altman DG, Doré CJ. Randomisation and baseline comparisons in clinical trials. *Lancet* 1990; **335**:149–153.
27. Assmann SF, Pocock SJ, Enos LE, *et al*. Subgroup analysis and other (mis)uses of baseline data in clinical trials. *Lancet* 2000; **355**:1064–1069.
28. Pocock SJ, Assmann SE, Enos LE, *et al*. Subgroup analysis, covariate adjustment and baseline comparisons in clinical trial reporting: current practice and problems. *Stat Med* 2002; **21**:2917–2930.

29. Brookes ST, Whitley E, Peters TJ, *et al.* Subgroup analyses in randomised controlled trials: quantifying the risks of false-positives and false-negatives. *Health Technol Assess* 2001; **5**:33.
30. The Canadian Cooperative Study Group. A randomized trial of aspirin and sulfinpyrazone in threatened stroke. *N Engl J Med* 1978; **299**:53–59.
31. Lewis S, Clarke M. Forest plots: trying to see the wood and the trees. *BMJ* 2001; **322**:1479–1480.
32. Matthews JNS, Altman DG. Statistics notes: Interaction 3: how to examine heterogeneity. *BMJ* 1996; **313**:862.
33. Antiplatelet Trialists' Collaboration. Collaborative overview of randomised trials of antiplatelet therapy. I: Prevention of death, myocardial infarction, and stroke by prolonged antiplatelet therapy in various categories of patients. *BMJ* 1994; **308**:81–106.
34. Clarke M, Halsey J. DICE 2: a further investigation of the effects of chance in life, death and subgroup analyses. *Int J Clin Pract* 2001; **55**:240–242.
35. ISIS-2 collaborative group. Randomised trial of intravenous streptokinase, oral aspirin, both, or neither among 17187 cases of suspected acute myocardial infarction: ISIS-2. *Lancet* 1988; **ii**:349–360.
36. Matthews JNS, Altman DG, Campbell MJ, *et al.* Analysis of serial measurements in medical research.*BMJ* 1990; **300**:230–235.
37. Kerry SM, Bland JM. Trials which randomize practices I: how should they be analysed? *Fam Pract* 1998; **15**:80–83.
38. Glynn RJ, Buring JE. Ways of measuring rates of recurrent events. *BMJ* 1996; **312**:364–367.
39. Cuervo LG, Clarke M. Balancing benefits and harms in health care — we need to get better evidence about harms. *BMJ* 2003; **327**:65–66.
40. Schulz KF, Grimes DA. Sample size slippages in randomised trials: exclusions and the lost and wayward. *Lancet* 2002; **359**:781–785.
41. Collins R, MacMahon S. Reliable assessment of the effects of treatment on mortality and major morbidity, I: clinical trials. *Lancet* 2001; **357**:373–380.
42. Bath F, Owen V, Bath P. Quality of full and final publications reporting acute stroke trials. A systematic review. *Stroke* 1998; **29**:2203–2210.

INDEX